Interpreting the Quantum World

This is a book about the interpretation of quantum mechanics, in particular about how to resolve the measurement problem introduced by the orthodox interpretation of the theory.

The heart of the book is a new result that shows how to construct all possible 'no collapse' interpretations, subject to certain natural constraints and the limitations imposed by the hidden variable theorems. From this perspective one sees precisely where things have gone awry and what the options are. Various interpretations, including Bohm's causal interpretation, Bohr's complementarity interpretation, and the modal interpretation are shown to be special cases of this result, for different choices of a 'preferred' observable. A feature of the book is a novel treatment of the main hidden variable theorems, and an extended critique of contemporary 'decoherence' theories of measurement. The discussion is self-contained and organized so that the technical portions may be skipped without losing the argument.

This book will be of interest to advanced undergraduates and graduate students in philosophy of science, physics, and mathematics with an interest in foundational problems in quantum physics. General readers with some technical sophistication will also find the book of value.

JEFFREY BUB received his PhD in mathematical physics from London University in 1966, where he studied physics with David Bohm at Birkbeck College and philosophy of science with Karl Popper and Imre Lakatos at the London School of Economics. With Bohm he published a 'hidden variable' dynamical reduction theory of quantum mechanics – the first theory to propose an explicit dynamics for the reduction or 'collapse' of the quantum state on measurement as a solution to the measurement problem. His early book, *The Interpretation of Quantum Mechanics*, was influential in developing the concept of a 'quantum logic,' and his numerous publications on the measurement problem and interpretative issues have helped shape the debate on the conceptual foundations of quantum mechanics. After working with Herbert Feigl at the Minnesota Center for Philosophy of Science, he taught physics and philosophy at Yale and subsequently at the University of Western Ontario. He has held visiting positions at Tel Aviv University, Princeton, and the University of California, and is currently Professor of Philosophy at the University of Maryland. *Interpreting the Quantum World* was joint winner of the 1998 Lakatos Award.

Interpreting the Quantum World

JEFFREY BUB

CAMBRIDGE
UNIVERSITY PRESS

PUBLISHED BY THE PRESS SYNDICATE OF THE UNIVERSITY OF CAMBRIDGE
The Pitt Building, Trumpington Street, Cambridge, United Kingdom

CAMBRIDGE UNIVERSITY PRESS
The Edinburgh Building, Cambridge CB2 2RU, UK http://www.cup.cam.ac.uk
40 West 20th Street, New York, NY 10011-4211, USA http://www.cup.org
10 Stamford Road, Oakleigh, Melbourne 3166, Australia

First published 1997
First paperback edition (with corrections) 1999

Typeset in Times 10/13 pt [VN]

A catalogue record for this book is available from the British Library

Library of Congress Cataloguing in Publication data

Bub, Jeffrey.
Interpreting the quantum world / Jeffrey Bub.
p. cm.
Includes bibliographical references and index.
ISBN 0 521 56082 9 (hc)
1. Quantum theory. I. Title.
QC174.12.B83 1997
530.1′2–dc20 96-31563 CIP

ISBN 0 521 56082 9 hardback
ISBN 0 521 65386 x paperback

Transferred to digital printing 2004

For David Bohm

Contents

Preface to the revised edition

The appearance of a revised edition presents me with the opportunity to make some corrections and improvements suggested by readers.

The main improvement is a much simplified and strengthened proof of the central theorem of the book in chapter 4. The original proof involved a 'weak separability' assumption (introduced to avoid a dimensionality restriction) that required several preliminary definitions and considerably complicated the formulation of the theorem. Sheldon Goldstein pointed out (private communication) that the proof goes through without this assumption.

What the theorem says can now be formulated quite simply. We know from the 'no go' hidden variable theorems that we cannot generate the probabilities defined by a quantum state, for ranges of values of the observables of a quantum mechanical system, from a measure function on a probability space of elements representing all possible assignments of values to these observables, if the value assignments are required to satisfy certain constraints. What this means is that, if we accept the constraints as reasonable and require that *all* observables are assigned values, we cannot interpret the quantum probabilities as measures of ignorance of the actual unknown values of these observables. In fact, the 'no go' theorems show that there are no consistent value assignments at all to certain well-chosen finite sets of observables, quite apart from the question of generating the quantum probabilities as measures over possible value assignments.[a] We also know that if we consider any quantum state ψ and any single observable R, the probabilities defined by ψ for ranges of values of R *can* be represented in this way, essentially because the Hilbert space subspaces associated with the ranges of values of a single observable generate a Boolean algebra (or Boolean lattice). So the possibility of consistent value assignments, or the representation of quantum probabilities as measures over such value assignments, must fail somewhere between considering a single observable and all observables.

[a] Note that the 'irreducibility of quantum probabilities in this sense arises from certain structural features of Hilbert space, brought out for the first time by the 'no go' theorems. It does not follow from earlier considerations, such as Heisenberg's uncertainty principle, which refers only to a reciprocal relationship between the statistical distributions of certain observables for a given quantum state, and says nothing about hypothetical value assignments to observables.

The question at issue is this: beginning with an arbitrary quantum state ψ and the Boolean lattice generated by a single observable R, how large a set of observables can we add to R before things go wrong? More precisely, what is the maximal lattice extension $\mathcal{D}(\psi, R)$ of this Boolean lattice, generated by the subspaces associated with ranges of values of observables, on which we can represent the probabilities defined by ψ, for the ranges of values of $\mathcal{R}$ and these additional observables, in terms of a measure over 2-valued homomorphisms on $\mathcal{D}(\psi, R)$?[b] The theorem provides an answer to this question on the further assumption that $\mathcal{D}(\psi, R)$ is invariant under automorphisms of the lattice $\mathcal{L}$ of all subspaces of Hilbert space that preserve the ray representing the state ψ and the 'preferred observable' R.

This analysis, in terms of the lattice structure of finite-dimensional Hilbert spaces, has been generalized by Rob Clifton and co-workers to cover continuous observables and mixed states in the general framework of C^*-algebras. See Clifton (1999), Clifton and Zimba (1998), and Clifton and Halvorson (1999).

It turns out that the sublattice $\mathcal{D}(\psi, R) \subset \mathcal{L}$ is unique. In fact, it is a generalization of the orthodox sublattice obtained by taking all the subspaces asigned probability 1 or 0 by ψ as representing 'definite' or 'determinate' properties of the system in the state ψ, and all other properties as indeterminate (so that the propositions asserting that the system possesses these properties in the state ψ are neither true nor false or, as the physicist would say, are 'meaningless' in the state ψ). From the standpoint of the theorem, the orthodox sublattice is obtained by choosing R as the unit observable I, but this choice leads to the measurement problem as I show in sections 4.1 and 5.1. Other choices for R can be associated with various non-orthodox 'no collapse' interpretations, for example Bohmian mechanics and the modal interpretation.

The choice of some 'preferred observable' R other than I requires introducing a dynamics for the evolution of *actual values* of observables associated with the 'determinate sublattice' $\mathcal{D}(\psi,R)$, as this sublattice evolves over time with the unitary evolution of ψ as a solution to Schrödinger's equation of motion. Of course, this dynamics for actual values will have to mesh with the Schrödinger dynamics tracked by ψ. I sketch such a dynamics in chapter 5. It turns out to be a stochastic dynamics that reduces to the deterministic dynamics Bohm introduced for the actual values of position in configuration space in his 1952 hidden variable theory, if we take R as continuous position in configuration space. This question has now been investigated in full generality by Bacciagaluppi and Dickson (1999). Michael Dickson has suggested (private communication) that my analysis of 'environmental modelling' and decoherence in 5.4, insofar as it depends on the details of this stochastic dynamics, might require some fine-tuning. I leave it to future readers to develop a more sophisticated analysis based on the general framework developed by Bacciagaluppi and Dickson.

[b] A 2-valued homomorphism is a map that assigns 1's and 0's to the elements of the sublattice in a structure-preserving way, and so defines an assignment of values to the determinate observables associated with the sublattice.

The uniqueness theorem can be understood 'neutrally,' as a way of relating a variety of 'no collapse' interpretations. But I see the significance of the theorem as much more radical. Quantum mechanics arose as a non-commutative matrix mechanics. As Bohr and Heisenberg saw it, quantum mechanics is a 'rational generalization' of classical mechanics, incorporating the quantum of action and the correspondence principle. What Heisenberg did was to ask how the kinematics of classical mechanics could be modifed so as to yield Bohr's frequency condition for the radiation emitted by an atom, when an electron jumps between orbits, rather than the classical frequency condition. Guided by the correspondence principle, he arrived at a theory of motion in which the representatives of certain dynamical variables do not commute. Shortly afterwards, Schrödinger developed wave mechanics from the wave–particle duality idea of de Broglie and proved the equivalence of the two theories.

Initially, physicists took the wave theory as a new way of modelling the microworld and regarded Heisenberg's non-commutative mechanics as a formally equivalent version of wave mechanics without any special foundational significance. But it soon became clear that the wave version of quantum mechanics did not automatically resolve the conceptual problems implicit in the non-commutative theory. Replacing a commutative algebra of dynamical variables with a non-commutative algebra is equivalent to replacing the representation of dynamical properties by the subsets of a set with the representation of these properties by the subspaces of a vector space, that is, it is equivalent to the replacement of a Boolean algebra for the representation of properties by a non-Boolean algebra of a certain sort. In fact, the salient structural feature of the transition from classical to quantum mechanics, as von Neumann saw (see the discussion in the Coda), is the replacement of a set-theoretic or Boolean structure for modelling the properties of a mechanical system with a projective geometry, rather than wave–particle duality. This structural change introduces a new element, the *angle* between subspaces representing properties, that is not present in a set-theoretic representation. The angles are related to probabilities – in fact, by Gleasons's theorem, to the only way probabilities can be defined on a non-Boolean structure of this sort. In the light of the 'no go' theorems for the representation of these probabilities as measures over possible value assignments on the non-commutative algebra of observables, the fundamental question of interpretation for quantum mechanics is how to understand these probabilities.

Now, the transition from classical mechanics to general relativity can be understood as involving the discovery that geometry is not only empirical but *dynamical*. That is, we now realize that geometry is not *a priori*. It makes sense to ask: what is the geometry of the world? As it turns out, the geometry of our universe is not a fixed Euclidean geometry, but rather a non-Euclidean geometry that changes dynamically as the distribution of mass in the universe changes. The significance of the uniqueness theorem, as I see it, is that just as the transition from classical mechanics to relativistic mechanics involves the discovery that geometry is dynamical, so the transition from classical mechanics to quantum mechanics involves the discovery that possibility is dynamical: the possibility structure of our universe is not a fixed, Boolean structure, as we supposed classically, but is in fact a non-Boolean structure that changes dynamically. The unitary Schrödinger evolution of the quantum state in time tracks the evolution of this possibility structure as a changing sublattice $\mathscr{D}(\psi, R)$ in the lattice of all subspaces of Hilbert space. So the Schrödinger time-dependent equation characterizes the temporal evolution of what is *possible*, not what is *actual* at time t.

At any particular time t, what is actually the case is selected as a sub-structure of the 'determinate sublattice' $\mathscr{D}(\psi, R)$ at time t by a 2-valued homomorphism on $\mathscr{D}(\psi, R)$, just as what is actually the case in a classical world is selected by a 2-valued homomorphism on the fixed Boolean lattice of possibilities. What is actually the case at time t in a quantum world must change over time in a way that meshes with the evolving possibility structure. So, in a classical world, change is described by the evolution over time of what is actual, where what is actually the case at time t is selected by a 2-valued homomorphism – the classical state – as a temporally evolving substructure against the background of a *fixed* Boolean lattice of possibilities. In a quantum world, what is actually the case at time t is selected by a 2-valued homomorphism as a temporally evolving substructure on a *changing* background of possibilities. So in a quantum world there is a dual dynamics: the Schrödinger dynamics for the evolution of possibility, and a dynamics for how what is actually the case changes with time, which turns out to be a generalization of Bohmian dynamics. From this perspective, we can understand the phenomena of interference and entanglement – essentially quantum phenomena – as arising from the way in which what is actually the case at t changes from t to t' in such a way as to mesh with the change in possibility structure from t to t'.

Apart from the new proof of the uniqueness theorem and minor corrections, I have revised the discussion in 5.3 on non-ideal measurements. Guido Bacciagaluppi pointed out (in a review for *Nuncius*) that I had defined positive operator valued measures in a non-standard way. The argument goes through nonetheless, and the new formulation avoids reference to POVs. Bacciagaluppi also pointed out a flaw in the argument I attributed to Rob Clifton in the Coda, about how Lorentz invariance might be preserved in Bohmian modal interpretations, and I have accordingly dropped that argument.

Preface

It was Michael Whiteman,[1] applied mathematician and mystic at the University of Cape Town, who first introduced me to the Einstein–Podolsky–Rosen 'paradox' and other mysteries of quantum mechanics. That was in 1962, and I was hooked. The following year I went to London on a Jan Smuts Memorial Scholarship (a princely sum of £500 for two years, renewable for a third), with the idea of studying philosophy of physics under Popper at the London School of Economics. Popper was visiting the US at the time and, apart from attending Lakatos' fascinating evening seminar on mathematical discovery,[2] I soon withdrew from the program. The sudden immersion in courses on the history of philosophy was too much of a culture shock after four years as a mathematics and physics undergradate.

Whiteman had suggested that I look up his friend G.J. Whitrow[3] if things didn't work out at the LSE. Whitrow's advice was clear: if I was interested in foundational problems of quantum mechanics, the choice was between Bohm in London or Rosenfeld in Copenhagen. Since a move to Copenhagen seemed too daunting a prospect at the time, I telephoned Bohm at Birkbeck College. He agreed to a meeting, with the understanding that he wasn't accepting any new graduate students. Eventually, on the strength of a paper I had written on the Einstein–Podolsky–Rosen argument (with some embarrassingly critical comments on Bohm's 'hidden variable' theory), I persuaded him to change his mind.

The theoretical physics group at Birkbeck consisted of Bohm, Hiley, and about half a dozen graduate students. Bohm was interested in understanding algebraic topology as a process-based rather than object-based formal language for physics, and we worked through the section on polyhedral complexes in Hodge's (1952) book on harmonic integrals in his graduate seminar. I recall being mystified. On the days when he came into Birkbeck, Bohm would arrive in the morning, begin a discussion on a topic he'd been thinking about, and continue until late afternoon. We'd all troop down to the cafeteria for lunch, or sometimes (mercifully, since the food was pretty awful) to one of the restaurants on Charlotte Street nearby. There were lots of gems in those discussions

[1] Author of *Philosophy of Space and Time* (1967), and *The Mystical Life* (1961).

[2] The material was later published as a four part article in *The British Journal for the Philosophy of Science* **14**, 1963–4, and in an expanded version in Lakatos (1976).

[3] Author of *The Natural Philosophy of Time* (1961).

but invariably, after we had brainstormed with Hiley over a particularly intriguing or puzzling idea, Bohm had thought of some entirely new way of looking at the matter by the next session.

After a while, it began to seem increasingly unlikely to me that I'd ever manage to work on anything long enough to launch a PhD dissertation. I started to spend more time away from Birkbeck doing my own reading, and I also began to visit the LSE regularly, where I sat in on Popper's seminar after he returned. It was during this period that I stumbled on a paper by Margenau (1963a) on the measurement problem. I read all the references in the bibliography I could lay my hands on, including London and Bauer's *La Théorie de l'Observation en Mécanique Quantique* (1939), which I managed to obtain from the publisher in Paris and laboriously translated into English.

When I surfaced again at Birkbeck, Bohm asked me to give a seminar presentation to the group on what I had been doing. He seemed genuinely interested and suggested I take a look at papers by Wiener and Siegel (Wiener and Siegel, 1953, 1955; Siegel and Wiener, 1956) as a way of resolving the measurement problem. Bohm's thought was that one should be able to exploit the Wiener–Siegel 'differential space' approach to quantum mechanics to construct an explicit nonlinear dynamical 'collapse' theory for quantum measurement processes.[4]

After that, I knew I had my PhD topic. We eventually published the theory as 'A Proposed Solution of the Measurement Problem in Quantum Mechanics by a Hidden Variable Theory' (Bohm and Bub, 1966a), together with a critique of the Jauch and Piron 'no go' theorem for hidden variables (Bohm and Bub, 1966b). Both articles appeared in the same issue of *Reviews of Modern Physics* as Bell's seminal critique of 'no go' theorems (Bell, 1966). As I recall, the underlying ideas of the theory were Bohm's, but he left it to me to work out the details. Bohm was never very interested in mathematical rigour – he simply 'saw' that things would work out in a certain way, and his physical intuition was always on the right track (although it was sometimes a frustrating business to map out all the twists and turns in his thinking).

I was Bohm's graduate student from 1963 to 1965. After I left Birkbeck we corresponded regularly for a few years. Bohm continued to work on the 'collapse' theory during the late 1960s – I have a lengthy unpublished manuscript of his dating from that time – but eventually we both lost interest.[5] I discovered quantum logic which, for a while, I thought was the answer to all the conceptual puzzles of quantum mechanics. Bohm found my fascination with quantum logic incomprehensible, and our correspondence languished.

[4] Curiously, he did not bring up his own 1952 hidden variable theory in this connection, and I don't recall him discussing the theory while I was a student at Birkbeck.

[5] The manuscript ('On the Role of Hidden Variables in the Fundamental Structure of Physics') has now been published in *Foundations of Physics* **26**, 719–86 (1996). The Bohm–Bub theory has recently been resurrected by Ron Folman (1994, 1995), who has been looking for experimental confirmation in a possible deviation from the quantum mechanically predicted experimental distribution for the decay time of massive particles, specifically the tau lepton, in the very short decay time region. See also OPAL Collaboration (1996). For an account of some earlier experimental tests of the theory, see Belinfante (1973), chapter 4.

My first and most important intellectual debt is to Bohm, and this book is dedicated to his memory.

After I left Birkbeck, I had a one-year post-doctoral position in the Chemistry Department at the University of Minnesota. Alden Mead, a physical chemist in the department, visited Birkbeck on a sabbatical during my last year there and offered me a position as his assistant. I was supposed to work on fundamental length theories, but I don't think much became of that. Ford Hall, the home of the Minnesota Center for the Philosophy of Science, beckoned from just across the mall.

The Center was an exciting place. Herbert Feigl and Grover Maxwell were there permanently, and there were many visitors. Hilary Putnam passed through and gave a talk on quantum logic that profoundly influenced my thinking on the interpretation of quantum mechanics. I was captivated by von Neumann's (1939) notion of a non-Boolean logic for quantum systems, and the idea that what is conceptually puzzling about quantum mechanics relative to classical mechanics is that the properties of quantum systems 'fit together' in a non-Boolean way, and this is what we ought to try and understand.

At the Center I met Bill Demopoulos. We were both intrigued by quantum logic – after a brief joint flirtation with Whitehead's process philosophy it was like a breath of fresh air. Working through the Kochen and Specker papers (1965, 1967) together was the beginning of a long collaboration and friendship. The ideas on quantum logic in my book *The Interpretation of Quantum Mechanics* (1974) reflect this collaboration. I still think the essential difference between classical and quantum mechanics is captured by the insight that going from classical to quantum mechanics involves the transition from a Boolean to a non-Boolean possibility structure for the properties of a physical system (see Demopoulos, 1976).

Bohm's ideas on hidden variables, and von Neumann's concept of a quantum logic understood as a possibility structure for events, have always been the two main influences on my approach to the conceptual problems of quantum mechanics. In a sense, this book reconciles these two opposing themes, from the perspective of a 'modal' interpretation in the sense of van Fraassen (1973, 1974, 1981, 1991), although the implementation of this notion is very different from van Fraassen's. But my more immediate intellectual debt is to Rob Clifton, with whom I have enjoyed a lively and extremely productive email correspondence and collaboration for the past two years or so. Much of the book is an extended discussion of our joint paper on a uniqueness theorem for 'no collapse' interpretations of quantum mechanics (Bub and Clifton, 1996).

I had constructed a class of 'no collapse' interpretations (Bub, 1992a, b, 1993a, 1994b, 1996), which I presented variously as versions of the modal interpretation or as Bohmian interpretations (in the sense of Bohm's 1952 hidden variable theory). The basic idea came out of my analysis of Bohr's reply to the Einstein–Podolsky–Rosen argument (Bub, 1989, 1990). Clifton was working on modal interpretations that exploit the biorthogonal decomposition theorem and proved a result justifying the common framework of the Kochen and Dieks formulations as unique, subject to certain constraints (Clifton, 1995b). After I received a draft of this theorem, I proved a uniqueness theorem for the class of 'no collapse' interpretations I had constructed (Bub, 1994a, 1995c). Later, Clifton saw the possibility of replacing the assumptions in my original uniqueness theorem with fewer and more natural assumptions, which eventually led to our joint theorem. My account of the motivation for the theorem, and the significance of the theorem for the modal interpretation, Bohmian mechanics, and Bohr's complementarity interpretation draws on the analysis in our joint paper. Needless to say, I bear sole responsibility for any foolishness in this exposition.

Like many others working on problems in the foundations of quantum mechanics, I have found enlightenment and inspiration in the writings of John Bell and David Mermin, and over the years I have benefited from discussions on the interpretation problem and the measurement problem with Roger Cooke, Bas van Fraassen, R.I.G. Hughes, Allen Stairs, Itamar Pitowsky, Michael Redhead, Harvey Brown, Jeremy Butterfield and, more recently, David MacCallum, David Albert, Andrew Elby, Jeff Barrett, Ron Folman, Pekka Lahti, Bradley Monton, Michael Dickson, Guido Bacciagaluppi and Jitendra Subramanyam. Rob Clifton, Bradley Monton, David MacCallum, and Jo Clegg read the manuscript in various drafts, and I have incorporated many of their suggestions for improvements in both style and content.

Finally, I owe a special debt to my wife, Robin Shuster – muse extraordinaire and Socratic midwife to many of the ideas presented here. I doubt that I would have completed the book without her constant encouragement and support.

Introduction

This is a book about the interpretation of quantum mechanics, and about the measurement problem. The conceptual entanglements of the measurement problem have their source in the orthodox interpretation of 'entangled' states that arise in quantum mechanical measurement processes. The heart of the book is a uniqueness theorem (Bub and Clifton, 1996; see chapter 4) that characterizes alternative 'no collapse' interpretations of the theory, in particular observer-free interpretations that don't involve the measurement problem. From the perspective of the uniqueness theorem, one sees precisely where things have gone awry and what the options are.

One might wonder why, and in what sense, a fundamental theory of how physical systems move and change requires an interpretation. Quantum mechanics is an irreducibly statistical theory: there are no states of a quantum mechanical system in which all dynamical variables have determinate or 'sharp' values – no states that are 'dispersion-free' for all dynamical variables. Moreover, so-called 'no go' theorems exclude the possibility of defining new states in terms of 'hidden variables,' in which all dynamical variables – or even certain finite sets of dynamical variables – have determinate values, if we assume that the values assigned to functionally related dynamical variables by the new hidden variable states are subject to certain constraints, and we require that the quantum statistics can be recovered by averaging over these states. So it is standard practice to refer agnostically to 'observables' rather than dynamical variables (which suggest determinate values evolving in time), and to understand quantum mechanics as providing probabilities for the outcomes of measurements of observables under physically well-defined conditions.

This neutrality only goes so far. All standard treatments of quantum mechanics take an observable as having a determinate value if the quantum state is an eigenstate of that observable.[6] If the state is not an eigenstate of the observable, no determinate value is attributed to the observable. This principle – sometimes called the 'eigenvalue–eigenstate link'[7] – is explicitly endorsed by Dirac (1958, pp. 46–7) and von Neumann (1955, p. 253), and clearly identified as the 'usual' view by Einstein, Podolsky, and

[6] For an account of quantum states and their representation in Hilbert space see the appendix.
[7] The term is due to Arthur Fine (1973, p. 20).

Rosen (1935) in their classic argument for the incompleteness of quantum mechanics (see chapter 2). Since the dynamics of quantum mechanics described by Schrödinger's time-dependent equation of motion is linear, it follows immediately from this orthodox interpretation principle that, after an interaction between two quantum mechanical systems that can be interpreted as a measurement by one system on the other, the state of the composite system is not an eigenstate of the observable measured in the interaction, and not an eigenstate of the indicator observable functioning as a 'pointer.' So, on the othodox interpretation, neither the measured observable nor the pointer reading have determinate values, after a suitable interaction that correlates pointer readings with values of the measured observable. This is the measurement problem of quantum mechanics.

There are three possible ways of resolving the measurement problem: We adopt what Bell (1990) has termed a 'FAPP' ('for all practical purposes') solution, or we change the linear dynamics of the theory (which, as I see it, means changing the theory), or we change the orthodox Dirac–von Neumann interpretation principle.

FAPP solutions range from the Daneri–Loinger–Prosperi (1962, 1966) quantum ergodic theory of macrosystems[8] to the currently fashionable 'decoherence' theories. Essentially, the idea here is to exploit the fact that a macroscopic measuring instrument is an open system in virtually continuous interaction with its environment. Because of the typical sorts of interactions that take place in our world between such systems and their environments, it turns out that almost instantaneously after a measurement interaction, the 'reduced state' of the measured system and measuring instrument as a composite subsystem of the universe is, for all practical purposes, indistinguishable from a state that supposedly can be interpreted as representing a classical probability distribution over determinate but unknown values of the pointer observable. The information required to exhibit characteristic quantum interference effects between different pointer-reading states is almost immediately irretrievably lost in the many degrees of freedom of the environment. Since there are well-known difficulties with such an 'ignorance interpretation,' there is usually a further move involving an appeal to Everett's (1957, 1973) 'relative state' or 'many worlds' interpretation of quantum mechanics, where determinateness is only claimed in some relative sense. I discuss versions of this approach in chapter 8, where I argue that the measurement problem is not resolved by this manoeuvre.

The Bohm–Bub 'hidden variable' theory (1966a) modifies the linear dynamics by adding a nonlinear term to the Schrödinger equation that effectively 'collapses' or projects the state onto an eigenstate of the pointer reading and measured observable in a measurement process (the resulting eigenstate depending on the hidden variable). Currently the Ghirardi–Rimini–Weber theory (1986), with later contributions by Pearle (Ghirardi, Grassi, and Pearle, 1990, 1991; Pearle, 1989, 1990), is a much more

[8] For a critique, see Bub (1968).

sophisticated stochastic dynamical 'collapse' theory, formulated as a continuous spontaneous localization theory.

The remaining possibility is to adopt an alternative principle for selecting the set of observables that have determinate values in a given quantum state. This was Bohm's approach, and also – very differently – Bohr's. Bohm's 1952 hidden variable theory or 'causal' interpretation (Bohm, 1952a; Bohm and Hiley, 1993) takes the position of a system in configuration space[9] as determinate in every quantum state. Certain other observables can be taken as determinate at a given time together with this 'preferred' always-determinate observable, depending on the state at that time. Alternative formulations of Bohm's theory present different accounts of 'nonpreferred' observables such as spin. On the formulation proposed here, the theory is a 'modal' interpretation of quantum mechanics, in the broad sense of van Fraassen's notion (see chapter 6). For Bohr, an observable has a determinate value only in the context of a specific, classically describable experimental arrangement suitable for measuring the observable. Since the experimental arrangements suitable for locating a quantum system in space and time, and for the determination of momentum–energy values, turn out to be mutually exclusive, there is no unique description of the system in terms of the determinate properties associated with the determinate values of a fixed preferred observable. So which observables have determinate values is settled pragmatically by what we choose to observe, via the classically described measuring instruments we employ, and is not defined for the system alone. Bohr terms the relation between space–time and momentum–energy concepts 'complementary,' since both sets of concepts are required to be mutually applicable for the specification of the classical state of a system.

What is generally regarded as the 'Copenhagen interpretation' is some fairly loose synthesis of Bohr's complementarity interpretation and Heisenberg's ideas on the significance of the uncertainty principle. It is usual to pay lip service to the Copenhagen intepretation as the 'orthodox' interpretation of quantum mechanics, but the interpretative principle behind complementarity is very different from the Dirac–von Neumann principle. (I discuss the relationship in detail in sections 7.1 and 7.2). Unlike Dirac and von Neumann, Bohr never treats a measurement as an interaction between two quantum systems, and hence has no need for a special 'projection postulate' to replace the linear Schrödinger evolution of the quantum state during a measurement process. Both Dirac and von Neumann introduce such a postulate to describe the stochastic projection or 'collapse' of the state onto an eigenstate of the pointer reading and measured observable – a state in which these observables are determinate on their interpretation. (See Dirac, 1958, p. 36, and von Neumann, 1955, p. 351 and pp. 417–18.) The complementarity interpretation avoids the measurement problem by selecting as determinate an observable associated with an individual quantum 'phenomenon' manifested in a measurement interaction involving a specific classically describable experimental arrangement. Certain other observables, regarded as measured in the

[9] For an N-particle system, the configuration space of the system is a $3N$-dimensional space, coordinatized by the $3N$ position coordinates of the particles.

interaction, can be taken as determinate together with this observable and the quantum state.

Einstein viewed the Copenhagen interpretation as 'a gentle pillow for the true believer.'[10] For Einstein, a physical system has a 'being-thus,' a 'real state' that is independent of other systems or the means of observation (see the quotations in section 1.1 and section 6.1). He argued that realism about physical systems in this sense is incompatible with the assumption that the state descriptions of quantum mechanics are complete. What Einstein had in mind by a 'completion' of quantum mechanics is not entirely clear, but on one natural way of understanding this notion (as an observer-free 'no collapse' interpretation subject to certain physically plausible constraints), the possible completions of quantum mechanics are fully characterized by the uniqueness theorem in chapter 4.[11]

This book begins with a survey of the problem of interpretation, as it arises in the debate between Einstein and Bohr. Einstein's discomfort with quantum mechanics cannot be attributed to an aversion to indeterminism. He did not argue that quantum mechanics must be incomplete *because* 'God does not play dice with the universe.' Rather, as Pauli put it, Einstein's 'philosophical prejudice' was realism, not determinism (section 1.1). It is not that all indeterministic or stochastic theories were problematic for Einstein. What Einstein objected to were stochastic theories that violate certain realist principles; or rather, he objected to taking such theories as anything more than predictive instruments that would ultimately be replaced by a complete explanatory theory.

Chapter 1 continues with a discussion of the transition from classical to quantum mechanics, and a formulation of the measurement problem as a problem generated by the orthodox (Dirac–von Neumann) interpretation of the theory. My main aim here is to bring out the different ways in which dynamical variables and properties are represented in the two theories. In classical mechanics, the dynamical variables of a system are represented as real-valued functions on the phase space of the system and form a commutative algebra. The subalgebra of idempotent dynamical variables (the characteristic functions) represent the properties of the system and form a Boolean algebra, isomorphic to the Boolean algebra of (Borel) subsets of the phase space of the system. In quantum mechanics, the dynamical variables or 'observables' of a system are represented by a noncommutative algebra of operators on a Hilbert space, a linear vector space over the complex numbers, and the subalgebra of idempotent operators (the projection operators) representing the properties of the system is a non-Boolean algebra isomorphic to the lattice of subspaces of the Hilbert space. So the transition from classical to quantum mechanics involves the transition from a Boolean to a non-Boolean structure for the properties of a system.

There are restrictions on what sets of observables can be taken as simultaneously determinate without contradiction, if the attribution of determinate values to observ-

[10] In a letter to Schrödinger, dated May, 1928. Reprinted in Przibram (1967, p. 31).
[11] See Fine (1986), especially chapter 4, for a different interpretation of Einstein's view.

ables is required to satisfy certain constraints. The 'no go' theorems for hidden variables underlying the quantum statistics provide a series of such results that severely limit the options for a 'no collapse' interpretation of the theory.

In chapter 2, I present the Einstein–Podolsky–Rosen (1935) incompleteness argument, and several versions of Bell's extension of the argument to a 'no go' theorem demonstrating the inconsistency of stochastic and deterministic hidden variables, satisfying certain locality and separability constraints, with the quantum statistics.

Chapter 3 deals with the Kochen and Specker (1967) 'no go' theorem, showing the impossibility of assigning determinate values to certain finite sets of observables if the value assignments are required to preserve the functional relations holding among the observables. I present a new proof of the theorem for a set of 33 observables (1-dimensional projectors), based on a classical tautology that is quantum mechanically false proposed by the logician Kurt Schütte in an early (1965) unpublished letter to Specker.

Chapter 4 introduces the problem of interpretation, and contains the proof of the uniqueness theorem demonstrating that, subject to certain natural constraints, all 'no collapse' interpretations of quantum mechanics can be uniquely characterized and reduced to the choice of a particular preferred observable as determinate. The preferred observable and the quantum state at time t define a (non-Boolean) 'determinate' sublattice in the lattice $\mathscr{L}$ of all subspaces of Hilbert space – the sublattice of propositions that can be true or false at time t. The actual properties of the system at time t are selected by a 2-valued homomorphism (a yes–no map) on the determinate sublattice at time t, so the range of possibilities for the system at time t is defined by the set of 2-valued homomorphisms on the determinate sublattice. From this 'modal' perspective, the possibility structure of a quantum world is represented by a *dynamically evolving* (non-Boolean) sublattice in $\mathscr{L}$, while the possibility structure of a classical world is fixed for all time as the Boolean algebra $\mathscr{B}$ of subsets of a phase space. The dynamical evolution of the quantum state tracks the evolution of possibilities (and probabilities defined over these possibilities) through the evolution of the determinate sublattice, rather than actualities, while the dynamically evolving classical state defines the actual properties in a classical world as a 2-valued homomorphism on $\mathscr{B}$ and directly tracks the evolution of actual properties. In a quantum world, the dynamical state is distinct from the 'property state' (defined by a 2-valued homomorphism on the determinate sublattice), while the classical state doubles as a dynamical state and a property state.

Different choices for the preferred determinate observable correspond to different 'no collapse' interpretations of quantum mechanics. In chapter 5, I show how the orthodox (Dirac–von Neumann) interpretation without the projection postulate can be recovered from the theorem, and how the measurement problem is avoided in 'no collapse' interpretations by an appropriate choice of the preferred determinate observable. Property states must evolve in time so as to reproduce the quantum statistics over the determinate sublattices defined by the dynamical evolution of quantum states. Since the determinate sublattice at time t is uniquely defined by the quantum state at t and a preferred observable, it suffices to provide an evolution law for the actual values of the preferred determinate observable. Following a proposal by Bell (1987, pp. 176–7) and Vink (1993), I formulate a specific stochastic equation of motion for the case of a discrete preferred observable, which reduces to the deterministic evolution law of Bohm's theory in the continuum limit, if the preferred observable is continuous position in configuration space. It turns out that the interaction between a measuring instrument and its environment plays a crucial rôle in guaranteeing that the actual value of an appropriately chosen preferred determinate observable will evolve stochastically in time so that the observable functions as a stable pointer in ideal or non-ideal measurement interactions. In this respect, the choice of preferred observable is constrained by the dynamics of system–environment interactions in our world: if we want an interpretation of quantum mechanics to account for the measurement interactions that are possible in our world, we need to choose a preferred determinate observable for which measurement correlations persist under environmental 'monitoring.' It is not the phenomenon of decoherence as a loss of interference that is relevant here. Rather, the fact that measuring instruments are open systems interacting with environments with many degrees of freedom turns out to have a very different dynamical significance in an observer-free 'no collapse' interpretation with a fixed preferred determinate observable.

In chapter 6, I show how Bohm's causal interpretation (one natural way to develop an Einsteinian realism within quantum mechanics) and the modal interpretation (in a version generalized from earlier formulations by Kochen, 1985, and by Dieks, 1988, 1989a, 1994a,b) can be seen as two observer-free 'no collapse' interpretations in the sense of the theorem. Bohm's interpretation adopts position in configuration space as a fixed preferred determinate observable, while the modal interpretation can be understood as adopting a time-dependent preferred determinate observable derived from the quantum state.

I discuss Bohr's complementarity interpretation as a 'no collapse' interpretation in chapter 7, and show how this interpretation can be related to the orthodox (Dirac–von Neumann) interpretation from the perspective of the uniqueness theorem.

The 'new orthodoxy' appears to center now on the idea that the original Copenhagen interpretation has been vindicated by the recent technical results on environmental decoherence. Sophisticated versions of this view are formulated in terms of 'consistent histories' or 'decoherent histories,' and trade on features of Everett's 'relative state' interpretation of quantum mechanics as a solution to the measurement problem (popularly understood as a 'many worlds' theory, in some sense). In chapter 8, I argue that there is no real advance here with respect to Einstein's qualms about the Copenhagen interpretation. It is still a 'gentle pillow for the true believer,' perhaps now with the added attraction of a rather fancy goose-down comforter.

The coda concludes with a review of the main themes of the argument, and its significance for the debate on the interpretation of quantum mechanics and the measurement problem.

In the appendix, I develop some mathematical machinery dealing with the structure of Hilbert space and the representation of states, probabilities, and observables in quantum mechanics. The discussion, which is intended to be self-contained for a reader with some minimal mathematical competence, covers the 'entangled' states of quantum systems that arise in measurement interactions and situations of the Einstein–Podolsky–Rosen type, and the formalism for some illustrative examples dealing with spin. No particular formal background is assumed, beyond a passing familiarity with the basic concepts of vector spaces, complex numbers, and probability theory.

1

From classical to quantum mechanics

> There is indeed much talk of 'observables' in quantum theory books. And from some popular presentations the general public could get the impression that the very existence of the cosmos depends on our being here to observe the observables. I do not know that this is wrong. I am inclined to hope that we are indeed that important. But I see no evidence that it is so in the success of contemporary quantum theory.
>
> *J.S. Bell (1987, p. 170)*

1.1 The ideal of the detached observer

If you ask physicists what is so strange about quantum mechanics relative to classical mechanics, you might get different answers, but they would all boil down to the irreducibly probabilistic character of state descriptions in quantum mechanics: the dynamical variables of a quantum mechanical system (position, momentum, energy, spin, etc.) cannot all have determinate or 'sharp' values simultaneously – some dynamical variables must be indeterminate.

The standard argument is that interference, or 'wave–particle duality,' requires an irreducibly probabilistic account of phenomena, because the attribution of determinate values to the dynamical variables of the theory (or even determinate ranges of values) is limited by the uncertainty principle, once you take the physics of measurement at the microlevel seriously. Since the properties of a quantum system appear only under specific physical conditions defined by appropriate measuring instruments, and the theory tells us that some of these physical conditions are mutually exclusive, it seems that *what there is* depends in some essential way on *how we look*. We used to think of the physical universe as composed of separable systems with objective properties that change in certain ways under the action of forces. Now, the story goes, we see that the very notion of a system having a property in the absence of a specific interaction with a measuring instrument is untenable for quantum phenomena.

Bohr put it this way (1961, p. 74):

The recognition that the interaction between the measuring tools and the physical systems under investigation constitutes an integral part of quantum phenomena has not only revealed an unsuspected limitation of the mechanical conception of nature, as characterized by attribution of separate properties to physical systems, but has forced us, in the ordering of experience, to pay proper attention to the conditions of observation.

Einstein emphatically rejected this view, the 'Copenhagen interpretation' of quantum mechanics. In a letter to Schrödinger dated December 22, 1950 (reprinted in Przibram, 1967, p. 39), he writes:

You are the only contemporary physicist, besides Laue, who sees that one cannot get around the assumption of reality – if only one is honest. Most of them simply do not see what sort of risky game they are playing with reality – reality as something independent of what is experimentally established. They somehow believe that the quantum theory provides a description of reality, and even a *complete* description; this interpretation is, however, refuted, most elegantly by your system of radioactive atom + Geiger counter + amplifier + charge of gun powder + cat in a box, in which the ψ-function of the system contains the cat both alive and blown to bits. Is the state of the cat to be created only when a physicist investigates the situation at some definite time? Nobody really doubts that the presence or absence of the cat is something independent of the act of observation. But then the description by means of the ψ-function is certainly incomplete, and there must be a more complete description. If one wants to consider the quantum theory as final (in principle), then one must believe that a more complete description would be useless because there would be no laws for it. If that were so then physics could only claim the interest of shopkeepers and engineers; the whole thing would be a wretched bungle.

Schrödinger's cat in a box[12] illustrates the measurement problem of quantum mechanics: after the decay products of the radioactive atom have interacted with the Geiger counter, the whole system is completely described by a quantum state that is a linear superposition over live and dead states of the cat. The measurement problem and the interpretation of superpositions is the focus of much of this book. Here I want to emphasize Einstein's point: 'Nobody really doubts that the presence or absence of the cat is something independent of the act of observation.' Surely, this is true. But the standard view, expressed with a straight face, is that quantum mechanics requires us to regard any question concerning the status of the cat as (strictly speaking) 'meaningless' until we establish an observational relationship with the cat by opening the box. This might be construed as a radically instrumentalist thesis about physics – but then

[12] Schrödinger's own formulation refers to hydrocyanic acid, not gunpowder:

A cat is penned up in a steel chamber, along with the following diabolical device (which must be secured against direct interference by the cat): in a Geiger counter there is a tiny amount of radioactive substance, *so* small, that *perhaps* in the course of one hour one of the atoms decays, but also, with equal probability, perhaps none; if it happens, the counter tube discharges and through a relay releases a hammer which shatters a small flask of hydrocyanic acid. If one has left this entire system to itself for an hour, one would say that the cat still lives *if* meanwhile no atom has decayed. The first atomic decay would have poisoned it. The ψ-function of the entire system would express this by having in it the living and the dead cat (pardon the expression) mixed or smeared out in equal parts.

(Schrödinger, 1935b; reprinted in Wheeler and Zurek, 1983, p. 157)

'physics could only claim the interest of shopkeepers and engineers.' If quantum mechanics is not merely a useful and unusually accurate predictive device for what we are likely to observe under specified experimental conditions, surely the question we ought to consider is: what must the universe be like – even if there were no observers – for the theory to be true?

Einstein had an extensive correspondence with Max Born, in which he attempted to convey his objections to the Copenhagen interpretation of quantum mechanics, without much success. Born reproduces some 'caustic marginal comments' that Einstein wrote in a reply to one of Born's letters dated March 18, 1948 (Born, 1971, p. 164). Aside from 'Ugh!' and 'Blush, Born, Blush!' there is a succinct summary of Einstein's view:

I just want to explain what I mean when I say that we should try to hold on to physical reality. We all of us have some idea of what the basic axioms in physics will turn out to be. The quantum or the particle will surely not be amongst them; the field, in Faraday's and Maxwell's sense, could possibly be, but it is not certain. But whatever we regard as existing (real) should somehow be localized in time and space. That is, the real in part of space *A* should (in theory) somehow 'exist' independently of what is thought of as real in space *B*. When a system in physics extends over the parts of space *A and B*, then that which exists in *B* should somehow exist independently of that which exists in *A*. That which really exists in *B* should therefore not depend on what kind of measurement is carried out in part of space *A*; it should also be independent of whether or not any measurement at all is carried out in space *A*. If one adheres to this programme, one can hardly consider the quantum-theoretical description as a complete representation of the physically real. If one tries to do so in spite of this, one has to assume that the physically real in *B* suffers a sudden change as a result of a measurement in *A*. My instinct for physics bristles at this. However, if one abandons the assumption that what exists in different parts of space has its own, independent, real existence, then I simply cannot see what it is that physics is meant to describe. For what is thought to be a 'system' is, after all, just a convention, and I cannot see how one could divide the world objectively in such a way that one could make statements about parts of it.

A core feature of Einstein's realism is the notion that 'what exists in different parts of space has its own, independent, real existence,' an existence that is 'independent of whether or not any measurement at all is carried out.' Pauli tried to explain Einstein's position to Born in a letter dated March 30, 1954 (Born, 1971, p. 218):

Now from my conversations with Einstein I have seen that he takes exception to the assumption, essential to quantum mechanics, that the *state of a system is defined only by specification of an experimental arrangement. Einstein wants to know nothing of this.* If one were able to measure with sufficient accuracy, this would of course be as true for small macroscopic spheres as for electrons. It is, of course, demonstrable by specifying thought experiments, and I presume that you have mentioned and discussed some of these in your correspondence with Einstein. But Einstein has the philosophical prejudice that (for macroscopic bodies) a state (termed 'real') can be defined 'objectively' under *any* circumstances, that is, *without* specification of the experimental arrangement used to examine the system (of the macro-bodies), or to which the system is being 'subjected.' It seems to me that the discussion with Einstein can be reduced to this hypothesis of his, which I have called the idea (or the 'ideal') of the 'detached observer.' But to me and other representatives of quantum mechanics, it seems that there is sufficient experimental and theoretical evidence against the practicability of this ideal.

Born seems to have conflated Einstein's realist thesis with determinism. Pauli clears up this confusion in a letter to Born dated March 31, 1954 (Born, 1971, p. 221):

Also, Einstein gave me your manuscript to read; he was *not at all* annoyed with you, but only said you were a person who will not listen. This agrees with the impression I have formed myself insofar as I was unable to recognise Einstein whenever you talked about him in either your letter or your manuscript. It seemed to me as if you had erected some dummy Einstein for yourself, which you then knocked down with great pomp. In particular, Einstein does not consider the concept of 'determinism' to be as fundamental as it is frequently held to be (as he told me emphatically many times), and he denied energetically that he had ever put up a postulate such as (your letter, para. 3): 'the sequence of such conditions must also be objective and real, that is, automatic, machine-like, deterministic.' In the same way, he *disputes* that he uses as criterion for the admissibility of a theory the question: 'Is it rigorously deterministic?'

Einstein's point of departure is 'realistic' rather than 'deterministic,' which means that his philosophical prejudice is a different one.

So the issue for Einstein is not determinism, but realism. Quantum mechanics, as a fundamental theory of motion, is apparently at odds with some of our most basic assumptions about physical reality. Einstein concluded that quantum mechanics cannot be regarded as a complete theory, as 'the whole story,' in principle, about the motion of physical systems. But if quantum mechanics *is* 'the whole story,' how do we make sense of a quantum mechanical universe?

It is really quite astonishing that mainstream contemporary physics has rejected this question as misguided. Aage Petersen (1963, p. 12) quotes Bohr as saying:

There is no quantum world. There is only an abstract quantum physical description. It is wrong to think that the task of physics is to find out how nature *is*. Physics concerns what we can *say* about nature.

Physicists seem loath to acknowledge the existence of 'facts' in a quantum mechanical universe, or that it makes any sense to talk about an event 'occurring' or a possibility being 'realized' or the corresponding proposition being 'true' because, supposedly, quantum mechanics doesn't sanction such notions, or only sanctions them in a relative sense.

For example, Omnès' ninth thesis, one of twenty-one theses summarizing what he calls (1994, p. xiii) 'the interpretation of quantum mechanics, not an interpretation,' is the statement (1994, p. 507):

The theory is unable to give an account of the existence of facts, as opposed by their uniqueness to the multiplicity of possible phenomena. This impossibility could mean that quantum mechanics has reached an ultimate limit in the agreement between a mathematical theory and physical reality. It might also be the underlying reason for the probabilistic character of the theory.

Gell-Mann and Hartle (1990) conclude what they take to be (p. 430) 'an attempt at extension, clarification, and completion of the Everett interpretation' with the comment (p. 455):

The problem with the 'local realism' that Einstein would have liked is not the locality but the *realism.* Quantum mechanics describes *alternative* decohering histories and one cannot assign 'reality' simultaneously to different alternatives because they are contradictory. Everett and others have described this situation, not incorrectly, but in a way that has confused some, by saying that the histories are all 'equally real' (meaning only that quantum mechanics prefers none over another except via probabilities) and by referring to 'many worlds' instead of 'many histories.'

As Wheeler (1983, p. 194) sees it, the occurrence of events requires our participation as observers:

What we have the right to say of past spacetime, and past events, is decided by choices – of what measurements to carry out – made in the near past and now. The phenomena called into being by these decisions reach backward in time in their consequences..., back even to the earliest days of the universe. Registering equipment operating in the here and now has an undeniable part in bringing about that which appears to have happened. Useful as it is under everyday circumstances to say that the world exists 'out there' independent of us, that view can no longer be upheld. There is a strange sense in which this is a 'participatory universe.'

How, one might ask, could the transition from classical to quantum mechanics have resulted in such a radical shift in the conception of the aim of physics? How could 'the ideal of the detached observer,' the objectivity of the external world, be rejected as mistaken?

Einstein expressed his opinion about the Copenhagen interpretation rather more forcefully in private correspondence than in his published writings. In a letter to Schrödinger dated May 1928 (reprinted in Przibram, 1967, p. 31), he writes:

The Heisenberg–Bohr tranquilizing philosophy – or religion? – is so delicately contrived that, for the time being, it provides a gentle pillow for the true believer from which he cannot very easily be aroused. So let him lie there.

The salient difference between classical and quantum mechanics for the problem of interpretation is that the properties of a classical mechanical system form a Boolean algebra while the properties of a quantum mechanical system form a non-Boolean algebra, not embeddable into a Boolean algebra. What is novel about quantum systems relative to classical systems is that the properties of a quantum system 'fit together' in an essentially non-Boolean way, and it is this difference that is the source of the conceptually puzzling features of quantum mechanics. The rejection of 'the ideal of the detached observer' is the Copenhagen response to non-Booleanity. The following two sections address this issue. I consider contemporary reformulations of the Copenhagen interpretation in chapter 8.

1.2 States and properties in classical mechanics

Classical mechanics, the theory of motion developed by Newton and reformulated in the eighteenth and nineteenth centuries by Euler, Lagrange, Hamilton, and others, is an account of how physical systems move and change. The theory is applied to a model classical mechanical universe, which might be a system of n particles, or a single particle constrained to move on the surface of a sphere, or a rigid body such as a pendulum in a gravitational field of force. The analysis of the model is applicable to some aspect of the motion of a real physical system, such as the orbit of the earth around the sun, or the period of a real pendulum, to the extent that the model incorporates all relevant dynamical features of the system. So to show, on the basis of classical mechanics, that the orbit of the earth around the sun is an ellipse, it suffices to consider a model universe consisting of two point particles representing the sun and the earth, with masses m_s and m_e, moving under the action of a mutually attractive gravitational force proportional to the masses and inversely proportional to the square of their distance apart.

The *state* of a model classical mechanical universe is specified by an assignment of values to a certain set of dynamical variables, the generalized position and momentum variables. In the case of a universe consisting of n particles, an appropriate choice for these variables would be the $3n$ Cartesian position coordinates and corresponding momentum coordinates of the n particles. For a rigid body such as a pendulum, or a particle constrained to move on the surface of a sphere, an appropriate choice would be an angle variable and the corresponding angular momentum. All other dynamical variables are represented as functions of these positions and momenta, and so their values are determined by the state. The equations of motion yield rates of change for positions and momenta in terms of partial derivatives of the Hamiltonian H, a function of positions and momenta, and generate a possible history of the universe from the position and momentum values specified at some particular time. If the potential energy V is independent of velocity, and the transformation equations that define the generalized position coordinates in terms of Cartesian coordinates do not depend explicitly on time, then H is equal to the total energy, $T + V$, where T is the kinetic energy.

The states of subsystems of a classical universe are similarly specified by assignments of values to the generalized positions and momenta of the subsystems. So the subsystems in a classical universe are separable as classical systems with local states that are partial states of the global state of the universe, and we can understand a classical universe as consisting of separable classical systems interacting under the influence of the forces encoded in the Hamiltonian. The evolution of the universe over time – a history of the universe – is given by a particular dynamical evolution of the state.

The state of a classical universe plays two distinct rôles: a *diachronic* rôle in tracking dynamical change over time, and a *synchronic* rôle as the specification of a 'possible world' at a particular time. A 'possible world' is characterized by the selection of a set of determinate values for the dynamical variables as one of the possible sets of values for these variables, or equivalently by the selection of a maximal set of properties for the systems in the universe as one of the possible maximal sets of properties. In its diachronic rôle, a classical state is a *dynamical state*: it tracks dynamical change in the universe through its temporal evolution. In its synchronic rôle, a classical state is a *property state*:[13] it specifies the properties of the universe as one of the possible maximal sets of properties. Think of a property state as a maximal 'list' of properties, a catalogue of all the properties of the universe at a particular time.

For example, a classical particle confined to move along a line always has determinate position and momentum properties, represented by assigning specific ranges of values to the dynamical variables Q and P, denoting position and momentum respectively. It always makes sense to say that the particle is *somewhere* on the line, say between two points indicated by different values of Q, and has a momentum lying between certain values of P. If the position is between 5.2 and 6.8, then it is certainly between 5 and 7, so the particle also has this position property, but not the property associated with the range of values between 2 and 5, and not necessarily the property associated with the range of values between 2 and 6. The particle also has other properties, associated with ranges of values of other dynamical variables, such as the energy E, for example (which might be represented as a function of position and momentum as $E(Q,P) = P^2/2m + V(Q)$, with $P^2/2m$ the kinetic energy and $V(Q)$ the potential energy, where the negative gradient of the potential function represents the force on the particle).

The state of the particle is represented geometrically by a point in the 2-dimensional state space or *phase space* of the particle, with orthogonal Cartesian coordinates defined by Q-values and P-values. Hamilton's equations describe how the state changes in phase space, and hence how the dynamical properties of the particle change over time under the influence of the potential $V(Q)$ (the Hamiltonian H being equal to the energy E in this case):

$$\frac{dQ}{dt} = \frac{\partial H}{\partial P}$$

$$\frac{dP}{dt} = -\frac{\partial H}{\partial Q}$$

A system of n classical particles moving in 3-dimensional real space is represented on a $6n$-dimensional phase space, where the different dimensions represent the positions and momenta of the n particles, with three position coordinates and three corresponding momentum coordinates for each particle. So the state of a classical system is represented by a point in the phase space of the system, dynamical variables are represented by real-valued functions on the phase space, and dynamical properties are represented by subsets of phase space. The property that the energy $E(Q,P)$ of a particle

[13] Van Fraassen (1991) uses the term 'value state,' having in mind the assignment of values to dynamical variables, or truth values to the corresponding propositions.

on a line lies in a certain range, for example, is associated with the subset of the 2-dimensional phase space for which the Q-values and P-values yield an E-value lying in the specified range. If the particle has mass m and performs a simple harmonic motion about the origin with frequency $\omega/2\pi$, then $E = P^2/2m + m\omega^2Q^2/2$ and constant values of E correspond to ellipses in phase space about the origin. The proposition that E lies in a certain range corresponds to the set of points between two ellipses about the origin.

Ideally, every subset of phase space corresponds to a property of the system, but for both formal and physical reasons we would want to exclude certain subsets (such as the subset of phase space for which the value of the energy E lies in the set of rational numbers).[14] So the subsets representing dynamical properties are usually restricted to the Borel subsets of phase space (the subsets generated by the open, closed, or half-open intervals of values under union, intersection, and complement).

Each Borel subset F of phase space X is associated with a characteristic function P_F, defined by:

$$P_F(x) = 1, \text{ if } x \in F$$
$$P_F(x) = 0, \text{ if } x \notin F$$

where x is a variable ranging over the elements (points) of phase space. The characteristic functions are *idempotent* dynamical variables: $P_F{}^2 = P_F$ (where $P_F{}^2(x) = P_F(x) \cdot P_F(x)$).[15] To say that the system has a certain property is to say that the characteristic function P of the corresponding subset takes the value 1, which means that the state of the system is represented by a point lying in the subset. To say that the system lacks the property is to say that P takes the value 0, which means that the state is represented by a point outside this subset.

Consider again the particle confined to move along a line, and the two properties: 'left of zero,' denoted by f, and 'at zero or right of zero,' denoted by g. In the phase space of the particle, these properties are represented by two subsets that partition the space into two mutually exclusive and collectively exhaustive regions. Together with the null property 0, represented by the empty set (which the particle never has), and the universal property 1, represented by the whole phase space (which the particle always has), these two properties generate a *Boolean algebra* $\mathcal{B}_2$ containing four elements,[16] which can be represented in terms of a 'Hasse diagram' as shown in figure 1.1. The lines here indicate partial ordering relations ('$\leq$') that hold between pairs of properties,

[14] Note that if we regard this subset of phase space as corresponding to a property of the system, then we are also committed to the complementary property, corresponding to the energy taking a value in the set of irrational numbers.

[15] The letters P and Q are traditionally used to refer to momentum and position, respectively. I also use P, with or without a subscript, to denote the characteristic function of a Borel subset of phase space, because the characteristic functions are the counterparts of projection operators on Hilbert space, the idempotent elements of the algebra of dynamical variables of quantum mechanics. Characteristic functions and projection operators represent properties or propositions – hence the letter P (for projection operator, or property, or proposition).

[16] The subscript '2' refers to the number of atoms. See below.

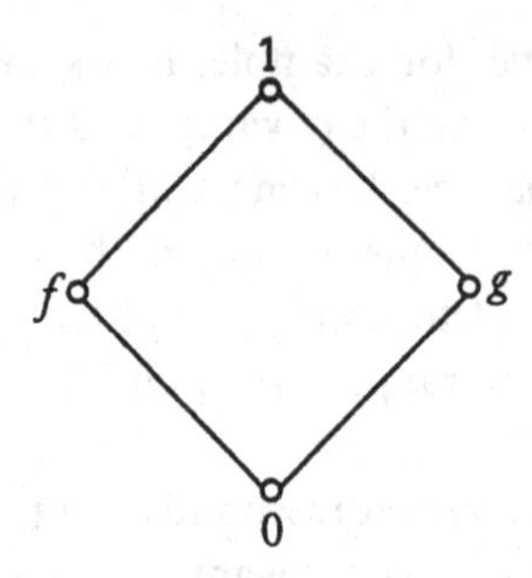

Figure 1.1. Hasse diagram of the 4-element Boolean algebra $\mathscr{B}_2$.

corresponding to the set inclusion relations between their representative subsets (which are trivial in this case). So, for example, $0 \leq f$, which represents the fact that the null set is included in the subset representing the property 'left of zero' or, equivalently, the statement that if the system has the property 0, then it also has the property 'left of zero' (which is vacuously true as a conditional statement, because the system never has the property 0). Similarly, $f \leq 1$. But f is not ordered with respect to g: the subset representing 'left of zero' neither includes the subset representing 'at zero or right of zero,' nor is included in this subset. The elements 0 and 1 of the algebra are referred to as the 'zero' and 'unit' elements of the algebra (or the minimum and maximum elements of the algebra).

If we consider three position properties:

f: to the left of -1
m: in the closed interval between -1 and $+1$
g: to the right of $+1$

then we generate an 8-element Boolean algebra $\mathscr{B}_3$, represented by figure 1.2.

The ordering relations are more complicated now. The element g' denotes the property represented by the subset that is the set-theoretic complement of the subset representing g. It is equivalent to the set-theoretic union of the subsets representing the properties f and m. Similarly, for m' and f', where m' is equivalent to the union of the subsets representing the properties f and g, and f' is equivalent to the union of the subsets representing the properties m and g. On the diagram, the symbol '$\vee$' denotes the *least upper bound* (*lub* or *supremum*) of the related elements. For example, $f \vee m$ denotes the (unique) element above f and m (with respect to the partial ordering) that is also below all other elements that are above f and m. This can be read directly off the diagram. There are several elements above f and m. For example, g', m', and 1 are all upper bounds of f, and g', f', and 1 are all upper bounds of m. (A partial ordering is transitive, so that if $f \leq g'$ and $g' \leq 1$, then $f \leq 1$.) There are two elements above both f and m: g' and 1. But only g' is the *least* upper bound of f and m, because $g' \leq 1$.

Just as $g' = f \vee m$, the least upper bound of f and m, so m can be expressed as the greatest lower bound of g' and f', $m = g' \wedge f'$, where the symbol '$\wedge$' denotes the *greatest*

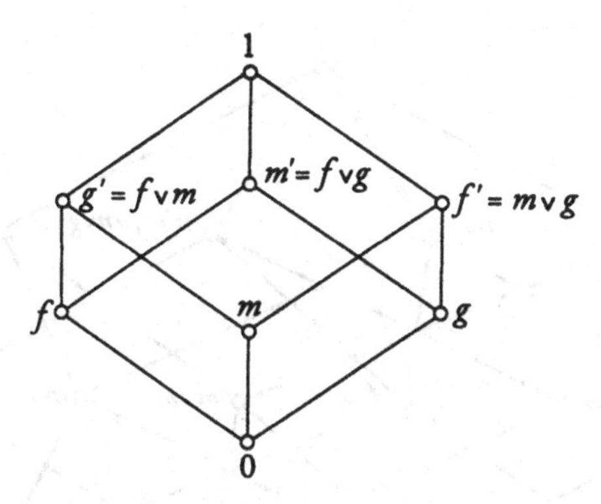

Figure 1.2. Hasse diagram of the 8-element Boolean algebra $\mathcal{B}_3$.

lower bound (*glb* or *infimum*) of the related elements: $g' \wedge f$ is the (unique) element below g' and m' (with respect to the partial ordering) that is also above all other elements that are below g' and f. Similarly, $f = g' \wedge m'$ and $g = m' \wedge f$. Again, these relations can be read directly off the diagram. Set-theoretically, $f = g' \wedge m'$ corresponds to the fact that the subset representing the property f is the set-theoretic intersection of the complement of the set representing the property g, and the complement of the set representing the property m.

Partitioning the phase space into four mutually exclusive and collectively exhaustive subsets, corresponding to four position properties f, m, n, g for the particle on a line, yields the 16-element Boolean algebra $\mathcal{B}_4$ shown in figure 1.3.

Notice that g', the property represented by the set-theoretic complement of the set representing the property g, is also the property represented by the set-theoretic union of the sets representing the properties f, m, and n. On the diagram, this is indicated by the element $f \vee m \vee n$, which can be read either as $(f \vee m) \vee n$, or as $f \vee (m \vee n)$, the least upper bound of f, m, and n. In other words, the operation of forming the least upper bound is associative: $(f \vee m) \vee n = f \vee (m \vee n)$, for all f, m, n. Similarly, the operation of forming the greatest lower bound is associative.

The *complement*, x', of an element x in a Boolean algebra (the algebraic counterpart of set-theoretic complementation) is defined by the conditions $x \wedge x' = 0$ and $x \vee x' = 1$. So again, the complement of f, say, can be read directly off the diagram. We want the element, f', such that the greatest lower bound of f and f' is the minimum element in the algebra, and the least upper bound of f and f' is the maximum element in the algebra. This is the element $m \vee n \vee g$ in $\mathcal{B}_4$, $m \vee g$ in $\mathcal{B}_3$, and g in $\mathcal{B}_2$.

A classical state of the particle, in its synchronic rôle as a property state, selects a subset of the elements in a Boolean algebra. In the case of $\mathcal{B}_2$, this is either the set $\{f, 1\}$ or the set $\{g, 1\}$. That is, there are two possible classical states: the state in which f is true, and the state in which g is true (1 is always true). For the Boolean algebra $\mathcal{B}_3$, a classical state selects one of the three sets $\{f, f \vee m, f \vee g, 1\}$, $\{m, f \vee m, m \vee g, 1\}$, or $\{g, f \vee g, m \vee g, 1\}$. And for $\mathcal{B}_4$, the state selects one of the four sets $\{f, f \vee m, f \vee n, f \vee g, f \vee m \vee n, f \vee m \vee g,$

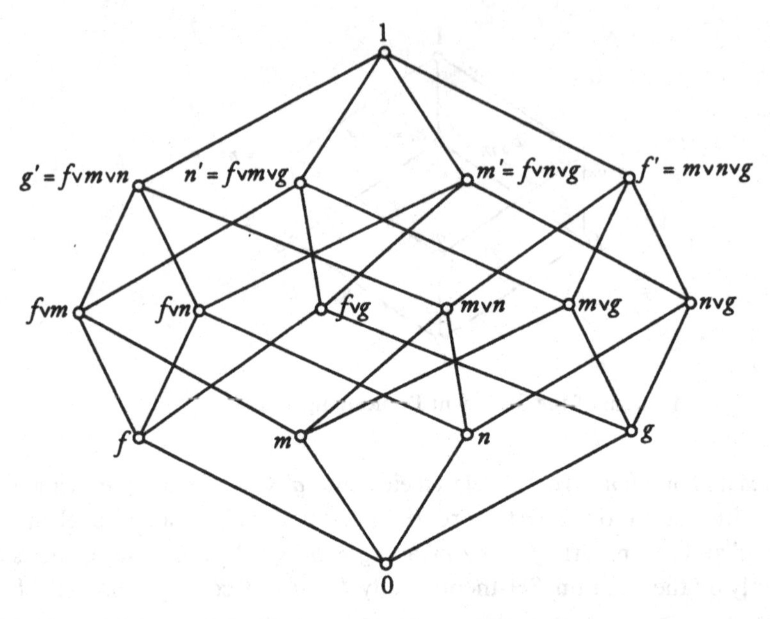

Figure 1.3. Hasse diagram of the 16-element Boolean algebra $\mathscr{B}_4$.

$f \vee n \vee g$, 1}, {$m, f \vee m, m \vee n, m \vee g, f \vee m \vee n, f \vee m \vee g, m \vee n \vee g$, 1}, {$n, f \vee n, m \vee n, n \vee g, f \vee m \vee n, f \vee n \vee g, m \vee n \vee g$, 1}, or {$g, f \vee g, m \vee g, n \vee g, f \vee m \vee g, f \vee n \vee g, m \vee n \vee g$, 1}. This is shown in figure 1.4. Notice that, in each case, the state selects a set of elements that is above one of the atoms f, m, n, or g in the algebra, where an *atom* is a minimal element above 0: an element e such that there is no element between 0 and e (with respect to the ordering). Such a set is a maximal set of properties that can co-obtain for the system (in this restricted set of properties), and the set of states includes all such maximal sets of co-obtaining properties or 'possible worlds.' Equivalently, the set of states includes all the maximal sets of propositions that can be true of the system simultaneously.

If instead of considering only two, three, or four position properties, we considered n position properties, for some large n, we would have a Boolean algebra with n atoms and 2^n elements. And if we included momentum, energy, and other properties, the Boolean algebra would be larger still. But structurally these large Boolean algebras of properties would be similar to the elementary cases considered above, as would Boolean algebras with an infinite number of atoms (although the infinite case, not surprisingly, introduces new complexities).

Formally, a Boolean algebra (or Boolean lattice)[17] is a complemented distributive lattice, where a *lattice* is a partially ordered set, in which a supremum and infimum exists for every pair of elements.

[17] I use these terms interchangeably.

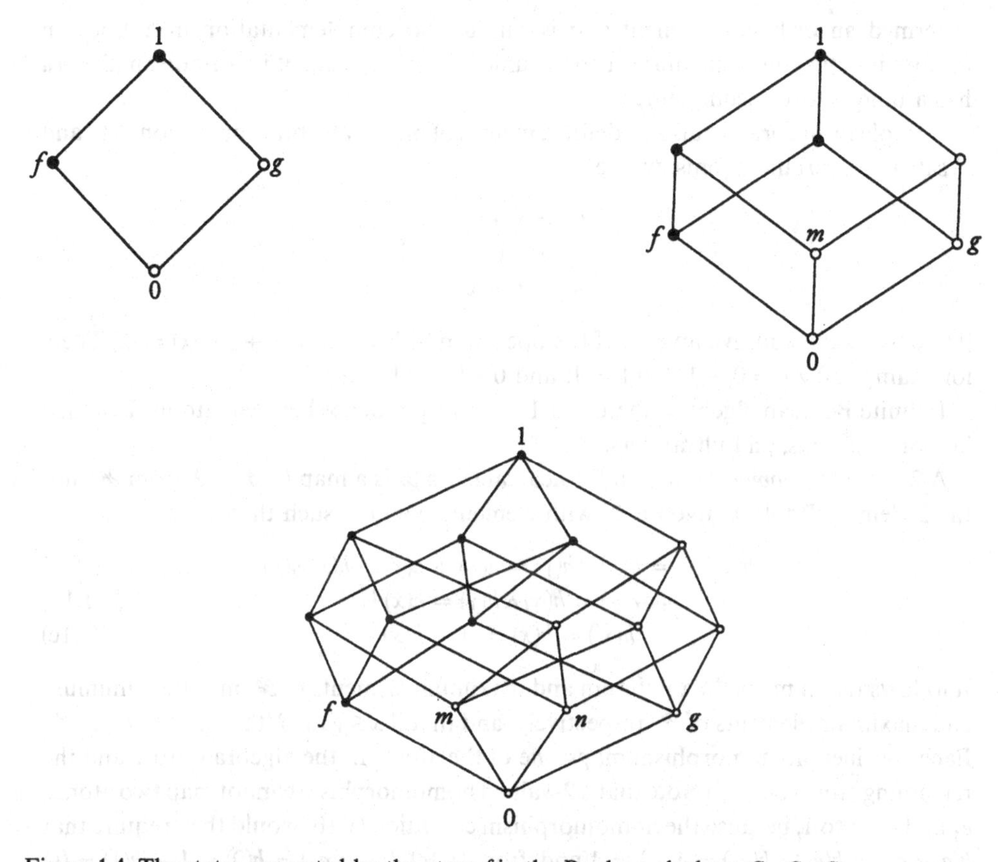

Figure 1.4. The states generated by the atom f in the Boolean algebras $\mathscr{B}_2$, $\mathscr{B}_3$, $\mathscr{B}_4$.

A lattice is distributive if and only if every triple of elements satisfies the distributive law:

$$x \wedge (y \vee z) = (x \wedge y) \vee (x \wedge z)$$
$$x \vee (y \wedge z) = (x \vee y) \wedge (x \vee z)$$

A lattice is complemented if it has a maximum element, 1, and a minimum element, 0, and each element has a complement (that is, for each element x, there exists at least one element c such that $x \wedge c = 0$ and $x \vee c = 1$). The complement of a lattice element need not be unique. In a distributive lattice, complementation is unique, and the unique complement x' of an element x satisfies two further conditions: $(x')' = x$, and $y' \leq x'$ if $x \leq y$. An element x' satisfying the four conditions:

(i) $x \wedge x' = 0$
(ii) $x \vee x' = 1$
(iii) $(x')' = x$
(iv) $y' \leq x'$ if $x \leq y$

is termed an orthocomplement and is unique. So complementation in a Boolean algebra is orthocomplementation and is unique – every element in a Boolean algebra has a unique (ortho)complement.

A Boolean algebra can also be defined in term of binary algebraic operations '+' and '·' satisfying certain axioms, where:

$$\begin{aligned} x \vee y &= x + y - x \cdot y \\ x \wedge y &= x \cdot y \\ x' &= 1 - x \end{aligned}$$

(Here '−' is the additive inverse of the operation '+': $x - x = x + (-x) = 0$.) Then, for example, $0 \vee 1 = 0 + 1 - 0 \cdot 1 = 1$, and $0 \wedge 1 = 0 \cdot 1 = 0$.

In finite Boolean algebras, there is a 1–1 correspondence between atoms, 2-valued homomorphisms, and ultrafilters.

A *2-valued homomorphism* on a Boolean algebra $\mathscr{B}$ is a map $h: \mathscr{B} \to \mathscr{Z}$ from $\mathscr{B}$ onto the 2-element Boolean algebra $\mathscr{Z}$ with elements 0 and 1, such that

$$h(x \vee y) = h(x) \vee h(y) = h(x) + h(y) - h(x) \cdot h(y) \quad (1.1a)$$
$$h(x \wedge y) = h(x) \wedge h(y) = h(x) \cdot h(y) \quad (1.1b)$$
$$h(x') = h(x)' = 1 - h(x) \quad (1.1c)$$

It follows that h maps the minimum and maximum elements of $\mathscr{B}$ onto the minimum and maximum elements of $\mathscr{Z}$, respectively, and that if $x \le y$ in $\mathscr{B}$, then $h(x) \le h(y)$ in $\mathscr{Z}$. Each 2-valued homomorphism maps one of the atoms in the algebra onto 1 and the remaining atoms onto 0. (Note that a 2-valued homomorphism cannot map two atoms, e_1 and e_2 onto 1, because the homomorphism condition (1.1b) would then require that $h(e_1 \wedge e_2) = h(e_1) \wedge h(e_2) = 1 \wedge 1 = 1$ and, from (1.1c), $h(e_1 \wedge e_2) = h(0) = 1 - h(1) = 0$.)

A 2-valued homomorphism on a Boolean algebra is a classical truth-value assignment – with 0 representing 'false' and 1 representing 'true' – to the elements of the algebra regarded as propositions (more precisely, as equivalence classes of logically equivalent propositions). So, for example, $h(x \vee y) = 1$ if and only if $x = 1$ or $y = 1$, and $h(x \wedge y) = 1$ if and only if $x = 1$ and $y = 1$.

The equivalence classes of logically equivalent propositions of a logic form what is called the *Lindenbaum algebra*[18] of the logic, under the partial ordering defined by implication. That is, the set of propositions logically equivalent to a proposition x is taken as 'less than or equal to ($\le$)' the set of propositions logically equivalent to a proposition y, if x implies y in the sense that the conditional $x \to y$ ('if x then y') is a theorem of logic. If the logic is sound and complete, this coincides with the relation of semantic entailment between x and y. To say that x *semantically entails* y is to say that y is true in all cases in which x is true. It can easily be shown that the relation '$\le$' defined in this way between equivalence classes of logically equivalent propositions is a partial ordering relation. The Lindenbaum algebra of classical logic is a Boolean algebra, and

[18] See Bell and Slomsen (1969, p. 42).

the partial ordering corresponds to the partial ordering by set inclusion of the subsets representing the propositions. Loosely, if we identify logically equivalent propositions, the propositions of classical logic form a Boolean algebra. It is easy to see that '$\vee$' corresponds to disjunction, '$\wedge$' corresponds to conjunction, and the complement corresponds to negation. For example, the proposition x implies the disjunction $x \vee y$ (if x is true, then $x \vee y$ must be true as well), and y implies $x \vee y$, so the disjunction $x \vee y$ is an upper bound of both x and y. And if x and y each imply a proposition z, then the disjunction $x \vee y$ implies z, so the disjunction $x \vee y$ is the least upper bound or supremum of x and y.

A *filter* in a lattice $\mathscr{L}$ is a non-empty subset $\mathscr{f}$ of $\mathscr{L}$ satisfying the conditions:

(i) if $x \in \mathscr{f}$ and $y \in \mathscr{f}$, then $x \wedge y \in \mathscr{f}$
(ii) if $x \in \mathscr{f}$ and $x \leq y$, then $y \in \mathscr{f}$

Filters can be partially ordered by set-inclusion. An *ultrafilter* in a Boolean algebra is a maximal filter with respect to this ordering. A filter $\mathscr{f}$ in a Boolean algebra $\mathscr{B}$ is an ultrafilter if and only if, for each element $x \in \mathscr{B}$, either $x \in \mathscr{f}$ or $x' \in \mathscr{f}$. So a filter in a Boolean algebra corresponds to a (not necessarily maximal) set of true propositions, containing all conjunctions of propositions in the set, and all propositions implied by any proposition in the set. An ultrafilter corresponds to a 'possible world,' a maximal such set of true propositions.

In a finite Boolean algebra $\mathscr{B}$, each ultrafilter is generated by an atom in $\mathscr{B}$, and the elements in the ultrafilter will be just those elements mapped onto 1 by the 2-valued homomorphism corresponding to the atom. In an infinite Boolean algebra, there can also be ultrafilters that are not generated by atoms.

Every nonzero element in a Boolean algebra is contained in some ultrafilter. (This is the algebraic counterpart of the completeness theorem for classical sentential logic, that a sentence is satisfiable, in the sense that it can be assigned a truth value 'true' in a 'possible world', if and only if it is consistent, where consistency here means that a contradiction cannot be derived from the sentence in the proof theory of the logic.) Every Boolean algebra $\mathscr{B}$ is isomorphic to a *field* $\mathscr{F}$ of subsets of the set X of ultrafilters in $\mathscr{B}$ (a set of subsets of X that contains X and is closed under unions, intersections, and complements). This is Stone's (1936) representation theorem. The field $\mathscr{F}$ is called the 'Stone space' of the Boolean algebra $\mathscr{B}$.

In a classical mechanical universe, the algebra of dynamical variables is a commutative algebra: the commutative algebra of real-valued functions on phase space (where the product of two functions at a point in phase space is defined as the product of their values at the point). The algebra of properties is a subalgebra of this algebra, the algebra of idempotent dynamical variables or characteristic functions corresponding to the Borel subsets of phase space. Each idempotent dynamical variable P can take one of two possible values, 0 and 1, and these can be associated with the truth values 'false' and 'true,' respectively, for the proposition asserting that the associated property obtains (that the value a of some dynamical variable A lies in a certain range of values E: '$a \in E$').

For example, a dynamical variable A, with k distinct possible values $a_1, a_2, \ldots, a_k$, can be represented as a sum of idempotent dynamical variables or properties $P_{a_1}, P_{a_2}, \ldots, P_{a_k}$:

$$A = \sum_{i=1}^{k} a_i P_{a_i}, \tag{1.2}$$

where P_{a_i} maps every point in the Borel subset of phase space associated with the value a_i onto 1, and every point outside this subset onto 0. The subsets corresponding to the characteristic functions P_{a_i}, $i = 1, \ldots, k$, partition the phase space into k non-overlapping regions. So if x is a point in phase space – an n-tuple of position and momentum values – belonging to the jth region, then $A(x) = \Sigma_{i=1}^{k} a_i P_{a_i}(x) = a_j$ (because $P_{a_i}(x) = 1$ and $P_{a_i}(x) = 0$ *if* $i \neq j$). Each range of values of A is associated with an idempotent dynamical variable. The range of values in the set $E = \{a_i: i \in I\}$ is associated with the idempotent variable $\Sigma_{i \in I} P_{a_i}$, the characteristic function that takes the value 1 on the union of the phase space subsets associated with the P_{a_i}, for i in some index set I, and 0 elsewhere. The A-properties form a Boolean algebra of 2^k elements generated from the k atoms associated with the k possible values of A. If the possible values of A form a continuum, then A can be represented as an integral:

$$A = \int_{-\infty}^{\infty} a dP_A(a), \tag{1.3}$$

where each continuous range of values E is associated with a characteristic function $P_A(E)$ on phase space, and $P_A(a)$ denotes the characteristic function $P_A((-\infty, a])$.

The algebra of properties or propositions of a classical system with dynamical variables $A, B, C, \ldots$ forms a Boolean algebra, isomorphic to the σ-field[19] of Borel subsets of phase space. Classical states correspond to atoms in the Boolean algebra of properties: minimal nonzero elements of the algebra representing points – singleton subsets – of phase space. A classical state defines a 2-valued homomorphism on the Boolean algebra of properties that assigns to each idempotent dynamical variable one of its two possible values, 0 or 1, which can be identified with the minimum and maximum elements of a 2-element Boolean algebra. So a classical state selects an ultrafilter of properties (the properties mapped onto 1) as the collection of properties belonging to the state in the synchronic sense, the properties represented by subsets of phase space containing the state as a phase point.

1.3 States, properties, and probabilities in quantum mechanics

Quantum mechanics is derived as a generalization of classical mechanics in which commutation relations are imposed on the dynamical variables. The Poisson bracket,

[19] A σ-field is closed under countable unions (and hence under countable complements).

$\{U,V\}$, of two classical dynamical variables, U and V, is defined as:

$$\{U,V\} = \sum_i \left(\frac{\partial U}{\partial Q_i} \frac{\partial V}{\partial P_i} - \frac{\partial U}{\partial P_i} \frac{\partial V}{\partial Q_i} \right)$$

and it is required that:

$$[U, V] = i\hbar\{U,V\},$$

where $[U, V]$ denotes the commutator $UV - VU$, $i = \sqrt{-1}$, and $\hbar = \frac{h}{2\pi}$, where h is Planck's constant. So, for example, for the position Q and momentum P of a particle moving along a line, $QP - PQ = i\hbar$. For the Cartesian components of angular momentum, J_x, J_y, and J_z, $[J_x, J_y] = J_z$ (and cyclic permutations). We obtain a noncommutative algebra of dynamical variables, with an associated non-Boolean algebra of properties or idempotent dynamical variables, that is representable as an operator algebra on a Hilbert space, a linear space over the complex numbers.[20] This is the import of Schrödinger's proof of the equivalence of his wave mechanics formulation of quantum theory and Heisenberg's matrix mechanics formulation of the theory, and von Neumann's Hilbert space representation of matrix and wave mechanics as a general theory of quantum mechanics.[21]

A finite-dimensional Hilbert space is just a finite-dimensional Euclidean space over the complex numbers. Vectors in the space can be multiplied by complex numbers and added or 'superposed' to form new vectors in the space, and the scalar or 'inner' product of two vectors is a complex number. Just as a classical dynamical variable with k distinct possible values can be represented as a sum of idempotent dynamical variables or characteristic functions on the state space of classical mechanics (where the characteristic functions are associated with disjoint subsets of classical states that exhaust the space), so a quantum dynamical variable A with distinct possible values a_1, $a_2, \ldots, a_k$ (the 'eigenvalues' of A) can be represented uniquely as a sum of projection operators:

$$A = \sum_{i=1}^{k} a_i P_{a_i} \tag{1.4}$$

on an n-dimensional Hilbert space, where the P_{a_i} here are projection operators onto $k \leq n$ orthogonal subspaces that span the space (the 'eigenspaces' of A). Compare with equation (1.2) in the classical case. The general form of this representation in an infinite-dimensional Hilbert space for dynamical variables with a continuum of values involves an integral instead of a sum

$$A = \int_{-\infty}^{\infty} a dP_A(a), \tag{1.5}$$

[20] For an account of Hilbert space, see the appendix, section A.1.

[21] See von Neumann (1955, pp. 17–33).

where $P_A(a)$ is the projection operator $P_A((-\infty,a])$ onto the subspace of Hilbert space associated with the range $(-\infty,a]$ of A. The existence of such a representation in quantum mechanics is guaranteed by the spectral representation theorem. Compare with equation (1.3) in the classical case.[22]

The projection operators represent idempotent dynamical variables, with two possible values, 0 and 1. That is, they represent properties or propositions in quantum mechanics. In the noncommutative algebra of dynamical variables of a quantum mechanical system, the subalgebra of idempotent dynamical variables is isomorphic to the lattice of subspaces of Hilbert space, a non-Boolean lattice. In the commutative algebra of dynamical variables of a classical mechanical system, the subalgebra of idempotent dynamical variables is isomorphic to the lattice of subspaces of phase space, a Boolean lattice.

In quantum mechanics, the proposition asserting that the value of a dynamical variable lies in a certain range is represented by a *subspace* of Hilbert space, or the corresponding projection operator onto the subspace. In classical mechanics, the proposition asserting that the value of a dynamical variable lies in a certain range is represented by a *subset* of phase space, or the corresponding characteristic function. The points in the subset represent classical states for which the proposition is true. Whether we can make an analogous statement for quantum propositions represented by subspaces of Hilbert space depends on how we understand the notion of state in the synchronic sense in quantum mechanics, the notion of a quantum property state.

Before considering this question, note that the algebra of dynamical variables of a mechanical system can be generated from the subalgebra of idempotent dynamical variables. This is obvious in classical mechanics, where every dynamical variable (real-valued function on phase space) can be represented as a sum or integral of idempotent dynamical variables represented by characteristic functions on a phase space. In quantum mechanics, it follows from the possibility of representing each dynamical variable as a sum or integral of projection operators on a Hilbert space. So the transition from classical to quantum mechanics can be characterized as the transition from a Boolean property structure, represented by the subsets of a set (the subsets of phase space), to a non-Boolean property structure, represented by the subspaces of a linear space (the subspaces of Hilbert space).

For example, the property structure of a classical system characterized by two distinct dynamical variables, A with two possible values a_1, a_2, and B with two possible values b_1, b_2, is represented by a 16-element Boolean algebra or Boolean lattice generated by four atoms representing the properties associated with the four possible pairs of values of A and B: $a_1b_1, a_1b_2, a_2b_1, a_2b_2$. (The value a_1 for A might represent, say, a particle being in a particular region of the universe – perhaps somewhere in a laboratory apparatus – and the value a_2 for A would then represent the particle being somewhere in the rest of the universe. Similarly, the values b_1, b_2 for B might

[22] For details, see the appendix, section A.1.

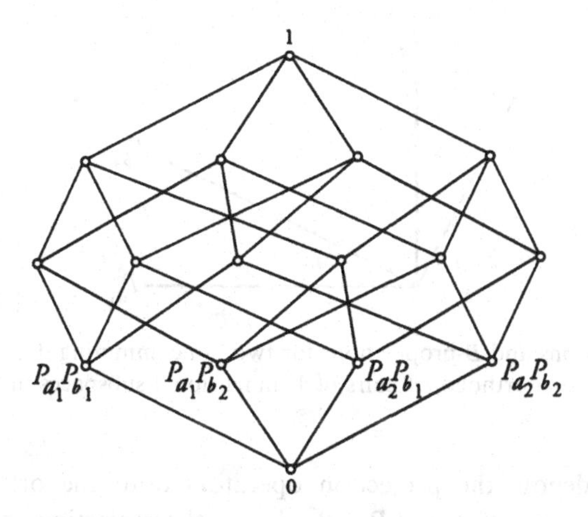

Figure 1.5. Boolean property structure of a classical system generated by the atoms associated with the possible values a_1, a_2 and b_1, b_2 of two distinct dynamical variables, A and B.

correspond to two complementary ranges of values of momentum. Or a_1, a_2 and b_1, b_2 might correspond to complementary ranges of values of components of angular momentum in different directions.)

If A and B are represented in terms of idempotent dynamical variables or properties P_{a_1}, P_{a_2} and P_{b_1}, P_{b_2}:

$$\left.\begin{aligned} A &= a_1 P_{a_1} + a_2 P_{a_2} \\ B &= b_1 P_{b_1} + b_2 P_{b_2} \end{aligned}\right\} \tag{1.6}$$

the four atoms can be represented as $P_{a_1}P_{b_1}, P_{a_1}P_{b_2}, P_{a_2}P_{b_1}, P_{a_2}P_{b_2}$.[23] The phase space is a set of four points in this case, corresponding to the four possible pairs of values of A and B, or the four properties $P_{a_1}P_{b_1}, P_{a_1}P_{b_2}, P_{a_2}P_{b_1}, P_{a_2}P_{b_2}$. This is shown in figure 1.5.

For the analogous quantum system, where A and B do not commute, A might denote spin in some direction **a**, and B spin in a different direction **b**, for a spin-$\frac{1}{2}$ system, where each spin has two possible values.[24] The two possible A-propositions, P_{a_1} and P_{a_2}, would then be represented by two orthogonal 1-dimensional subspaces in a 2-dimensional Hilbert space, $\mathscr{H}_2$ (or the corresponding projection operators onto these subspaces), and the two possible B-propositions, P_{b_1} and P_{b_2}, by a different pair of orthogonal 1- dimensional subspaces (or projection operators) in this space, as shown in figure 1.6.

Just as in classical mechanics, A and B can be represented in terms of idempotent dynamical variables or properties P_{a_1}, P_{a_2} and P_{b_1}, P_{b_2} via equations (1.6), where in this

[23] The symbol $P_{a_i}P_{b_j}$ here represents the product of the characteristic functions representing the properties. Alternatively, we could write $P_{a_i} \wedge P_{b_j}$ for the conjunctive properties, or simply $a_i \wedge b_j$.

[24] For a discussion of spin, see the appendix, section A.4.

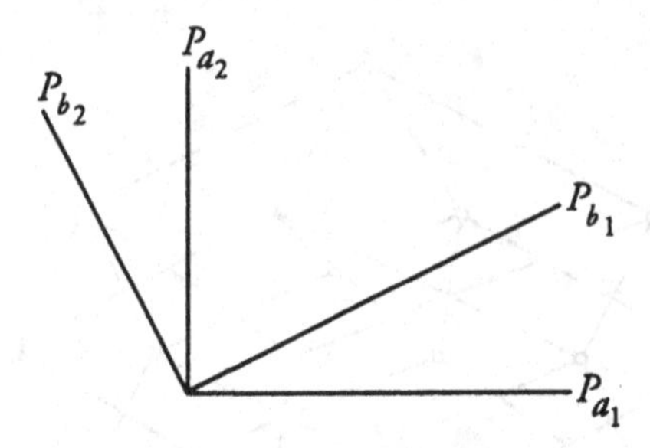

Figure 1.6. *A*-propositions and *B*-propositions for two noncommuting dynamical variables *A* and *B*, represented by two orthogonal pairs of 1-dimensional subspaces in $\mathscr{H}_2$.

case P_{a_1} and P_{a_2} denote the projection operators onto the orthogonal pair of 1-dimensional eigenspaces of *A*, and P_{b_1}, P_{b_2} denote the projection operators onto the orthogonal pair of 1-dimensional eigenspaces of *B*. Since the projection operators P_{a_i} do not commute with the projection operators P_{b_j}, for any i, j, the operators *A* and *B* do not commute. The property structure of this quantum system is represented by a 6-element non-Boolean lattice with four atoms P_{a_1}, P_{a_2} and P_{b_1}, P_{b_2}, shown in figure 1.7.

If we consider the set of spins in all directions, the lattice of properties is isomorphic to the lattice of all subspaces of $\mathscr{H}_2$ (figure 1.8).

If the dynamical variables each have three possible values, a_1,a_2,a_3 and b_1,b_2,b_3:

$$\left.\begin{aligned} A &= a_1 P_{a_1} + a_2 P_{a_2} + a_3 P_{a_3} \\ B &= b_1 P_{b_1} + b_2 P_{b_2} + b_3 P_{b_3} \end{aligned}\right\} \tag{1.7}$$

the property structure in the classical case is represented by a 512-element Boolean algebra, generated by the nine atoms representing the properties $P_{a_i}P_{b_j}$ associated with the nine possible pairs of values $a_i b_j$ $(i = 1, 2, 3; j = 1, 2, 3)$ of *A* and *B*. If *A* and *B* do not commute, the property structure of the analogous quantum system is represented by a 23-element non-Boolean lattice with six atoms, representing the properties P_{a_i}, P_{b_j} associated with the three possible *A*-values, a_1, a_2, a_3, and the three possible *B*-values, b_1, b_2, b_3 (figure 1.9).

For example, *A* and *B* might represent the spins in two different directions for a spin-1 system, where each spin has three possible values. The six atoms represent two orthogonal triples of 1-dimensional subspaces or rays, with no rays in common. The fifteen elements above the atoms represent the planes spanned by these rays.

If *A* and *B* are related dynamical variables that can be represented in terms of idempotent variables P_{a_1}, P_{a_2}, P_{a_3} and P_{b_1}, P_{b_2}, P_{b_3}, with $P_{a_1} = P_{b_1} = P$, as:

$$\begin{aligned} A &= a_1 P + a_2 P_{a_2} + a_3 P_{a_3} \\ B &= b_1 P + b_2 P_{b_2} + b_3 P_{b_3} \end{aligned}$$

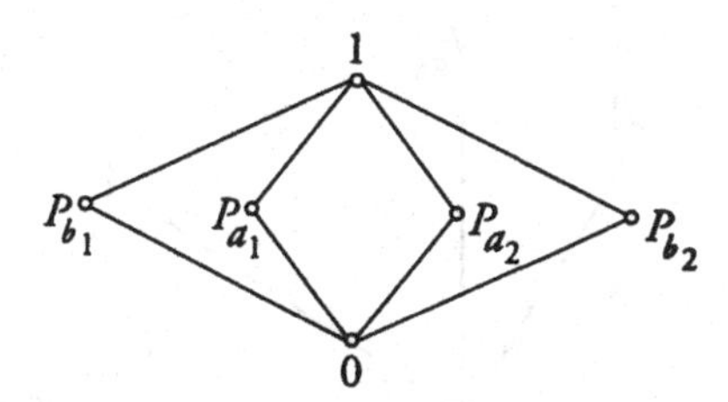

Figure 1.7. Non-Boolean property structure generated by atoms associated with the propositions of Figure 1.6.

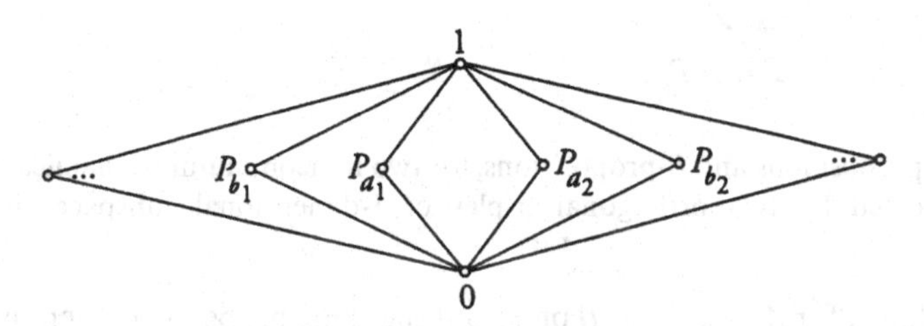

Figure 1.8. Non-Boolean property structure of a spin-$\frac{1}{2}$ quantum system.

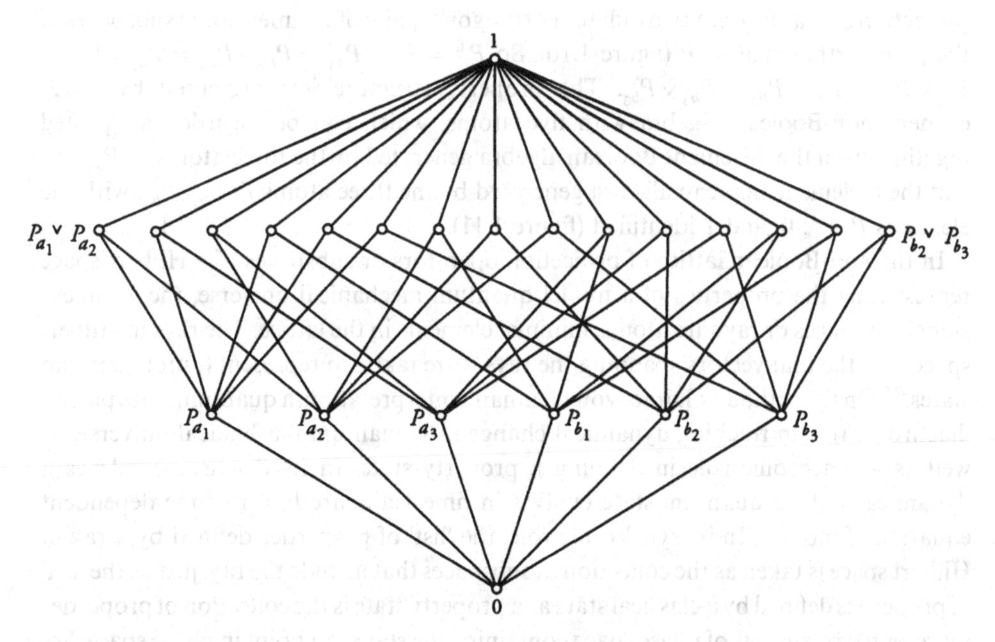

Figure 1.9. Non-Boolean property structure of a quantum system generated by the atoms associated with the possible values a_1, a_2, a_3 and b_1, b_2, b_3 of two noncommuting dynamical variables, A and B.

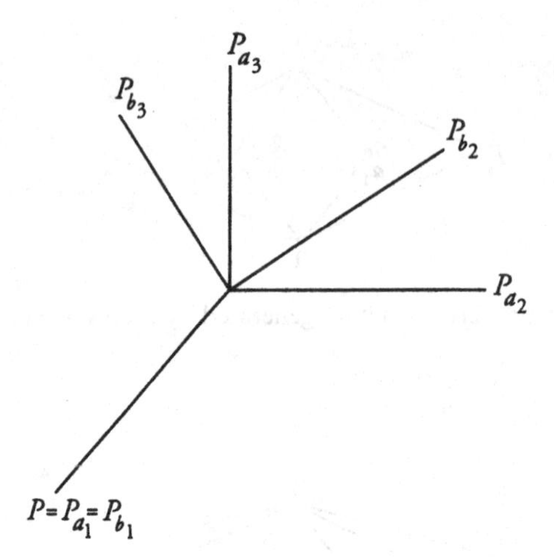

Figure 1.10. A-propositions and B-propositions, for two noncommuting dynamical variables A and B, represented by two orthogonal triples of 1-dimensional subspaces in $\mathscr{H}_3$, with $P_{a_1} = P_{b_1} = P$.

so that the value a_1 for A and b_1 for B picks out the same property, P, then the property structure in the classical case is represented by a 32-element Boolean algebra generated by the five atoms representing the properties $P, P_{a_2}P_{b_2}, P_{a_2}P_{b_3}, P_{a_3}P_{b_2}, P_{a_3}P_{b_3}$ associated with the five distinct values of A and B. In the quantum case, P_{a_2}, P_{a_3} and P_{b_2}, P_{b_3} are projection operators onto two different orthogonal pairs of 1-dimensional subspaces in the plane orthogonal to P (figure 1.10). So $P^{\perp} = P_{a_2} \vee P_{a_3} = P_{b_2} \vee P_{b_3} = P_{a_2} \vee P_{b_2} = P_{a_2} \vee P_{b_3} = P_{a_3} \vee P_{b_2} = P_{a_3} \vee P_{b_3}$. The property structure is represented by a 12-element non-Boolean algebra with five atoms, which can be regarded as 'pasted together' from the 8-element Boolean algebra generated by the three atoms P, P_{a_2}, P_{a_3} and the 8-element Boolean algebra generated by the three atoms P, P_{b_2}, P_{b_3}, with the elements P, $P^{\perp}$, 0, and 1 identified (figure 1.11).

In the non-Boolean lattice of projection operators or subspaces of a Hilbert space representing the properties of a model quantum mechanical universe, the 1-dimensional subspaces or rays are atoms, minimal elements in the lattice. The rays in Hilbert space – or the unit vectors spanning the rays – are taken to represent (pure) quantum states.[25] On the orthodox Dirac–von Neumann interpretation, a quantum state plays a diachronic rôle in tracking dynamical change in a quantum mechanical universe, as well as a synchronic rôle in defining a property state. In its diachronic rôle as a dynamical state, a quantum state evolves in time via Schrödinger's time-dependent equation of motion. In its synchronic rôle, the 'list' of properties defined by a ray in Hilbert space is taken as the collection of subspaces that include the ray, just as the 'list' of properties defined by a classical state as a property state is the collection of properties represented by subsets of phase space containing the state as a point in phase space. So,

[25] For a discussion of states in quantum mechanics, and the distinction between 'pure' and 'mixed' states, see the appendix, section A.2.

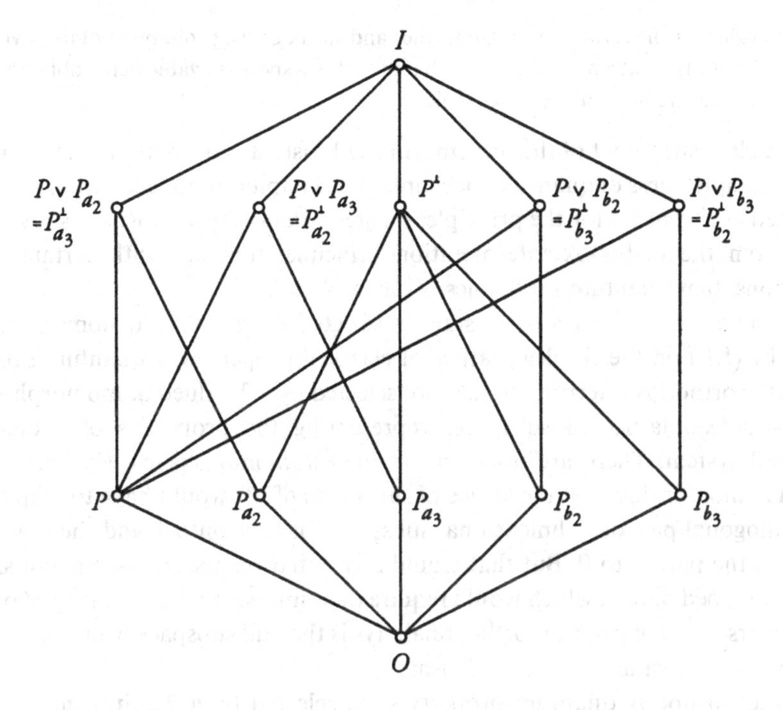

Figure 1.11. Non-Boolean property structure generated by the atoms associated with the propositions of Figure 1.10.

a subspace of Hilbert space representing a quantum proposition includes all the rays representing quantum states for which the proposition is true, just as a subset of phase space representing a classical proposition contains all the points representing classical states for which the proposition is true. In terms of dynamical variables, the quantum state defined by a ray in Hilbert space selects the set of determinate dynamical variables – the set of dynamical variables that have determinate values – as those dynamical variables for which the state is an eigenstate (that is, for which the state lies in one of the eigenspaces of the dynamical variable). The actual value of the dynamical variable is the value corresponding to the label of the eigenspace containing the state.

This orthodox interpretation principle (the 'eigenvalue–eigenstate link') is explicitly proposed by both von Neumann (1955, p. 253)[26] and Dirac. In Dirac's formulation (1958, pp. 46–7):

> The expression that an observable 'has a particular value' for a particular state is permissible in quantum mechanics in the special case when a measurement of the observable is certain to lead to the particular value, so that the state is an eigenstate of the observable.... In the general case we cannot speak of an observable having a value for a particular state, but we can speak of its having

[26] Von Neumann formulates this principle for properties represented by projection operators (see principle β on p. 253), but this is of course equivalent to the formulation in terms of general observables.

an average value for the state. We can go further and speak of the probability of its having any specified value for the state, meaning the probability of this specified value being obtained when one makes a measurement of the observable.

There is a clear statement of the principle in the Einstein–Podolsky–Rosen argument for the incompleteness of quantum mechanics (see chapter 2). Indeed, the argument is formulated as a *reductio* for the principle: Einstein, Podolsky, and Rosen show that it follows from the orthodox interpretation principle, together with certain realist assumptions, that quantum mechanics is incomplete.

Now, while a classical property state is selected by a 2-valued homomorphism (defined by (1.1)) on the Boolean lattice of classical properties, a quantum property state, on the orthodox interpretation, is not selected by a 2-valued homomorphism on the non-Boolean lattice of subspaces representing the properties of a quantum mechanical system. There are no such 2-valued homomorphisms. For example, a 2-valued homomorphism on the lattice of subspaces of $\mathscr{H}_2$ would have to map one of every orthogonal pair of 1-dimensional subspaces or rays onto 1 and the remaining member of the pair onto 0. But that would mean that distinct non-orthogonal rays would be mapped onto 1, which would require their intersection to be mapped onto 1. But the intersection of two non-orthogonal rays is the null subspace, which is mapped onto 0 by every 2-valued homomorphism.

Nor is an orthodox quantum property state selected by a 2-valued map on the non-Boolean lattice that reduces to a 2-valued homomorphism on each Boolean sublattice of properties (where each Boolean sublattice is generated by a family of properties corresponding to the assignment of different ranges of values E to a dynamical variable A, for all Borel subsets of values E). I shall refer to such maps as 'Boolean homomorphisms.' There are no Boolean homomorphisms on the lattice of subspaces of a Hilbert space, except in the case of a 2-dimensional Hilbert space, by the Kochen and Specker theorem (see chapter 3), and even in this case the collection of subspaces containing a ray – the property state according to the orthodox interpretation – does not correspond to a collection of subspaces selected by a Boolean homomorphism.[27]

The orthodox decision to take the ray representing a quantum state as a property state in the above sense has the consequence that we can no longer apply quantum mechanics to a quantum mechanical universe in the same way that classical mechanics is applied to a classical universe. According to the orthodox interpretation, at any particular time t the ray representing the quantum state specifies the property state as the collection of properties represented by the subspaces containing the ray. We can understand the properties in this collection as obtaining in the universe at time t – as actual properties of the universe at time t – but since the property state does not correspond to a Boolean homomorphism on the lattice of properties, we can't

[27] The orthodox property state in the case of a 2-dimensional Hilbert space $\mathscr{H}_2$ is the ultrafilter generated by the ray representing the state and contains the property represented by the ray and the universal property represented by $\mathscr{H}_2$. A Boolean homomorphism on the lattice of subspaces selects one ray in each orthogonal pair of rays in $\mathscr{H}_2$.

understand the properties that are not in this collection as not obtaining. That is, we can take the propositions corresponding to the properties in the property state as true, but we can't take the propositions that correspond to all the properties that are not in the property state as false, for this will involve a contradiction. There will be infinitely many sets of propositions of the form 'the value of A lies in the range E,' for all Borel sets E, that would *all* have to be taken as false, because no proposition asserting that A lies in some range belongs to the property state. We could take the propositions represented by subspaces orthogonal to the ray as false, but this leaves the propositions represented by subspaces that are neither orthogonal to the ray nor contain the ray indeterminate – neither true nor false.

On the orthodox interpretation, following Born (1926), the ray representing the quantum state assigns *probabilities* to *all* the propositions represented by subspaces of the Hilbert space: probability 1 to propositions in the property state defined by the ray, probability 0 to propositions represented by subspaces orthogonal to the ray, and nonzero probabilities to all other propositions. It follows that these probabilities cannot be represented on a standard (Kolmogorov) probability space, defined over the different possible property states or 'possible worlds' specified by the orthodox interpretation, as measures over these property states (as in the case of a classical probability theory).[28] So a nonzero probability assigned to a proposition asserting that the value of a dynamical variable A lies in a certain range is said to be 'the probability of finding the proposition to be true on measurement' or 'the probability of finding the value of the dynamical variable in the specified range on measurement.' This means that quantum mechanics, on the orthodox interpretation, can only provide probabilities for the results of measurements on a system by some agent or device *external* to the system. To mark this distinction between classical and quantum mechanics, dynamical variables in quantum mechanics are referred to as 'observables,' and an observable is understood to have no determinate value unless the ray representing the quantum state lies in one of the eigenspaces of the observable.

This notion of 'measurement' is undefined dynamically, as is the notion of an observable having no determinate value at one time and coming to have a determinate value at some other time as the outcome of a 'measurement' of the observable in question (or a 'measurement' of some observable that commutes with the observable in question). One would like, as Bell put it (1987, p.41), to have an interpretation of quantum mechanics, our most fundamental theory of motion, in terms of 'beables' rather than 'observables.'

> It would be foolish to expect that the next basic development in theoretical physics will yield an accurate and final theory. But it is interesting to speculate on the possibility that a future theory will not be *intrinsically* ambiguous and approximate. Such a theory could not be fundamentally about 'measurements,' for that would again imply incompleteness of the system and unanalyzed interventions from outside. Rather it should again become possible to say of a system not that

[28] The question of whether the probabilities specified by quantum states can be represented on a Kolmogorov probability space defined over *non-orthodox* property states is, of course, still open. This question is addressed in chapters 3 and 4.

such and such may be *observed* to be so but that such and such *be* so. The theory would not be about '*observ*ables' but about '*be*ables.'

The inability of the orthodox interpretation to provide an *internal* account of measurement (in terms of the dynamics of interacting systems) is what is known as the measurement problem of quantum mechanics. The following section considers what goes wrong with an internal account of measurement on the orthodox interpretation, and how Dirac and von Neumann proposed to resolve this problem.

1.4 Schrödinger's cat and quantum jumps

The measurement problem arises formally as a direct consequence of the orthodox (Dirac–von Neumann) interpretation and the linearity of the quantum dynamics, as given by Schrödinger's equation of motion for the time evolution of quantum states. In this section, I outline the problem as a formal problem for ideal 'first kind' measurements, in a terminology introduced by Pauli (1933). I take up the question of non-ideal measurements and the interaction between a measuring instrument and its environment in chapter 5.

Pauli distinguished between 'first kind' and 'second kind' measurements. 'First kind' measurements yield the same result on immediate repetition, which requires, in particular, that the measured system preserves its identity and is not destroyed in the measurement interaction. The measurement of a spin component of an electron by a Stern–Gerlach magnet is a 'first kind' measurement. By contrast, an energy measurement of a photon by a photon spectrometer, which absorbs and so destroys the photon, or a measurement of the momentum of a neutron by observing a recoil proton, which alters the momentum of the neutron in the collision, are examples of 'second kind' measurements.

Consider a model quantum mechanical universe consisting of two systems S and M, with M designated as the measuring instrument and S as the measured system. The pure states of $S + M$ are represented by rays or unit vectors on the tensor product[29] of the Hilbert spaces $\mathcal{H}_S$ and $\mathcal{H}_M$ of S and M, respectively. Suppose that the quantum state of $S + M$ takes an initial product form represented by a unit vector $|a_i\rangle|r_0\rangle$, where $|a_i\rangle$ is an eigenstate of some observable A of S – that is, a unit vector in the ith eigenspace of A – and $|r_0\rangle$ is an eigenstate of some 'pointer' or indicator observable R of M (so that A initially has the value a_i and R initially has a zero or 'ready' value r_0, according to the orthodox interpretation).

For each eigenstate $|a_i\rangle$ of the measured observable A, there exists a unitary evolution:[30]

$$|a_i\rangle|r_0\rangle \rightarrow |a_i\rangle|r_i\rangle$$

[29] See the appendix, section A.5.
[30] See the appendix, section A.2.

that correlates the pointer state with the eigenstate of A, and hence can be taken as representing an ideal 'first kind' measurement of A in this model universe. (If we enlarge the universe by a second measuring instrument M' with pointer observable R' – or, equivalently, take the pointer observable of M as a pair of commuting observables capable of registering consecutive measurement results – then immediate repetition of the measurement yields the same result: $|a_i\rangle|r_0\rangle|r'_0\rangle \rightarrow |a_i\rangle|r_i\rangle|r'_0\rangle \rightarrow |a_i\rangle|r_i\rangle|r'_i\rangle$.) In this case, each subsystem in the universe is associated with its own quantum state, both before and after the measurement interaction represented by the unitary evolution of the composite state, as in a classical mechanical universe. According to the orthodox interpretation, the property state after the measurement interaction is defined by the collection of properties represented by subspaces that contain the final product state $|a_i\rangle|r_i\rangle$, and this includes the property associated with the value a_i of A and the correlated R-property.

Since the dynamics is linear, it now follows that

$$|\alpha\rangle|r_0\rangle = \sum_i c_i|a_i\rangle|r_0\rangle \rightarrow \sum_i c_i|a_i\rangle|r_i\rangle \tag{1.9}$$

for an initial state in which the subsystem S has no determinate A-value and is represented by a state $|\alpha\rangle = \Sigma_i c_i|a_i\rangle$ in the Hilbert space of S that is a linear superposition of A-eigenstates. In this case, each subsystem is no longer associated with its own state after the measurement interaction represented by the unitary evolution of the composite state: the final state is the 'entangled' state $\Sigma_i c_i|a_i\rangle|r_i\rangle$, a linear superposition of product states. According to the orthodox interpretation, the property state is defined by the collection of properties represented by subspaces that contain the final entangled state $\Sigma_i c_i|a_i\rangle|r_i\rangle$. But no A-property belongs to this collection, and no R-property belongs to this collection. So neither the pointer reading nor the observable measured has a determinate value after the measurement interaction. (Note that an immediate repetition of the measurement in this case, for an enlarged universe consisting of S, M, and M', yields the final state $\Sigma_i c_i|a_i\rangle|r_i\rangle|r'_i\rangle$, with a similar correlation between A-values and pointer readings. So the measurement is repeatable, in the sense that if we could interpret the first instrument as registering a determinate outcome, the second measuring instrument would register the same outcome.)

If the pointer observable R is nonmaximal (degenerate),[31] as would be the case if M is a macroscopic system, the initial 'ready' state of M should be represented more properly as $|r_0, u\rangle$, where u is a degeneracy index. That is:

$$r = \sum_i r_i P_{r_i},$$

where

$$P_{r_i} = \sum_u |r_i, u\rangle\langle u, r_i|$$

[31] See the appendix, section A.1.

is the projection operator onto the eigenspace of R corresponding to the eigenvalue r_i. An ideal measurement transition would then be represented as:

$$|\alpha\rangle|r_0, u\rangle = \sum_i c_i|a_i\rangle|r_0, u\rangle \rightarrow \sum_i c_i|a_i\rangle|r_i, v_{i,u}\rangle \tag{1.10}$$

where $v_{i,u}$ is the value of the degeneracy index after the transition, depending on i and on the initial valve, u. Evidently, this does not affect the measurement problem in any way.

As Schrödinger (1935b) pointed out, if M represents a cat and R takes two possible values, associated with the cat being alive and the cat being dead, and the cat interacts with a microsystem S, such as an atom that can either decay or not decay in a certain time (where these events are associated with the two possible values of A), and the decay event triggers a device that kills the cat, then the cat will be neither alive nor dead after the measurement interaction, according to the orthodox interpretation.

The upshot of this analysis is that 'measurement,' in the sense required by the orthodox interpretation of the dynamical state as yielding probabilities of measurement outcomes, can't be understood dynamically. It follows that an observable that has no determinate value cannot come to have a determinate value as the result of a measurement understood as a dynamical interaction between a measured system and a measuring instrument, and so the requirement of the orthodox interpretation that observables come to have determinate values when measured has no dynamical justification. For some proofs that this measurement problem is formally insoluble, under very general assumptions about measurement in quantum mechanics, see Fine (1970), Shimony (1974), and Brown (1986).

Von Neumann's response to this problem was to propose two modes of evolution for a quantum system S: a unitary evolution described by Schrödinger's equation of motion between measurements, and a non-unitary, stochastic, entropy-increasing evolution induced by measurements (1955, p. 351 and pp. 417–18). For a (pure or mixed) state of S represented by a density operator W,[32] the unitary evolution (von Neumann's 'process 2') is:

$$W \rightarrow W(t) = U(t)WU^{-1}(t), \tag{1.11}$$

where $U(t) = \exp(-iHt/\hbar)$ if H is time independent. The evolution induced by the measurement of a maximal observable A with eigenvectors $|a_i\rangle$ (von Neumann's 'process 1') is:

$$W \rightarrow W' = \sum_i \mathrm{tr}(WP_{a_i})P_{a_i}, \tag{1.12}$$

where P_{a_i} is the projection operator onto the eigenvector $|a_i\rangle$.[33] The density operator W' represents a mixture of pure states $|a_i\rangle$, with weights equal to the probabilities

[32] For an account of the representation of pure and mixed quantum states by density operators, see the appendix, section 2.

[33] I write P_{a_i} for the projection operator onto the vector $|a_i\rangle$, rather than $P_{|a_i\rangle}$.

assigned by W to the corresponding eigenvalues. So if the state of the system S immediately before the measurement of A is a pure state represented by the vector $|\alpha\rangle = \Sigma_i c_i |a_i\rangle$, or the density operator $W = P_\alpha$, the measurement results in the projection of $|\alpha\rangle$ onto one of the eigenvectors $|a_i\rangle$ of A with probability $\mathrm{tr}(WP_{a_i}) = |c_i|^2$.

An observable is maximal (or nondegenerate) if it has n distinct eigenvalues in an n-dimensional Hilbert space. If A is maximal,

$$\sum_i \mathrm{tr}(WP_{a_i})P_{a_i} = \sum_i P_{a_i} W P_{a_i},$$

and 'process 1' can be represented as the transition:

$$W \to W' = \sum_i P_{a_i} W P_{a_i} \tag{1.13}$$

If A is nonmaximal with $k < n$ eigenspaces $\mathscr{H}_{a_i}$ and $W = P_\alpha$, the operator

$$\frac{P_{a_i} P_\alpha P_{ai}}{\mathrm{tr}(P_{a_i} P_\alpha P_{ai})}$$

is the projection operator onto the ray spanned by the normalized projection of $|\alpha\rangle$ onto the eigenspace $\mathscr{H}_{a_i}$ (the range of the projection operator P_{a_i}).[34] So $\Sigma_i P_{a_i} P_\alpha P_{a_i}$ represents a weighted sum of the normalized projections of the pure state represented by $|\alpha\rangle$ onto the eigenspaces $\mathscr{H}_{a_i}$ of A:

$$\sum_i P_{a_i} P_\alpha P_{a_i} = \sum_i \mathrm{tr}(P_{a_i} P_\alpha P_{a_i}) \frac{P_{a_i} P_\alpha P_{a_i}}{\mathrm{tr}(P_{a_i} P_\alpha P_{a_i})}$$

The proposal to take the generalized form of 'process 1' for measurements of maximal and nonmaximal observables as the transition (1.13) is due to Lüders (1951) and is known as the 'Lüders rule.' Notice that

$$\sum_i P_{a_i} P_\alpha P_{a_i} \neq \sum_i \mathrm{tr}(P_\alpha P_{a_i}) P_{a_i}$$

if A is nonmaximal. The density operator $\Sigma_i P_{a_i} P_\alpha P_{a_i}$ represents a weighted sum of the normalized projections of the state $|\alpha\rangle$ onto the eigenspaces $\mathscr{H}_{a_i}$, with weights $\mathrm{tr}(P_{a_i} P_\alpha P_{a_i}) = \mathrm{tr}(P_\alpha P_{a_i})$, while $\Sigma_i \mathrm{tr}(P_\alpha P_{a_i}) P_{a_i}$ represents a weighted sum of the projection operators P_{a_i}, with the same weights. The Lüders rule characterizes a 'first kind' minimally disturbing measurement of an observable, in which an initial pure state of the measured system is projected orthogonally onto one of the eigenspaces of the observable (with a probability equal to the square of the length of this projection). By contrast, the transition (1.12) for a degenerate observable A characterizes a maximally

[34] See the appendix, section A.2.

disturbing measurement, in which an initial pure state can be projected onto any vector in one of the eigenspaces of A (with the same probability).[35]

Dirac makes a similar proposal (1958, p. 36):

When we measure a real dynamical variable ξ, the disturbance involved in the act of measurement causes a jump in the state of the dynamical system. From physical continuity, if we make a second measurement of the same dynamical variable ξ immediately after the first, the result of the second measurement must be the same as that of the first. Thus after the first measurement has been made, there is no indeterminacy in the result of the second. Hence, after the first measurement has been made, the system is in an eigenstate of the dynamical variable ξ, the eigenvalue it belongs to being equal to the result of the first measurement. This conclusion must still hold if the second measurement is not actually made. In this way we see that a measurement always causes the system to jump into an eigenstate of the dynamical variable that is being measured, the eigenvalue this eigenstate belongs to being equal to the result of the measurement.

Dirac's 'quantum jumps' and von Neumann's 'projection postulate'[36] introduce a consistency problem into the theory. We can think of a measurement process as a dynamical interaction between a system S and a measuring instrument M_1. As such, the composite system $S + M_1$ ought to evolve unitarily. Suppose we measure an observable A of S by an interaction that correlates eigenvalues of A with eigenvalues of some pointer or indicator observable R of M_1. We can also measure simultaneously the observables $A \otimes I_{M_1}$ and $I_S \otimes R$ of $S + M_1$ by a second measuring instrument M_2.[37] Then we ought to get the same final state for S whether we apply the projection postulate directly to S or to the system $S + M_1$ after the unitary evolution characterizing the measurement interaction. Von Neumann shows that this is indeed the case.

If S is in the pure state $|\alpha\rangle = \Sigma_i c_i |a_i\rangle$, or $W = P_\alpha$, immediately prior to the measurement of A, applying the projection postulate directly to S yields the mixed state represented by the density operator $W' = \Sigma_i |c_i|^2 P_{a_i}$ for the state of S after the measurement. Von Neumann argues that there exists a measurement interaction in the formal sense, characterized by a unitary evolution of the state of $S + M_1$:

$$|\psi_0\rangle = |\alpha\rangle|r_0\rangle \rightarrow |\psi\rangle = \sum_i c_i |a_i\rangle|r_i\rangle$$

for which process 1 and process 2 are consistent. The transition from $|\psi_0\rangle$ to $|\psi\rangle$ can be taken as representing a measurement of A by the pointer R of the instrument M_1, because the state $|\psi\rangle$ correlates eigenvalues of A with distinct pointer readings, in the

[35] Notice that even a 'maximally disturbing' measurement in this sense does not alter the value of the measured observable, as in a 'second kind' measurement. If the initial pure state of the measured system is an eigenstate of the measured observable A for a certain degenerate eigenvalue, it remains in the corresponding eigenspace under the transition defined by von Neumann's 'process 1.'

[36] The term is due to Margenau (1936).

[37] $A \otimes I_{M_1}$ represents the tensor product of the observable A of the system S and the unit observable I_{M_1} in the Hilbert space of the instrument M_1. Similarly, $I_s \otimes R$ represents the tensor product of the unit observable in the Hilbert space of S and the observable R of M_1. See appendix, section A.5.

sense that $\text{prob}_\psi(a_i\&r_i) = |c_i|^2$ and $\text{prob}_\psi(a_i\&r_j) = 0$ if $i \neq j$, where these probabilities refer to potential measurements on $S + M_1$ by an instrument M_2. Applying the projection postulate to the state $|\psi\rangle$ of $S + M_1$ for a simultaneous measurement of $A \otimes I_{M_1}$ and $I_S \otimes R$ of $S + M_1$ by M_2 yields the mixture:

$$W'' = \sum_i |c_i|^2 P_{a_i} \otimes P_{r_i}$$

The reduced state[38] of S alone is $W''(S) = \Sigma_i |c_i|^2 P_{a_i} = W'$, the mixed state derived by applying the projection postulate directly to the system S. So, applying the projection postulate directly to the system S is consistent with its application to the system $S + M_1$ after a suitable interaction between S and M_1 governed by the equation of motion of the theory.

Von Neumann attempted to justify assuming two different modes of evolution for quantum states by the following argument (1955, pp. 418ff.):

Let us now compare these circumstances with those which actually exist in nature or in its observation. First, it is inherently entirely correct that the measurement or the related process of the subjective perception is a new entity relative to the physical environment and is not reducible to the latter. Indeed, subjective perception leads us into the intellectual inner life of the individual, which is extra-observational by its very nature (since it must be taken for granted by any conceivable observation or experiment).... Nevertheless, it is a fundamental requirement of the scientific viewpoint – the so-called principle of the psycho-physical parallelism – that it must be possible so to describe the extra-physical process of the subjective perception as if it were in reality in the physical world – i.e., to assign to its parts equivalent physical processes in the objective environment, in ordinary space. . . . In a simple example, these concepts might be applied about as follows: We wish to measure a temperature. If we want, we can pursue this process numerically until we have the temperature of the environment of the mercury container of the thermometer, and then say: this temperature is measured by the thermometer. But we can carry the calculation further, and from the properties of the mercury, which can be explained in kinetic and molecular terms, we can calculate its heating, expansion, and the resultant length of the mercury column, and then say: this length is seen by the observer. Going still further, and taking the light source into consideration, we could find out the reflection of the light quanta on the opaque mercury column, and the path of the remaining light quanta into the eye of the observer, their refraction in the eye lens, and the formation of an image on the retina, and then we would say: this image is registered by the retina of the observer. And were our physiological knowledge more precise than it is today, we could go still further, tracing the chemical reactions which produce the impression of this image on the retina, in the optic nerve tract and in the brain, and then in the end say: these chemical changes of his brain cells are perceived by the observer. But in any case, no matter how far we calculate – to the mercury vessel, to the scale of the thermometer, to the retina, or into the brain, at some time we must say: and this is perceived by the observer. That is, we must always divide the world into two parts, the one being the observed system, the other the observer. In the former, we can follow up all physical processes (in principle at least) arbitrarily precisely. In the latter, this is meaningless. The boundary between the two is

[38] The 'reduced state' of a subsystem of a composite system is defined in the appendix, section A.5.

arbitrary to a very large extent. . . . That this boundary can be pushed arbitrarily deeply into the interior of the body of the actual observer is the content of the principle of the psycho-physical parallelism – but this does not change the fact that in each method of description the boundary must be put somewhere, if the method is not to proceed vacuously, i.e., if a comparison with experiment is to be possible. Indeed, experience only makes statements of this type: an observer has made a certain (subjective) observation; and never any like this: a physical quantity has a certain value.

Now quantum mechanics describes the events which occur in the observed portions of the world, so long as they do not interact with the observing portion, with the aid of the process 2, but as soon as such an interaction occurs, i.e., a measurement, it requires the application of process 1. The dual form is therefore justified. However, the danger lies in the fact that the principle of the psycho-physical parallelism is violated, so long as it is not shown that the boundary between the observed system and the observer can be displaced arbitrarily in the sense given above.

The possibility of an arbitrary placement of the boundary between the observer and the observed system avoids a consistency problem. Von Neumann shows that if we divide the world into three parts – the observed system (S), the measuring instrument (M_1), and the observer (M_2) – then the boundary between the observer and the observed system can be drawn between S and $M_1 + M_2$ or between $S + M_1$ and M_2. In terms of his temperature measurement example, von Neumann points out that S could be taken as the system measured, M_1 as the thermometer, and M_2 as the light plus the observer. Or S could represent the system measured together with the thermometer, M_1 the light and the eye of the observer as a refracting device, and M_2 the observer interfacing with $S + M_1$ through the retina. Or S could represent everything up to the retina, M_1 the retina, optic nerve, and brain of the observer, and M_2 'his abstract "ego".'[39]

That this boundary can be shifted arbitrarily is shown by the equivalence *for* S of applying the projection postulate to S (in an observation of S by $M_1 + M_2$) and applying the projection postulate to $S + M_1$ (in an observation of $S + M_1$ by M_2), after an appropriate unitary dynamical interaction between S and M_1 characterizing the measurement. The boundary between S and $M_1 + M_2$ can be shifted to the boundary between $S + M_1$ and M_2.

Consistency aside, the introduction of the projection postulate still appears to be a blatantly *ad hoc* move to avoid a problem that has its source in the orthodox interpretation principle. The only argument for introducing a special mode of state transition characterizing measurement processes is that 'experience only makes statements of this type: an observer has made a certain (subjective) description; and never any like this: a physical quantity has a certain value.' This echoes Bohr's remark, quoted from Petersen (1963, p. 12) in section 1.1, that the task of physics is not 'to find out how nature *is*,' but rather 'what we can *say* about nature.' But where is the argument that quantum mechanics, as opposed to classical mechanics, is a theory about subjective descriptions or what we can say about nature? After all, quantum mechanics

[39] Von Neumann (1955, p. 421).

is, in effect, a non-Boolean generalization of classical mechanics, and classical mechanics manages quite nicely to be a theory of how nature *is*, in principle, by ascribing values to physical quantities.

For Einstein, the measurement problem simply points to the incompleteness of quantum mechanics. Recall Einstein's remarks on Schrödinger's cat in the letter to Schrödinger quoted in section 1.1: 'Nobody really doubts that the presence or absence of the cat is something independent of the act of observation. But then the description by means of the ψ-function is certainly incomplete, and there must be a more complete description.' The following two chapters consider the question of the completeness of quantum mechanics, beginning with the Einstein–Podolsky–Rosen (1935) incompleteness argument.

2

Bell's 'no go' theorem

But it seems certain to me that the fundamentally statistical character of the theory is simply a consequence of the incompleteness of the description. This says nothing about the deterministic character of the theory; that is a thoroughly nebulous concept anyway, so long as one does not know how much has to be given in order to determine the initial state ('cut').[40]

Einstein, from a letter to Schrödinger dated 22 December, 1950. See Przibram (1967, p. 40)

2.1 The Einstein–Podolsky–Rosen incompleteness argument

'Einstein Attacks Quantum Theory. Scientist and Two Colleagues Find It Is Not "Complete" Even Though "Correct".' So reports *The New York Times* on May 4, 1935, based on an interview with Boris Podolsky prior to the publication in *The Physical Review* of the landmark paper 'Can Quantum-Mechanical Description of Physical Reality be Considered Complete?' co-authored by Einstein, Podolsky, and Rosen (1935). Einstein was sufficiently annoyed by Podolsky's indiscretion in publicizing the dispute in the 'secular press' that he issued a statement dissociating himself from Podolsky's remarks (*The New York Times*, May 7, 1935, p. 21):

Any information upon which the article 'Einstein Attacks Quantum Theory' in your issue of May 6 [sic] is based was given to you without my authority. It is my invariable practice to discuss scientific matters only in the appropriate forum and I deprecate advance publication of any announcement in regard to such matters in the secular press.

The Einstein–Podolsky–Rosen (EPR) argument has been enormously influential in the debate on the foundations of quantum mechanics. While EPR argue for the incompleteness of quantum mechanics, Bell's 'no go' theorem, which is in a sense an extension of the EPR argument, appears to support the opposite conclusion. In this

[40] The reference here is to the boundary or 'cut' between what we take to be a separable physical system and the rest of the universe.

section, I present an analysis of the argument, as originally formulated by EPR. I consider various versions of Bell's theorem in subsequent sections of this chapter.

The motivation for the EPR argument is Einstein's realist philosophy of physics, characterized by two informal independence principles: a *separability* principle, and a *locality* principle. These are metaphysical principles, not part of the formal framework of quantum mechanics. The argument demonstrates an inconsistency between these realist principles and an associated 'criterion of physical reality,' and certain structural and interpretative principles of quantum mechanics, specifically the orthodox (Dirac–von Neumann) interpretation of the quantum state and (for the original version of the argument, as presented by EPR) the projection postulate as well.

In a short paper on the EPR argument entitled 'Quantum Mechanics and Reality' that Einstein sent to Born in a letter dated April 5, 1948 (reprinted in Born, 1971, p. 170), Einstein repeats the point made in the March 18 letter, quoted in section 1.1 ('But whatever we regard as existing (real) should somehow be localized in time and space. That is, the real in part of space *A* should (in theory) somehow "exist" independently of what is thought of as real in space *B*.'). He argues (in the April 5 letter) that physical objects are thought of as arranged in a space–time continuum and for this it is essential that, at a specific time, they 'claim an existence independent of one another, insofar as these things "lie in different parts of space".' He refers to this assumption of mutually independent existence as '(the "being-thus") of spatially distant things.'[41]

What Einstein appears to have in mind here is a separability principle that can be formulated as follows: If two dynamical systems are spatially separated,[42] then each system can be characterized by its own properties, independently of the properties of the other system. That is, each system separately has a 'being-thus,' a characterization in terms of certain properties intrinsic to the system, insofar as the systems are separable as dynamical systems.

The locality principle is the requirement that no influence on a dynamical system can directly affect another system that is spatially separated from the first system. In particular, measurement performed on a system cannot alter any properties of another system that is spatially separated from the first system. In the March 18 letter: 'That which really exists in *B* should therefore not depend on what kind of measurement is carried out in part of space *A*; it should also be independent of whether or not any measurement at all is carried out in space *A*.'

The EPR argument – apparently Podolsky was mostly responsible for the formulation of the published version – begins with a necessary condition for a complete theory (EPR, p. 777):[43]

Every element of the physical reality must have a counterpart in the physical theory.

[41] I use the translation by Don Howard (1985, p. 187) rather than Born's translation in Born (1971) for the comments in the April 5 letter.

[42] In a relativistic setting: space-like separated.

[43] All page references to the EPR article in this section are to Einstein, Podolsky, and Rosen (1935).

A sufficient condition is proposed for the existence of an 'element of physical reality' (EPR, p. 777):

If, without in any way disturbing a system, we can predict with certainty (i.e., with probability equal to unity) the value of a physical quantity, then there exists an element of physical reality corresponding to this physical quantity.

This is the EPR 'criterion of physical reality.' Evidently, the existence of an element of physical reality corresponds to the existence of a determinate value for a physical quantity or dynamical variable of a system; that is, to the system having a determinate property. The application of the criterion of reality is to a pair of well-separated systems, where the outcome of a measurement on one system allows – through the orthodox interpretation of the quantum state and the projection postulate – the prediction with certainty of a measurement outcome on the second system.[44] Since the two systems are well separated, separability requires that each system has its own elements of reality, and locality requires that elements of reality of the second system cannot be disturbed by measurements on the first system. So the prediction with certainty in such a case, on the basis of a measurement on the first system, can be associated with an element of reality possessed by the second system.

EPR review the quantum mechanical description of the motion of a particle with a single degree of freedom, and point out that if the quantum state of the system is an eigenstate of a physical quantity (dynamical variable, observable) A with eigenvalue a, then the system has the value a with certainty. So, according to the criterion of reality, there is an element of physical reality corresponding to A. If the state of the system is not an eigenstate of A then (EPR, p. 778) 'we can no longer speak of the physical quantity A having a particular value.' This, of course, is the orthodox (Dirac–von Neumann) interpretative principle. They illustrate this principle for a state of the particle given by a plane wave, in which the momentum has a definite value with certainty, and is therefore real by the reality criterion. In this case, a definite position value cannot be predicted with certainty. EPR remark (EPR, p. 778):

The usual conclusion from this in quantum mechanics is that *when the momentum of a particle is known, its coordinate has no physical reality.*

The next step in the argument is to establish the following disjunction D: Either (1) the quantum mechanical description of reality given by the quantum state is incomplete, or (2) quantities represented by noncommuting Hilbert space operators cannot have simultaneous reality (that is, if A and B do not commute, the system cannot have both an A-property and a B-property).

This follows, they argue, because if quantities represented by noncommuting operators *do* have simultaneous reality – simultaneously definite or determinate values – and the quantum state *does* provide a complete description of reality (that is, if the

[44] In the following section, I formulate a version of the EPR argument that does not depend on the projection postulate.

conjunction of the negation of (1) and the negation of (2) – symbolically, $\neg(1) \wedge \neg(2)$ – is true), then the quantum state would contain the values of quantities represented by noncommuting operators (by the condition of completeness), in the sense that these values would be predictable from the state. (The corresponding properties would belong to the property state.) Since this is not the case, the disjunction D, $(1) \vee (2)$, follows. In other words, if we assume that a quantum state represents the 'whole story' about a physical system, the fact that the quantum state does *not* contain the values of quantities represented by noncommuting operators entails, via the condition of completeness, the negation of $\neg(1) \wedge \neg(2)$, which is logically equivalent to D. So D follows from the condition of completeness and the fact that the quantum state does not contain the values of noncommuting quantities.

The first section of the paper concludes with the observation that the usual assumption, in the application of quantum mechanics to physical systems, is that the quantum state provides a complete description of physical reality. EPR state that the aim of the second section is to demonstrate that the conjunction of this assumption and the criterion of reality leads to a contradiction.

In the second section, EPR consider a composite system consisting of two subsystems, S_1 and S_2, in a pure state $|\psi\rangle$ at time T that can be represented in two different ways:

$$|\psi\rangle = \sum_i c_i |a_i\rangle |u_i\rangle = \sum_j d_j |b_j\rangle |v_j\rangle$$

It is assumed that the 'entangled' state $|\psi\rangle$ is the result of an interaction between the two systems, but that after time T the systems are separated and no longer interact. Suppose we measure the quantity U of S_2 after time T, with eigenvalues u_i and eigenvectors $|u_i\rangle$, and obtain the value u_k. Then we could conclude, on the basis of the projection postulate of the orthodox interpretation, that the measurement has resulted in the projection or collapse of the state $|\psi\rangle$ onto the state $|a_k\rangle|u_k\rangle$, and hence that the state of S_1 is $|a_k\rangle$. Alternatively, we could measure the quantity V of S_2, with eigenvalues v_j and eigenvectors $|v_j\rangle$. If we obtained the value v_l, we would conclude that the measurement has resulted in the collapse of $|\psi\rangle$ to the state $|b_l\rangle|v_l\rangle$, and hence that the state of S_1 is $|b_l\rangle$. So, the authors point out, it is possible to assign two different quantum states to the same system at the same time. That is, S_1 can be left in two different quantum states as a consequence of two different measurements performed on S_2. But, since the two systems no longer interact after time T, no 'real change' can take place in S_1 as a consequence of anything we might do to S_2 (EPR, p. 779). The inference here is evidently mediated by Einstein's separability and locality principles. Since the two systems are no longer capable of interacting, their elements of reality are separable: each system has its own 'being-thus,' and the elements of reality of one system cannot be altered by any manipulation of the elements of reality of the other system.

The eigenstates $|a_i\rangle$ and $|b_j\rangle$ can be eigenstates of noncommuting operators A and B.

EPR illustrate this with the example of position and momentum. Following Bohm (1951, pp. 615ff.), it is usual to consider the spin in two different directions of a spin-$\frac{1}{2}$ system in the singlet spin state. The singlet state of a pair of coupled spin-$\frac{1}{2}$ systems can be represented as:

$$|\psi\rangle = \tfrac{1}{\sqrt{2}}|+\rangle|-\rangle - \tfrac{1}{\sqrt{2}}|-\rangle|+\rangle$$

where $|+\rangle$ and $|-\rangle$ represent the two eigenstates of spin in any particular direction, corresponding to the eigenvalues ± 1 in units of $2\,\hbar$.[45] So:

$$\begin{aligned}|\psi\rangle &= \tfrac{1}{\sqrt{2}}|x+\rangle|x-\rangle - \tfrac{1}{\sqrt{2}}|x-\rangle|x+\rangle \\ &= \tfrac{1}{\sqrt{2}}|y+\rangle|y-\rangle - \tfrac{1}{\sqrt{2}}|y-\rangle|y+\rangle\end{aligned}$$

for the eigenvectors $|x+\rangle = |\sigma_x = +1\rangle$, $|x-\rangle = |\sigma_x = -1\rangle$ and $|y+\rangle = |\sigma_y = +1\rangle$, $|y-\rangle = |\sigma_y = -1\rangle$ of the spin quantities in the x- and y-directions, represented by the Pauli spin operators σ_x and σ_y, respectively. By measuring σ_x or σ_y on S_2, we can predict with certainty, and without in any way disturbing S_1, either the value of σ_x or the value of σ_y for S_1. Whatever the outcome of the spin measurement on S_2, we can predict with certainty that the outcome of a measurement of spin in the same direction on S_1 will be opposite in value. According to the criterion of reality, in the first case there is an element of reality of S_1 corresponding to σ_x (that is, an x-spin property obtains for S_1), while in the second case there is an element of reality of S_1 corresponding to σ_y (that is, a y-spin property obtains for S_1). But both the x-spin property and the y-spin property must obtain for S_1, since the spin measurements on S_2 cannot affect the 'being-thus' of S_1.

EPR now remark that they have established that it follows from the assumption that the quantum state provides a complete description of physical reality (the assumption $\neg$(1)), that two physical quantities represented by noncommuting operators can have simultaneous reality. That is, $\neg$(2) follows from $\neg$(1), given the criterion of reality as applied to a specific quantum mechanical system. They conclude, therefore, from $\neg$(2) and the disjunction D, (1) $\vee$ (2), that (1) follows: the description of reality given by the quantum state is incomplete.

This is not really how the argument goes in the paper. Rather, EPR apply the criterion of reality to a specific quantum mechanical system and establish $\neg$(2) directly (without $\neg$(1)). From (1) $\vee$ (2), and $\neg$(2), the alternative disjunct (1) follows immediately.

What they claim they *propose* to argue (at the end of section 1) is that $\neg$(1) together with the criterion of reality leads to a contradiction, and hence (assuming the criterion of reality) that (1) follows. What they claim to *have argued* (towards the end of section 2) is that $\neg$(1) entails $\neg$(2), given the criterion of reality as applied to a specific quantum system. From $\neg$(2) and the disjunction D, (1) follows, so presumably the contradiction they have in mind is that (1) follows from $\neg$(1). This reasoning is rather more convoluted than they need, and does not correctly characterize the argument they provide.

[45] See the appendix, sections A.4 and A.5.

EPR conclude by considering and rejecting an alternative criterion of reality: that two or more physical quantities correspond to simultaneous elements of reality only if they can be measured simultaneously, or values can be predicted with certainty for the two quantities simultaneously (EPR, p. 780):

Indeed, one would not arrive at our conclusion if one insisted that two or more physical quantities can be regarded as simultaneous elements of reality *only when they can be simultaneously measured or predicted.*

This is a *necessary* condition on *simultaneous* reality, while the original criterion of reality is proposed as a *sufficient* condition for physical reality. From the context, it appears that this is to be understood as a proposed additional qualification on the criterion of reality (not as an alternative criterion). For two quantities represented by noncommuting operators, like spin in the x-direction and spin in the y-direction for S_1, the EPR argument does not go through with this modified criterion, since the quantities cannot be measured simultaneously and only one or the other, but not both simultaneously, can be predicted with certainty by a measurement on S_2. But, EPR object, this would make the spin elements of reality at S_1 – the objectivity of spin in the x-direction and spin in the y-direction for S_1 – depend on what measurements are performed on S_2. As EPR see it (EPR, p. 780): 'No reasonable definition of reality could be expected to permit this.'

Here EPR invoke both separability and locality conditions implicitly. The assumption is that S_1 has its own elements of reality, its own 'being-thus,' if S_1 and S_2 are spatially separated and non-interacting. The 'being-thus' of S_1 – which quantities have values, and what these values are – cannot depend on what we do to S_2. In particular, measurements on S_2 cannot change the 'being-thus' of S_1.

The Einstein–Podolsky–Rosen argument appears to show that the state descriptions of quantum mechanics are incomplete. This raises the question of what would count as a 'completion' of the theory. Several authors, notably Fine (1986, chapter 4) and Jammer (1974, p. 254), have suggested that Einstein had something other than hidden variables in mind here. Einstein's negative reaction to Bohm's (1952a) hidden variable theory in correspondence with Renninger, Born, and others is often cited in support of this view. (For example, in a letter to Born dated May 12, 1952, reprinted in Born, 1971, p. 192, Einstein dismissed the theory as 'too cheap for me.') But Einstein's lack of enthusiasm for Bohm's theory, as this theory is usually formulated, should not be construed as a blanket rejection of 'completions' of quantum mechanics in the sense of hidden variable reconstructions of quantum statistics. Shimony (1971, p. 192) regards Einstein as 'the most profound advocate of hidden variables,' and this assessment is endorsed by Bell (1987, p. 89).

In a seminal review article, Bell (1966) presented a unified critical analysis of the main 'no go' theorems for hidden variables underlying the quantum statistics – theorems purporting to show that the probabilistic state descriptions of quantum mechanics cannot be recovered from distributions over the values of dynamical variables that are not taken into account in quantum mechanics, and are in this sense 'hidden' from the perspective of the theory. Bell clarified the significance of the assumptions involved in several of these theorems and showed that they exclude only certain rather restricted classes of hidden variable reconstructions of the quantum statistics. In particular, the earliest and most influential of these results by von Neumann (1955, chapter 4) turns out to exclude a completely trivial class of hidden variable theories. Bell, and later Mermin, characterized von Neumann's impossibility proof for hidden variables as 'silly' (Mermin, 1993, p. 805, cites an interview with Bell in *Omni*, May, 1988, p. 88). Remarkably, von Neumann's authority was such that this 'silly' proof helped entrench the consensus in the physics community against the very idea of hidden variables in quantum mechanics. Bohm felt that it lay behind the chilly reception to his 1952 hidden variable theory,[46] which demonstrated explicitly the possibility of reproducing all the empirical predictions of quantum mechanics via a hidden variable construction.

Bell concluded his analysis by observing (1966, p. 452) that the equations of motion in Bohm's theory 'have in general a grossly non-local character,' so that 'in this theory an explicit causal mechanism exists whereby the disposition of one piece of apparatus affects the results obtained with a distant piece.' He remarked that 'the Einstein–Podolsky–Rosen paradox is resolved in the way which Einstein would have liked least' and raised the question whether one could prove 'that *any* hidden variable account of quantum mechanics *must* have this extraordinary character.' Bell did succeed in proving that no hidden variable theory satisfying a locality constraint could reproduce the quantum statistics. This proof and its later elaborations is now known as 'Bell's theorem,' and was published in the first issue of a now defunct journal, *Physics*, in 1964 – predating the long-delayed publication of the 1966 review article!

Bell's theorem has spawned an avalanche of theoretical and experimental research on foundational questions in quantum mechanics related to the locality issue, and has even been characterized by Henry Stapp (1975, p. 271), quite seriously, as 'the most profound discovery of science.' If we consider a 'completion' of quantum mechanics as a 'no collapse' interpretation of the theory, then the central theorem of this book, proved in chapter 4, goes beyond Bell's result by characterizing all possible 'completions' of quantum mechanics, subject to certain natural constraints. As I show in chapter 6, Bohm's theory can be formulated as one such 'completion.'

Harvey Brown (1992) and Mermin (1993) refer to 'two theorems' of John Bell. One of these theorems is the locality theorem already mentioned. Bell's 'other theorem,' proved in the 1966 review article, follows as a corollary to Gleason's theorem (Gleason, 1957),[47] and establishes the impossibility of assigning values to all the observables of a system represented on a Hilbert space of three or more dimensions, if the values assigned to mutually compatible observables, represented by pairwise commuting self-adjoint operators, are required to preserve the functional relations holding among

[46] Private communication. See also the comment on von Neumann's theorem in Bohm and Hiley (1993, p. 116). The theorem first appeared in the German edition of von Neumann (1955), published in 1932.
[47] See the appendix, section A.2.

these observables. Bell's proof exploits a continuum of observables in $\mathscr{H}_3$, but the proof can be reformulated for a finite number of observables (see Mermin, 1993). This theorem is now more usually known as the 'Kochen and Specker theorem' (or sometimes as the 'Bell–Kochen–Specker theorem'), after Simon Kochen and E.P. Specker, who independently published a finitary version of the theorem in 1967.[48] I shall present a new proof of the Kochen and Specker theorem in chapter 3, following a suggestion in a letter (1965) by the logician Kurt Schütte to Specker.

In sections 2.3 and 2.4, I prove versions of Bell's locality theorem by Clauser and Horne (1974), and by Mermin (1981a,b), for stochastic and deterministic hidden variables, respectively. The strategy of these Bell 'no go' theorems is to derive an inequality for the correlation statistics of certain 2-valued measurement results on pairs of systems in 'entangled' quantum states, from the assumption that either the probabilities of measurement results, or the actual results themselves, are determined by the values of hidden variables that satisfy certain constraints that ensure the separability of the systems and the non-existence of nonlocal influences between the separate systems. It turns out that this inequality is violated by the correlation statistics predicted by quantum mechanics and confirmed by experiment.

In a later development, Greenberger, Horne, and Zeilinger (1989) proved the impossibility of assigning values to the spin component observables of a composite system consisting of four spin-$\frac{1}{2}$ systems, in such a way that the value assignment is consistent with the orthodox (Dirac–von Neumann) interpretation of the quantum state of the system and the separability and locality constraints of the EPR criterion of reality. The proof involves spin component observables over a continuous range of directions. Following a reformulation of the proof by Clifton, Redhead, and Butterfield (1991a,b) for a finite number of directions, Mermin (1990c, 1993) constructed an elegantly simplified version for a composite three-particle system in $\mathscr{H}_8$.[49] This is a 'Bell theorem without inequalities.' One might say that Bell's theorem for a composite two-particle system in $\mathscr{H}_4$ turns the EPR incompleteness argument on its head to draw the opposite conclusion, by extending the EPR analysis to additional pairs of observables, where each pair involves an observable measured on S_1 that is not the same as the observable measured on S_2 (for certain appropriately chosen pairs). The Greenberger–Horne–Zeilinger result in Mermin's version leads to a similar conclusion for a composite three-particle system in $\mathscr{H}_8$.[50]

In the following section, I consider the Mermin version of the Greenberger–Horne–Zeilinger (GHZ) result as a counter-argument to the EPR incompleteness argument.

[48] According to Kochen (private communication to Brad Monton), the proof itself dates from around 1961. Specker refers to 'an elementary geometrical argument' establishing such a result in Specker (1960, p. 246).

[49] The Clifton, Redhead, and Butterfield paper was published after Mermin's paper. Mermin acknowledges the rôle of a preprint of this paper in his proof.

[50] Heywood and Redhead (1983) and Stairs (1983) showed how to exploit the Kochen and Specker theorem for two spin-1 particles in $\mathscr{H}_9$ to draw a similar conclusion. For a review of these and related results, and an illuminating discussion of the relation between Bell's locality theorem and his 'other' theorem (the Kochen and Specker theorem), see Brown (1992).

2.2 The Greenberger–Horne–Zeilinger counter-argument

The EPR argument for a pair of spin-$\frac{1}{2}$ particles, S_1 and S_2, can be put this way: The singlet state:

$$|\psi\rangle = \tfrac{1}{\sqrt{2}}\,|x+\rangle|x-\rangle - \tfrac{1}{\sqrt{2}}\,|x-\rangle|x+\rangle$$

is an eigenstate of the observable $\sigma_x^1 \otimes \sigma_x^2$ with eigenvalue -1 (where the superscripts here refer to the corresponding Hilbert space or particle). Since the three observables:

$$\sigma_x^1 \otimes \sigma_x^2,\ \sigma_x^1 \otimes I_2,\ I_1 \otimes \sigma_x^2$$

commute pairwise and satisfy the functional relation:

$$\sigma_x^1 \otimes \sigma_x^2 = (\sigma_x^1 \otimes I_2)\cdot(I_1 \otimes \sigma_x^2)$$

their eigenvalues must satisfy the same functional relation (that is, any eigenvalue of $\sigma_x^1 \otimes \sigma_x^2$ must be expressible as a product of eigenvalues of $\sigma_x^1 \otimes I_2$ and $I_1 \otimes \sigma_x^2$), and a simultaneous measurement of the three observables must yield a set of three eigenvalues as measured values, $v(\sigma_x^1 \otimes \sigma_x^2)$, $v(\sigma_x^1 \otimes I_2)$, $v(I_1 \otimes \sigma_x^2)$, that satisfy this functional relation:[51]

$$v(\sigma_x^1 \otimes \sigma_x^2) = v(\sigma_x^1 \otimes I_2)v(I_1 \otimes \sigma_x^2) = v(\sigma_x^1)v(\sigma_x^2)$$

By the orthodox interpretation of the quantum state, the observable $\sigma_x^1 \otimes \sigma_x^2$ of the 2-particle system has the determinate value $v(\sigma_x^1 \otimes \sigma_x^2) = -1$ in the singlet state $|\psi\rangle$, so the measured values of σ_x^1 and σ_x^2 must satisfy the relation $v(\sigma_x^1)v(\sigma_x^2) = -1$. It follows that if we have two well-separated systems in the state $|\psi\rangle$, and we measure σ_x^1 on S_1, we can predict with certainty the outcome of a measurement of σ_x^2 on S_2, without disturbing S_2, and conversely, since $v(\sigma_x^1) = \pm 1$ and $v(\sigma_x^2) = \pm 1$, and the product of these two measured values must be equal to -1 in the state $|\psi\rangle$. By the EPR criterion of reality, therefore, σ_x^1 and σ_x^2 are elements of reality: the composite system in state $|\psi\rangle$ must have a determinate value for σ_x^1 and a determinate value for σ_x^2. But on the orthodox interpretation of the quantum state $|\psi\rangle$, neither σ_x^1 nor σ_x^2 have determinate values in the state $|\psi\rangle$ and so, by the condition of completeness ('every element of the physical reality must have a counterpart in the physical theory'), $|\psi\rangle$ is not a complete state description of the system.

On this formulation of the argument, it is unnecessary to add that the singlet state is also an eigenstate of $\sigma_y^1 \otimes \sigma_y^2$ with eigenvalue -1:

$$|\psi\rangle = \tfrac{1}{\sqrt{2}}\,|y+\rangle|y-\rangle - \tfrac{1}{\sqrt{2}}\,|y-\rangle|y+\rangle$$

and therefore, by the same reasoning, σ_y^1 and σ_y^2 must also have determinate values in the state $|\psi\rangle$. But the fact that noncommuting observables like σ_x^1, σ_y^1 and σ_x^2, σ_y^2 must all

[51] This functional relationship constraint on the simultaneously measurable eigenvalues of pairwise commuting observables is discussed further as a general constraint on value assignments in a hidden variable theory in section 3.1.

be determinate in the single state is certainly a more shocking conclusion of the argument. Notice that, while the original version of the EPR argument exploits the orthodox interpretation of the quantum state as well as the projection postulate, this formulation does not depend on the projection postulate.

The Greenberger–Horne–Zeilinger result (in Mermin's version)[52] provides a counter-argument to the EPR argument. Consider a composite system of three spin-$\frac{1}{2}$ particles, S_1, S_2, S_3, represented on a Hilbert space $\mathscr{H}(S_1) \otimes \mathscr{H}(S_2) \otimes \mathscr{H}(S_3)$, and the ten observables:

$$
\begin{aligned}
&\sigma_x^1 \otimes I_2 \otimes I_3,\ \sigma_y^1 \otimes I_2 \otimes I_3\\
&I_1 \otimes \sigma_x^2 \otimes I_3,\ I_1 \otimes \sigma_y^2 \otimes I_3\\
&I_1 \otimes I_2 \otimes \sigma_x^3,\ I_1 \otimes I_2 \otimes \sigma_y^3\\
&\sigma_x^1 \otimes \sigma_y^2 \otimes \sigma_y^3\\
&\sigma_y^1 \otimes \sigma_x^2 \otimes \sigma_y^3\\
&\sigma_y^1 \otimes \sigma_y^2 \otimes \sigma_x^3\\
&\sigma_x^1 \otimes \sigma_x^2 \otimes \sigma_x^3
\end{aligned}
$$

or in abbreviated form:

$$
\begin{aligned}
&\sigma_x^1,\ \sigma_y^1\\
&\sigma_x^2,\ \sigma_y^2\\
&\sigma_x^3,\ \sigma_y^3\\
&\sigma_x^1\sigma_y^2\sigma_y^3\\
&\sigma_y^1\sigma_x^2\sigma_y^3\\
&\sigma_y^1\sigma_y^2\sigma_x^3\\
&\sigma_x^1\sigma_x^2\sigma_x^3
\end{aligned}
$$

Since all ten observables are products of commuting observables with eigenvalues ± 1, they all have eigenvalues ± 1. Formally, if the eigenvalues of three observables, A on $\mathscr{H}(S_1)$, B on $\mathscr{H}(S_2)$, and C on $\mathscr{H}(S_3)$ are ± 1, then the eigenvalues of $A \otimes I_2 \otimes I_3$, $I_1 \otimes B \otimes I_3$, and $I_1 \otimes I_2 \otimes C$ are ± 1. The four commuting observables, $A \otimes B \otimes C$, $A \otimes I_2 \otimes I_3$, $I_1 \otimes B \otimes I_3$, and $I_1 \otimes I_2 \otimes C$ satisfy the functional relation:

$$A \otimes B \otimes C = (A \otimes I_2 \otimes I_3) \cdot (I_1 \otimes B \otimes I_3) \cdot (I_1 \otimes I_2 \otimes C)$$

and so their eigenvalues, $v(A \otimes B \otimes C)$, $v(A \otimes I_2 \otimes I_3)$, $v(I_1 \otimes B \otimes I_3)$, $v(I_1 \otimes I_2 \otimes C)$ satisfy the same functional relation:

$$v(A \otimes B \otimes C) = v(A \otimes I_2 \otimes I_3)v(I_1 \otimes B \otimes I_3)v(I_1 \otimes I_2 \otimes C)$$

It follows that each eigenvalue of $A \otimes B \otimes C$ is the product of eigenvalues of $A \otimes I_2 \otimes I_3$, $I_1 \otimes B \otimes I_3$, and $I_1 \otimes I_2 \otimes C$, and hence also equal to ± 1.

[52] Mermin (1990c, 1993).

Now, since

$$\sigma_x \sigma_y = i\sigma_z$$
$$\sigma_y \sigma_x = -i\sigma_z$$

for a spin-$\frac{1}{2}$ particle (and cyclic permutations), and

$$(\sigma_x)^2 = (\sigma_y)^2 = (\sigma_z)^2 = I,$$

the three observables

$$\sigma_x^1\sigma_y^2\sigma_y^3,\ \sigma_y^1\sigma_x^2\sigma_y^3,\ \sigma_y^1\sigma_y^2\sigma_x^3$$

commute pairwise, and

$$(\sigma_x^1\sigma_y^2\sigma_y^3)\cdot(\sigma_y^1\sigma_x^2\sigma_y^3)\cdot(\sigma_y^1\sigma_y^2\sigma_x^3) = -\sigma_x^1\sigma_x^2\sigma_x^3$$

Consider a simultaneous eigenstate, $|\phi\rangle$, of the three commuting observables $\sigma_x^1\sigma_y^2\sigma_y^3$, $\sigma_y^1\sigma_x^2\sigma_y^3$, $\sigma_y^1\sigma_y^2\sigma_x^3$ in which all three observables have the eigenvalue 1. The state $|\phi\rangle$ is also an eigenstate of $\sigma_x^1\sigma_x^2\sigma_x^3$ with eigenvalue -1. Suppose we have three mutually well-separated systems in the state $|\phi\rangle$. A measurement of σ_y on any two of the systems will enable us to predict with certainty the outcome of a measurement of σ_x on the third system, without disturbing this system, because the product of all three measured values must equal 1. (Alternatively, we could predict with certainty the outcome of a σ_x-measurement on a system, without disturbing the system, by measuring σ_x on the other two systems, because the product of all three measured values must be equal to -1.) Similarly, a measurement of σ_x on one system and σ_y on a second system will enable us to predict with certainty the outcome of a σ_y-measurement on the third system, without disturbing this system. In other words, we can predict with certainty the result of a σ_x-measurement or a σ_y-measurement on a system, without disturbing the system, by appropriate measurements on two other distant systems.

As Mermin puts it (1990c, p. 3375):

> If (like Einstein) one is afflicted with a strong antipathy toward nonlocal influences, then one is impelled to conclude that the results of measuring either component of any of the three particles must have already been specified prior to any of the measurements – i.e., that any particular system in the state ϕ must be characterized by numbers $m_x^1, m_y^1, m_x^2, m_y^2, m_x^3, m_y^3$ which specify the results of whichever of the four different sets (xyy, yxy, yyx, or xxx) of three single-particle spin measurements one might choose to make on the three far apart particles.

But now, since $|\phi\rangle$ is a simultaneous eigenstate of $\sigma_x^1\sigma_y^2\sigma_y^3$, $\sigma_y^1\sigma_x^2\sigma_y^3$, $\sigma_y^1\sigma_y^2\sigma_x^3$, and $\sigma_x^1\sigma_x^2\sigma_x^3$ with eigenvalues 1, 1, 1, and -1, respectively, the numbers $m_x^1, m_y^1, m_x^2, m_y^2, m_x^3$, m_y^3 must satisfy the relations:

$$m_x^1 m_y^2 m_y^3 = 1$$
$$m_y^1 m_x^2 m_y^3 = 1$$
$$m_y^1 m_y^2 m_x^3 = 1$$
$$m_x^1 m_x^2 m_x^3 = -1$$

This is impossible, because each m-term takes the value ± 1 and appears twice on the left of the equations, and so the product of the four left sides is 1 while the product of the right sides is -1.

What the counter-argument shows is that, for the well-separated spin-$\frac{1}{2}$ systems in the state $|\phi\rangle$, the assignment of determinate values to a certain finite set of observables by the orthodox (Dirac–von Neumann) interpretation of the state $|\phi\rangle$ is inconsistent with the assignment of values to these observables by the separability and locality constraints implicit in the application of the EPR criterion of reality. So one or the other (or both) of these principles has to go.

As Mermin points out, this counter-argument fails in $\mathscr{H}_4$. For a pair of spin-$\frac{1}{2}$ systems in the singlet state

$$|\psi\rangle = \tfrac{1}{\sqrt{2}}|+\rangle|-\rangle - \tfrac{1}{\sqrt{2}}|-\rangle|+\rangle$$

it follows from the orthodox interpretation of $|\psi\rangle$ that the observables $\sigma_x\sigma_x$ and $\sigma_y\sigma_y$ both have determinate values -1 in the state $|\psi\rangle$ (because $|\psi\rangle$ is a simultaneous eigenstate of both $\sigma_x\sigma_x$ and $\sigma_y\sigma_y$). By EPR locality and separability we can argue on the basis of the criterion of reality, as above, that

$$\sigma_x^1,\ \sigma_y^1,\ \sigma_x^2,\ \sigma_y^2$$

all have determinate values. Furthermore, if we assume that a measurement simply reveals these pre-existing determinate values as the numbers $m_x^1, m_y^1, m_x^2, m_y^2$, then these numbers must satisfy the constraints:

$$m_x^1 m_x^2 = -1$$
$$m_y^1 m_y^2 = -1$$

The observables $\sigma_x^1\sigma_y^2$ and $\sigma_y^1\sigma_x^2$ commute and their product is $\sigma_z^1\sigma_z^2$. The singlet state $|\psi\rangle$ is also an eigenstate of $\sigma_z^1\sigma_z^2$ with eigenvalue -1, so on the orthodox interpretation the observable $\sigma_z^1\sigma_z^2$ has the determinate value -1 in the state $|\psi\rangle$. A simultaneous measurement of $\sigma_x^1\sigma_y^2$ and $\sigma_y^1\sigma_x^2$ on the two-particle system in the state $|\psi\rangle$ must therefore yield opposite values for these product observables. If we could argue that $\sigma_x^1\sigma_y^2$ and $\sigma_y^1\sigma_x^2$ have determinate values (prior to the measurement), then these values would have to be $m_x^1 m_y^2$ and $m_y^1 m_x^2$, respectively, and satisfy the condition that:

$$(m_x^1 m_y^2)(m_y^1 m_x^2) = -1$$

in the state $|\psi\rangle$.

The three constraints

$$m_x^1 m_x^2 = 1$$
$$m_y^1 m_y^2 = 1$$
$$m_x^1 m_y^2 m_y^1 m_x^2 = -1$$

are inconsistent, since the product of the right sides of the equations is -1, while the product of the left sides is $+1$ (since each m-term takes the value ± 1, and appears twice). But in this two-particle example, there are no longer any grounds for arguing that the product observables $\sigma_x^1\sigma_y^2$ and $\sigma_y^1\sigma_x^2$ have determinate values, either on the basis of the orthodox interpretation of the quantum state $|\psi\rangle$ (since $|\psi\rangle$ is not an eigenstate of $\sigma_x^1\sigma_y^2$ or $\sigma_y^1\sigma_x^2$), or on the basis of the EPR criterion of reality (since the commuting observables $\sigma_x^1\sigma_y^2$ and $\sigma_y^1\sigma_x^2$ are nonlocal and so, while a measurement of $\sigma_x^1\sigma_y^2$ or $\sigma_y^1\sigma_x^2$ will allow us to predict with certainty the outcome of a measurement of $\sigma_y^1\sigma_x^2$ or $\sigma_x^1\sigma_y^2$, respectively, such a measurement will also necessarily disturb both systems).

2.3 Stochastic hidden variables

I shall derive a general version of Bell's inequality, the Clauser–Horne (1974) inequality, applicable to stochastic hidden variables – hidden variables that determine only the probabilities of measurement outcomes.

Consider two quantum systems moving in opposite directions from a source along a line (perhaps two particles resulting from the decay of a single particle) towards two detectors, M_L to the left of the source and M_R to the right of the source, each capable of measuring a variety of 2-valued observables on each system (say the spin component observables in any direction, if the particles are spin-$\frac{1}{2}$ particles).

Let $A, A', A'', \ldots$ denote the observables of the left system, S_L, measured by M_L, and B, $B', B'', \ldots$ the observables of the right system, S_R, measured by M_R. For the spin case, we suppose that M_L can be set to measure spin in the $\mathbf{a}, \mathbf{a}', \mathbf{a}'', \ldots$ directions and that M_R can be set to measure spin in the $\mathbf{b}, \mathbf{b}', \mathbf{b}'', \ldots$ directions. Denote the eigenvalues of $A, A', A'', \ldots$ by $a_+, a_-, a'_+, a'_-, a''_+, a''_-, \ldots$, and similarly for the eigenvalues of the observables of S_R. Let $p(a)$ represent the probability of detecting S_L as having a value a for the observable A (where a is a variable with two possible values, a_+ and a_-). Similarly, $p(b)$ represents the probability of detecting S_R as having a value b for the observable B, and $p(a\&b)$ represents the joint probability of detecting the values a and b for the observables A and B.

Now suppose there are statistical correlations between M_L-measurement outcomes and M_R-measurement outcomes, so that:

$$p(a\&b) \neq p(a)p(b) \tag{2.1}$$

and similarly for $p(a\&b')$, $p(a'\&b)$, $p(a'\&b')$, etc. Suppose, also, that these correlations have a *common cause* represented by a stochastic hidden variable λ, so that:

$$p(a\&b|\lambda) = p(a|\lambda)p(b|\lambda), \tag{2.2}$$

and similarly for the other joint probabilities, where the '|' represents conditionalization. That is, the joint probabilities are statistically independent, conditional on the common cause λ.

In section 2.5, I shall show that this assumption of conditional statistical independence defined by equation (2.2) is equivalent to the conjunction of two conditions that capture different aspects of separability and locality. To motivate the assumption here, consider a similar situation for large classical systems, say people. Suppose agents working for the covert operations branch of an intelligence agency are always sent out on missions in pairs, one agent of each pair travelling west to a country L, and the other agent travelling east to a distant country R. The missions might be something like assassinating a political figure, sabotaging an important military or industrial facility, and so on. Each type of mission in L is labelled $A, A', A'', \ldots$, and each type of mission in R is labelled $B, B', B'', \ldots$ Let a_+ represent the event (or proposition) of success on a mission of type A, and a_- represent failure, and similarly for other L-missions, and for R-missions. So $p(a_+)$ represents the probability of success for a mission of type A in the country L, and $p(a_+ \& b_-)$ represents the joint probability of success for a mission of type A in the country L and failure for a mission of type B in the country R, and similarly for other single and joint probabilities.

Suppose it turns out that

$$p(a_+) = p(a_-) = \tfrac{1}{2}$$
$$p(a'_+) = p(a'_-) = \tfrac{1}{2}, \text{ etc.}$$

and

$$p(b_+) = p(b_-) = \tfrac{1}{2}$$
$$p(b'_+) = p(b'_-) = \tfrac{1}{2}, \text{ etc.}$$

but

$$p(a_+ \& b_+) = 0.09$$
$$p(a'_+ \& b'_+) = 0.09, \text{ etc.}$$

In other words, there is a 50% chance of success for any type of mission in L and a 50% chance of success for any type of mission in R, but A and B missions carried out at the same time in L and R are both successful in only about 9 out of 100 cases on the average, and similarly for A' and B' missions, etc.

Imagine an internal investigation of this peculiar state of affairs. The very high correlation between success on a certain type of mission in L and failure on a similar mission in R (91 times out of 100, on the average) calls for an explanation, because we suppose that the actions of each agent, and the outcomes of these actions, are independent, and hence expect joint success in L and R for similar missions in about one quarter of the cases. (Similar considerations apply to the joint probabilities for A and B' missions, A' and B missions, etc., which we assume also differ from $\tfrac{1}{4}$). Suppose the investigation rules out any causal influence between the relevant events in L and R. Perhaps the two events are timed to occur virtually simultaneously, so that telephone communication between counter-intelligence agents of L and R after a mission in R, say, will be too late to affect the outcome of the mission in L. Since we can assume that the success of a particular agent on a particular mission depends solely on the intrinsic abilities of the agent, the level of difficulty of the type of mission, and uncontrollable local factors in the target area in the country in question, it follows that any correlation between events in L and R can only be due to a common causal factor originating in the agency itself, the source of the agents.

A common cause might, for example, be introduced by a 'mole' planted in the agency. Suppose the purpose of the mole is to thwart the success of as many missions as possible, but suppose also that the agents get their instructions only after they arrive in L and R, and that there is no communcation between the countries and anyone at the agency, the source of the agents, after the agents depart for their missions. What the mole is capable of influencing, one might suppose, is the choice of agents. The mole sees to it that when a top-notch agent like James Bond, with a probability of success of 0.9 for any type of mission, is sent to L, a totally inept agent, with a probability of success on any mission of 0.1, is sent to R, and conversely. (To avoid suspicion, the mole feels obliged to include one superior agent in each pair.) So, for half of all missions to L, James Bond types are selected while totally inept agents make up the other half, and similarly for missions to R.

If we assume, for simplicity, that half the agent pool consists of superior agents like James Bond who all have success rates of 0.9 for any mission, and the other half consists of inept agents, who all have success rates of 0.1 for any mission, this immediately explains the 50% success rate for any type of mission in L or R. Label the agent pairs, consisting of a competent agent paired with an inept agent, by a parameter λ (which can take values $\lambda_1, \lambda_2, \ldots$ for all such agent pairs in the agency) and suppose that each pair (that is, each value of λ) has equal *a priori* probability, $p(\lambda_i)$, of being selected for any coupled L-mission and R-mission, then:

$$\begin{aligned} p(a) &= \sum_i p(\lambda_i)p(a|\lambda_i) \\ &= \tfrac{1}{2}(0.9) + \tfrac{1}{2}(0.1) \\ &= \tfrac{1}{2} \end{aligned}$$

because for half the λ-values $p(a|\lambda_i) = 0.9$ and for the other half $p(a|\lambda_i) = 0.1$, and $\Sigma_i p(\lambda_i) = 1$. Also, $p(b) = \frac{1}{2}$, because the λ-values for which $p(b|\lambda_i) = 0.9$ are just the λ-values for which $p(b|\lambda_i) = 0.1$. Similarly, $p(a') = \frac{1}{2}$, etc., and $p(b') = \frac{1}{2}$, etc.[53]

Now, the joint probability of success for *A* and *B* missions at L and R is calculated as:

$$\begin{aligned} p(a\&b) &= \sum_i p(\lambda_i)p(a\&b|\lambda_i) \\ &= \sum_i p(\lambda_i)p(a|\lambda_i)p(b|\lambda_i) \\ &= \tfrac{1}{2}(0.9)(0.1) + \tfrac{1}{2}(0.1)(0.9) \\ &= 0.09 \end{aligned}$$

[53] For simplicity, I have chosen the λ-probabilities to be 0.9 and 0.1 for oppositely endowed pairs of agents, but it is not necessary that these numbers sum to 1, nor is it necessary that the λ-probabilities for a particular agent are the same for all types of mission to generate the correlations.

Why? Because we assume that, *once we fix the agent pair* λ_i, say, then the joint probability of success, $p(a\&b|\lambda_i)$, depends entirely on the intrinsic abilities of the agents in question (reflected in the λ-probabilities, 0.9 and 0.1), the mission types A and B,[54] and random or uncontrollable local factors on the ground, so to speak. In other words, there is no correlation between what happens in L and what happens in R for a particular agent pair – what happens in L or R for a particular agent pair is determined locally by factors present in L or R at the time of the events in question. So

$$p(a\&b|\lambda_i) = p(a|\lambda_i)p(b|\lambda_i)$$

that is, the joint probability of success, *given the agent pair*, is just the product of the probabilities of success for each agent separately.

This is conditional statistical independence. The common cause represented by the stochastic hidden variable λ 'screens off' events in L from events in R. Or, putting it somewhat paradoxically, conditionalizing on the common cause makes the correlations disappear. In fact, this is the distinguishing feature of a common cause of correlated events: an earlier event that 'screens off' the two events from each other, so that the joint conditional probability is just the product of the conditional probabilities of each event separately.

So we have correlations between success for A-missions in L and success for B-missions in R:

$$p(a_+\&b_+) = 0.09 \neq p(a_+)p(b_+) = \tfrac{1}{2} \cdot \tfrac{1}{2} = \tfrac{1}{4}$$

explained by a common cause λ taking values λ_i, such that conditional statistical independence holds for the probabilities of these events conditional on the common cause:

$$p(a_+\&b_+|\lambda_i) = p(a|\lambda_i)p(b|\lambda_i), \text{ for all } i,$$

and similarly for correlations between A' and B' missions, etc.

In the quantum mechanical context, conditional statistical independence is sometimes referred to as 'strong locality,' and also as 'factorizability.' The point of the example was to motivate this condition as intuitively compelling for the explanation of statistical correlations between events at spatially separated locations, when there are grounds to suppose that there can be no causal influence from an event at one location to an event at the distant location.

For a realist, a correlation between events at separated locations, L and R, requires a causal explanation, and if a causal influence from L to R or from R to L is ruled out, then – it would appear – there must be a common cause for the existence of the correlations. This common cause might be deterministic, so that the particular events about to occur at L and R are settled by the common cause, but at the very least there must be a stochastic common cause responsible for the correlations. That is, there must

[54] In this case, the λ-probabilities have been chosen to be independent of the type of mission for simplicity. So, in the set-up considered, the joint probability of success for any mission type in L and R is 0.09.

be two separate causal influences, originating in the common past of the two correlated events at L and R, moving towards L and R, respectively, where the two causal influences are related in virtue of their common origin and *separately* make it more or less likely (but not necessarily certain) that a particular event will occur at L and R at the time in question, depending on local conditions at L and R. The statistical correlation between events at the distant locations occurring at a particular time is then explained as nonmysterious in virtue of the relationship between the two causal influences *when these influences originated at a common location at an earlier time*. What the example shows is that a stochastic common cause is characterized by the statistical independence of events at L and R, conditional on the common cause.

The argument now proceeds to show that there cannot be a common cause for the statistical correlations in the quantum mechanical example. Since causal influences from L to R or R to L that could explain the correlation in the quantum mechanical set-up would require the superluminal transfer of information between L and R (because the two measurement events are outside each other's light cones), we suppose that such causal influences are ruled out by special relativity. So, what we apparently have in quantum mechanics is the existence of statistical correlations between distant events without any causal explanation. The really astonishing thing about the quantum mechanical correlations is just that they appear to be inconsistent with any explanation of the sort that would apply in the James Bond example.

Here's the argument: Consider the expression

$$\begin{aligned} K &= \alpha[\alpha'(1-\beta) + (1-\alpha')(1-\beta')] + (1-\alpha)[\alpha'\beta' + (1-\alpha')\beta] \\ &= \alpha + \beta + \alpha'\beta' - \alpha\beta - \alpha'\beta - \alpha\beta' \end{aligned}$$

If $0 \le \alpha \le 1$, $0 \le \alpha' \le 1$, $0 \le \beta \le 1$, $0 \le \beta' \le 1$, then K is a convex combination of convex combinations of terms lying between 0 and 1, and so:

$$0 \le K \le 1$$

(Each of the terms β, β', $(1-\beta)$, $(1-\beta')$ in the square brackets is a number between 0 and 1, so when fractions α' and $1-\alpha'$, respectively, of these terms are added they must sum to a number between 0 and 1. Each of the square bracket terms, therefore, lies between 0 and 1, so again when fractions α and $1-\alpha$, respectively, of these terms are added they must sum to a number between 0 and 1.)

Instead of $p(a|\lambda)$ or $p(a\&b|\lambda)$, etc., I shall write $p_\lambda(a)$, $p_\lambda(a\&b)$. The notation $p(a|\lambda)$ suggests that λ belongs to the same σ-field as the propositions or events on which the probabilities $p(\ )$ and $p(\,|\,)$ are defined. So, for example, $p(a|\lambda) = p(a\&\lambda)/p(\lambda)$, and the possible values of λ would have to form at most a countable set. More generally, we might want to consider a continuous parameter λ labelling a continuum of different possible common causes. In such a case

$$p(a) = \int p_\lambda(a)\rho(\lambda)\mathrm{d}\lambda,$$

where $\rho(\lambda)$ characterizes the distribution of λ-values.

Let

$$\alpha = p_\lambda(a),\ \alpha' = p_\lambda(a'),\ \beta = p_\lambda(b),\ \beta' = p_\lambda(b').$$

Then

$$K(\lambda) = p_\lambda(a) + p_\lambda(b) + p_\lambda(a')p_\lambda(b') - p_\lambda(a)p_\lambda(b) - p_\lambda(a')p_\lambda(b) - p_\lambda(a)p_\lambda(b')$$

Assuming conditional statistical independence:

$$p_\lambda(a)p_\lambda(b) = p_\lambda(a\&b), \text{ etc.} \tag{2.3}$$

averaging over λ yields:

$$\int K(\lambda)\rho(\lambda)\mathrm{d}\lambda = p(a) + p(b) + p(a'\&b') - p(a\&b) - p(a'\&b) - p(a\&b')$$

and so:[55]

$$0 \leq p(a) + p(b) + p(a'\&b') - p(a\&b) - p(a'\&b) - p(a\&b') \leq 1 \tag{2.4}$$

Similar inequalities to (2.4) follow if we exchange a' for a and/or b' for b, yielding four inequalities in all. These four inequalities are now known as the 'Clauser–Horne' inequalities (first derived by Clauser and Horne, 1974). What has been shown is that if there are correlations between M_L-measurement outcomes and M_R-measurement outcomes, and these correlations are the result of a common cause at the source of the system pairs, then the Clauser–Horne inequalities must hold for the single and joint probabilities.[56] The significance of this result is that the Clauser–Horne inequalities are violated by the probabilities generated by 'entangled' or nonseparable quantum states of composite systems consisting of widely separated particle pairs. Entangled states are pure states, like the singlet state in the tensor product Hilbert space of the composite EPR system, that are not expressible as product states. They induce statistical correlations of a specific sort between measurement outcomes on the distant particles. It follows that there are statistical correlations between measurement outcomes on separated quantum systems in certain states for which there can be no common cause.

This is the Clauser–Horne version of Bell's theorem. It is completely general, with no specific constraints on the relationship between the L-observables and the M-observables (other than that they are 2-valued), and no specific requirement on the nature of the correlations. The presumption is simply that there are correlations of some sort, and the theorem involves the derivation of inequalities that must be satisfied by certain combinations of single and joint probabilities, if these correlations have a common causal explanation.

The discussion of the Greenberger–Horne–Zeilinger argument in the previous section involved a system in a certain quantum state, showing the inconsistency between the assignment of values to observables by the orthodox (Dirac–von

[55] Averaging a number between 0 and 1 over λ with a distribution function $\rho(\lambda)$ that integrates to 1, or with weights $p(\lambda_i)$ that sum to 1, yields a number between 0 and 1.

[56] One can easily check that the inequalities hold in the case of the James Bond example.

Neumann) interpretation of the state and the EPR criterion of reality. Here we have an inconsistency between the probabilities generated by certain quantum states for the values of certain observables and a common causal explanation of the statistical correlations. What the EPR criterion of reality and the realist demand for a common causal explanation share is a commitment to Einstein's principles of separability and locality – the idea that physical systems can be separate in space (or space–time), that separate systems have their own elements of reality or 'being-thus,' and that all physical influence (by one element of reality on another) is local. The Clauser–Horne argument considers an extended stochastic notion of the 'being-thus' of a system, where a system is tagged with a 'being-thus' that has only a probabilistic influence on the outcome of certain physical interactions with measuring instruments. But the important point for the derivation of the inequality is that each physically separate system has its own *separate* 'being-thus,' whether this 'being-thus' is stochastic and imposes probabilistic constraints on events, or deterministic and settles definitively which events occur. (Note that James Bond was characterized above as having a stochastic 'being-thus,' a set of talents or intrinsic abilities encoded in a parameter λ that fixes only the probability of his success on any type of mission.)

If quantum mechanics is indeed 'incomplete,' these results impose severe limitations on any 'completion' of the theory, and hence on any 'no collapse' interpretation. Hidden variable theories that introduce stochastic hidden variables as common causes to explain the statistical correlations generated by certain entangled quantum states of composite systems cannot reproduce the quantum statistics. That is, no stochastic hidden variable theory can reproduce all the quantum statistics, if the hidden variables impose probabilistic constraints on property states of the separate systems that satisfy the separability and locality constraints encapsulated in conditional statistical independence.

The Greenberger–Horne–Zeilinger argument applies to a composite system in an entangled quantum state, and excludes deterministic hidden variables that extend the attribution of values to observables, defined by the orthodox interpretation of the state, to certain observables of the separate systems, on the basis of separability and locality constraints. Observer-free 'no collapse' interpretations of quantum mechanics of the sort considered in chapters 5 and 6, which replace the orthodox interpretation principle with different principles for attributing values to the observables of a system in a given quantum state, are subject to these results and will not, therefore, conform to all the requirements of Einstein's realism.

2.4 Deterministic hidden variables

Although the Clauser–Horne argument for stochastic hidden variables considered in the previous section also covers the deterministic case, where the hidden variable probabilities are all 0 or 1, I reproduce here a particularly simple and insightful derivation of a Bell inequality for deterministic hidden variables by Mermin (1981a, b), applied to the correlations generated by the singlet spin state.

Suppose the two quantum systems, S_L and S_M, moving in opposite directions are two spin-$\frac{1}{2}$ particles in the singlet state resulting, perhaps, from the decay of a spin-0 particle. Suppose the detectors M_R and M_L can each be set at one of three settings, for the measurement of the spin component in three alternative directions 120° apart, where the direction of the spin component observables A, A', A'' of S_L correspond to the directions of the spin component observables B, B', B'' of S_M.

In the singlet state.[57]

$$|\psi\rangle = \tfrac{1}{\sqrt{2}}\,|a_+\rangle|b_-\rangle - \tfrac{1}{\sqrt{2}}\,|a_-\rangle|b_+\rangle$$
$$= \tfrac{1}{\sqrt{2}}\,|a'_+\rangle|b'_-\rangle - \tfrac{1}{\sqrt{2}}\,|a'_-\rangle|b'_+\rangle, \text{ etc.,}$$

the measurement outcomes are perfectly oppositely correlated:

$$p_\psi(a_+\&b_+) = p_\psi(a_-\&b_-) = 0$$
$$p_\psi(a_+\&b_-) = p_\psi(a_-\&b_+) = \tfrac{1}{2}$$

and similarly for the pair A', B' and the pair A'', B''.

Perfect opposite correlation (or perfect correlation) and conditional statistical independence together restrict all common cause probabilities to 0 or 1:

$$p_\lambda(a_+) = 0 \text{ or } 1, \text{ etc.}$$

In other words, in the presence of perfect (opposite) correlation, conditional statistical independence requires that the common cause, λ, determines the outcome of any M_L-measurement and any M_R-measurement: for a particular choice of observables measured at L and R, λ tags the outcomes. So a particular value of λ encodes different individual particle pairs in the singlet state as pairs that will yield certain definite outcomes if subjected to one of the nine possible combinations of measurements: AB, AB', AB'', $A'B$, $A'B'$, $A'B''$, $A''B$, $A''B'$, $A''B''$. Since the singlet state does not distinguish these particle pairs in any way, the common cause λ is a deterministic hidden variable in this case.

To see that determinism – in the sense that all common cause probabilities are 0 or 1 – follows from conditional statistical independence and perfect (opposite) correlation, note that:

$$p(a_+\&b_+) = 0 = \overline{p_\lambda(a_+\&b_+)}$$
$$= \overline{p_\lambda(a_+)p_\lambda(b_+)}$$
$$= \int p_\lambda(a_+)p_\lambda(b_+)\rho(\lambda)\mathrm{d}\lambda,$$

where the '$\overline{}$' denotes averaging over the hidden variables λ, and similarly:

[57] See the appendix, section A.5.

Table 2.1. *Probabilities for the possible setting combinations of M_L and M_R*

	λ-probabilities						Setting combinations								
	a_+	a'_+	a''_+	b_+	b'_+	b''_+	AB	AB'	AB''	$A'B$	$A'B'$	$A'B''$	$A''B$	$A''B'$	$A''B''$
λ_1	1	1	1	0	0	0	d	d	d	d	d	d	d	d	d
λ_2	1	1	0	0	0	1	d	d	s	d	d	s	s	s	d
λ_3	1	0	1	0	1	0	d	s	d	s	d	s	d	s	d
λ_4	1	0	0	0	1	1	d	s	s	s	d	d	s	d	d
λ_5	0	1	1	1	0	0	d	s	s	s	d	d	s	d	d
λ_6	0	1	0	1	0	1	d	s	d	s	d	s	d	s	d
λ_7	0	0	1	1	1	0	d	d	s	d	d	s	s	s	d
λ_8	0	0	0	1	1	1	d	d	d	d	d	d	d	d	d

$$p(a_-\&b_-) = 0 = \int p_\lambda(a_-)p_\lambda(b_-)\rho(\lambda)\mathrm{d}\lambda$$
$$= \int(1 - p_\lambda(a_+))(1 - p_\lambda(b_+))\rho(\lambda)\mathrm{d}\lambda$$

Since $p_\lambda(a_+) \geq 0$, $p_\lambda(b_+) \geq 0$, we have for all λ:

$$p_\lambda(a_+)p_\lambda(b_+) = 0$$
$$1 - p_\lambda(a_+) - p_\lambda(b_+) + p_\lambda(a_+)p_\lambda(b_+) = 0,$$

and so

$$p_\lambda(a_+) + p_\lambda(b_+) = 1$$

It follows that either $p_\lambda(a_+) = 1$ and $p_\lambda(b_+) = 0$, or $p_\lambda(a_+) = 0$ and $p_\lambda(b_+) = 1$. Similarly for a', b', and a'', b''.

Consider an experiment in which the settings of the detectors M_L and M_R are chosen randomly after the particles move towards the detectors. To compute the probability of getting $++$ or $--$ (that is, 'same sign') on the assumption of a common cause for the perfect opposite correlations, note first that the common cause probabilities must all be either 0 or 1 by the above argument, for each of the three settings A, A', A'', and similarly for the three settings B, B', B'', with opposite values for A and B, A' and B', A'' and B''. So there are only $2 \times 2 \times 2 = 8$ possible combinations of 0, 1 probabilities for each of the nine possible setting combinations for M_L and M_R. Table 2.1 covers all these possibilities, with an entry of 's' or 'd' under the setting combination, depending on whether two results yield the same sign ($++$ or $--$) or different signs ($+-$ or $-+$).

The eight possible combinations of 0, 1 probabilities represent eight possible values for the deterministic common cause, or eight possible λ-types. Each λ-type is, in effect, an 'instruction set' coded into the particle pair that determines the outcome of any pair of measurements corresponding to any of the nine setting combinations. For example, the common cause λ_1 determines the outcome of an AB-measurement as $+-$, and we enter 'd' in the appropriate slot. The common cause λ_3 determines the outcome of an AB'-measurement as $++$, and we enter 's' in the appropriate slot.

Excluding the first and last rows in the table, the remaining rows all have five d's and four s's. That is, in each of the cases (for each of these λ-types) we expect to find 's' with a relative frequency of $\frac{4}{9}$ in a sufficiently long run of experiments, assuming a random choice of settings at M_L and M_R (which, we suppose, is under our control). So, for these λ-types:

$$p(\mathrm{s}) = \tfrac{4}{9}$$

If we make no assumption about the relative weights of the different λ-types in the distribution of particle pairs issuing from the source, including the types λ_1 and λ_8, then

$$p(\mathrm{s}) \leq \tfrac{4}{9} \approx 44.4\%$$

and $p(\mathrm{s}) = \frac{4}{9}$ only if types λ_1 and λ_8 are not present in the distribution. If we assume that both the setting combinations and the λ-types are chosen randomly, by the experimenter in the case of the setting combinations and by the uncontrollable conditions associated with the creation of the correlated particle pairs in the space–time region of the source, and also that these choices are statistically independent (that is, that there is no causal relationship between the choice of λ-type and the choice of setting combination, no 'conspiracy of nature' biasing the choice of setting combination in some way according to the choice of λ-type, or conversely), then we would expect that

$$p(\mathrm{s}) = \tfrac{1}{3}$$

since there are 24 s's out of 72 possible event types (a particular λ-type associated with a particular setting combination).

Rationally, then, one ought to give fair betting odds of no less than 5 to 4 of getting 'same sign' in a run of the experiment, and as high as 2 to 1, assuming randomness for the common causes as well as the setting combinations. But quantum mechanics predicts, and experiments confirm, that a 'same sign' outcome will occur half the time!

To compute the quantum mechanical probability of '$++$ or $--$' for a random choice of settings in many runs of the experiment, note that:

$$p_\psi(++ \text{ or} --|AB) = p_\psi(++ \text{ or } --|A'B') = p_\psi(++ \text{ or } --|A''B'') = 0$$

for the three settings $AB, A'B', A''B''$. The proposition '$++$ or $--$' for AB corresponds to a plane in the 4-dimensional Hilbert space of the two-particle system, the plane spanned by the vectors $|a_+\rangle|b_+\rangle$ and $|a_-\rangle|b_-\rangle$. The singlet state lies entirely in the plane spanned by the vectors $|a_+\rangle|b_-\rangle$ and $|a_-\rangle|b_+\rangle$. These two planes are orthogonal, represented by the projection operators $P_{++} \vee P_{--} = P_{++} + P_{--}$ and $P_{+-} \vee P_{-+} = P_{+-} + P_{-+}$, respectively. So the probability of '$++$ or $--$,' given by the square of the length of the projection of the singlet state onto the corresponding plane, is zero. Similarly for $A'B'$ and $A''B''$.

For the setting AB':

$$p_\psi(++ \text{ or } --|AB') = \sin^2 60° = \tfrac{3}{4}$$

because the angle between the M_L-setting A and the M_R-setting B' is 120°. Similarly for the five settings AB'', $A'B$, $A'B''$, $A''B$, $A''B'$.

A quick way to see this is to note that the probability $p_\psi(++ \text{ or } --|AB')$ is equal to the square of the length of the projection of the singlet state $|\psi\rangle$ onto the plane spanned by the vectors $|a_+\rangle|b'_+\rangle$ and $|a_-\rangle|b'_-\rangle$. If $P_{++'}$ and $P_{--'}$ represent the projection operators onto $|a_+\rangle|b'_+\rangle$ and $|a_-\rangle|b'_-\rangle$, respectively, then the projection operator onto this plane is:

$$P_{++'} \vee P_{--'} = P_{++'} + P_{--'}$$

What we want is the square of the length of the vector:

$$\begin{aligned}(P_{++'} + P_{--'})|\psi\rangle &= P_{++'}|\psi\rangle + P_{--'}|\psi\rangle \\ &= \tfrac{1}{\sqrt{2}}\langle b'_+|b_-\rangle|a_+\rangle|b'_+\rangle + \tfrac{1}{\sqrt{2}}\langle b'_-|b_+\rangle|a_-\rangle|b'_-\rangle\end{aligned}$$

where the first component represents the projection of $|\psi\rangle = \frac{1}{\sqrt{2}}\ |a_+\rangle|b_-\rangle - \frac{1}{\sqrt{2}}|a_-\rangle|b_+\rangle$ onto $|a_+\rangle|b'_+\rangle$, and the second component represents the projection of $|\psi\rangle$ onto $|a_-\rangle|b'_-\rangle$. (Note that $|a_-\rangle|b_+\rangle$ has zero projection onto $|a_+\rangle|b'_+\rangle$, and $|a_+\rangle|b_-\rangle$ has zero projection onto $|a_-\rangle|b'_-\rangle$.) Since the two components are orthogonal, the square of the length of this vector is just:

$$\tfrac{1}{2}|\langle b'_+|b_-\rangle|^2 + \tfrac{1}{2}|\langle b'_-|b_+\rangle|^2$$

Now, the relationship between the spin observables of a spin-$\frac{1}{2}$ particle is such that a spin observable B', oriented at an angle 120° in real space to a spin observable B, is associated with an orthogonal pair of eigenvectors $|b'_+\rangle, |b'_-\rangle$ oriented at an angle 60° in Hilbert space to the orthogonal pair of eigenvectors $|b_+\rangle, |b_-\rangle$ associated with B. So

$$\tfrac{1}{2}|\langle b'_+|b_-\rangle|^2 + \tfrac{1}{2}|\langle b'_-|b_+\rangle|^2 = \tfrac{1}{2}\sin^2 60° + \tfrac{1}{2}\sin^2 60° = \tfrac{3}{8} + \tfrac{3}{8} = \tfrac{3}{4}$$

and similarly for the five other setting pairs AB'', $A'B$, $A'B''$, $A''B$, $A''B'$ separated by an angle of 120°.

It follows that:

$$p_\psi(++ \text{ or } --|\text{random settings}) = \tfrac{1}{9}\,(3 \times 0 + 6 \times \tfrac{3}{4}) = \tfrac{1}{2}$$

One could design a sort of quantum mechanical roulette consisting of two separated 'wheels,' with a choice of three settings for each 'wheel' chosen randomly by the croupier, and one of eight possible common causes, also chosen randomly by the croupier, that can produce a '+' or a '−' at each 'wheel.' For example, a light might flash green for '+' or red for '−,' depending on the setting – say one of three types of slot (square, round, or triangular) at each 'wheel' – and the colour (green or red) of a bump (square, round, or triangular) on each of two 'dies,' one moving towards the left 'wheel,' the other (oppositely coloured to the left 'die' on each type of bump) moving towards the right 'wheel.' We suppose that if a left-moving 'die' falls onto a round slot of the left 'wheel,' then the colour of the round bump determines the colour of the light flash at the left 'wheel,' and similarly for the right 'wheel.' The gambler has to guess the outcome 'same colour' or 'different colour.' The empirically correct odds are even, and these are the odds recommended by quantum mechanics. But on the assumption that the game is not rigged, that there is no device under the table, biasing the setting choice of the 'wheels' in some way after the croupier randomly selects one of the eight possible pairs of oppositely coloured 'dice' (oppositely coloured with respect to each of the three types of bump), then even if the croupier cheats and selects only certain pairs of oppositely coloured 'dice' (so that certain pairs are never selected), there is no way the long-run relative frequency of 'same colour' can exceed $\frac{4}{9}$.

It would seem that such a quantum roulette must be producing the perfect opposite correlations (red–green or green–red) for the settings 'square–square,' 'round–round,' and 'triangular–triangular' without any common cause for the correlations – quite unlike the example of oppositely correlated covert operations in the James Bond example considered in the previous section.

This conclusion would, of course, be old news to physicists: the presumption has always been that there are no hidden variables underlying the quantum statistics, no common causes responsible for the statistical correlations of the theory. Rather, the peculiar features of the quantum statistics – interference, mysterious nonlocal correlations, and so forth – are taken as *sui generis*, explained by the empirical adequacy of the Hilbert space representation of quantum states. But note that, in the first place, this presumption requires justification in terms of structural features of the quantum description of events, and this is just what Bell's theorem provides. And, in the second place, since the singlet state $|\psi\rangle$ does not screen off M_L-outcomes from M_R-outcomes, that is to say:

$$p_{\psi}(a_+ \& b_+) \neq p_{\psi}(a_+)p_{\psi}(b_+),$$

$|\psi\rangle$ is not a common cause of the nonlocal statistical correlations, and so it is not clear in what sense quantum mechanics provides an *explanation* of the singlet state correlations, beyond a recipe for their calculation.

The general lesson of Bell's theorem is that a causal explanation of the statistical correlations generated by entangled quantum states for separated systems is not available cheaply: some feature of the separability and locality requirements that Einstein regarded as fundamental for a realist philosophy of physics will have to be given up. This price is paid by the observer-free 'no collapse' interpretations considered in chapters 5 and 6.

2.5 Locality and separability

The crucial assumption in the above proofs of Bell's theorem was the assumption of conditional statistical independence ('strong locality,' or 'factorizability'). I show here that this condition is equivalent to the conjunction of two independent conditions that capture different aspects of separability and locality.

Suppose that in a correlation experiment of the type considered above

$$p^{AB}(a\&b) = \int p_{\lambda}^{AB}(a\&b)\rho(\lambda)\mathrm{d}\lambda, \tag{2.5a}$$

where $\rho(\lambda)$ is the distribution of common causal factors or hidden variables yielding the joint probability of an outcome a (a_+ or a_-) for an A-measurement at M_L and an outcome b (b_+ or b_-) for a B-measurement at M_R (and similarly for other possible measurements AB', AB'', ..., $A'B$, $A'B'$, ..., etc.). If λ takes a finite or countable set of values (as in the singlet state example), then

$$p^{AB}(a\&b) = \sum_i p^{AB}(a\&b|\lambda_i)p(\lambda_i) \tag{2.5b}$$

and similarly for other pairs of observables.

The superscript AB on the 'surface' or measured probabilities – these are also the probabilities predicted by quantum mechanics – and the 'hidden' or common cause probabilities indicates that M_L is set to measure A and M_R is set to measure B.[58] This notation is redundant for joint probabilities (since the notation for the outcome refers explicitly to an AB-measurement), but not for the marginal probabilities:

$$p_\lambda^{AB}(a) \equiv p_\lambda^{AB}(a\&b_+) + p_\lambda^{AB}(a\&b_-)$$
$$p_\lambda^{AB}(b) \equiv p_\lambda^{AB}(a_+\&b) + p_\lambda^{AB}(a_-\&b)$$

(Note that a and b are variables over a-values and b-values, respectively. The symbol '$\equiv$' here denotes 'equal by definition.')

It is *not* assumed initially that

$$p_\lambda^{AB}(a) = p_\lambda^{AB'}(a) = p_\lambda^{AB''}(a) = \ \ldots$$

It may even be the case that all these probabilities are different and that none of these probabilities is equal to $p_\lambda^A(a)$, the probability of the outcome a for a measurement of A at M_R. A similar remark applies to the probabilities $p_\lambda^{AB}(b)$, $p_\lambda^{A'B}(b)$, $p_\lambda^{A''B}(b)$, etc., and $p_\lambda^B(b)$.

Suppose there are correlations in the surface probabilities. That is, suppose

$$p^{AB}(a\&b) \neq p^{AB}(a)\, p^{AB}(b)$$

and similarly for other observables.

Consider the following two conditions on the hidden probabilities:

Outcome independence (OI):

$$p_\lambda^{AB}(a|b) = p_\lambda^{AB}(a) \qquad (2.6a)$$
$$p_\lambda^{AB}(b|a) = p_\lambda^{AB}(b) \qquad (2.6b)$$

for all values a of A and b of B (and similarly for all pairs of observables).

Parameter independence (PI):

$$p_\lambda^{AB}(a) = p_\lambda^{A}(a) \qquad (2.7a)$$

for all values a of A, and similarly for B', B'', . . . , instead of B, and

$$p_\lambda^{AB}(b) = p_\lambda^{B}(b) \qquad (2.7b)$$

for all values b of B, and similarly for A', A'',. . . , instead of A (and similarly with A', A'',. . . , for A in (2.7a), and B', B'', . . . , for B in (2.7b)).

The conditions OI and PI, and the terms 'outcome independence' and 'parameter independence' were introduced by Shimony (1984a, b), following Jarrett's unpacking of conditional statistical independence as two conditions that he termed 'completeness'

[58] This handy terminological distinction between 'surface' and 'hidden' probabilities is due to van Fraassen (1982).

and 'locality' (Jarrett, 1984). Similar conditions were introduced earlier by Suppes and Zanotti (1976), and by van Fraassen (1982). Van Fraassen's 'causality' condition is a definitionally equivalent form of OI, and his 'hidden locality' condition is equivalent to PI. Jarrett's completeness and locality conditions correspond to Shimony's conditions OI and PI, respectively, but differ in that Jarrett explicitly considers the probabilistic dependence of outcomes at M_L and M_R on causal factors at M_L and M_R, in addition to the parameter settings at M_L and M_R (the particular observables measured). These are, in effect, hidden variables labelling hidden states of the measuring instruments that I have characterized below simply as 'independent random factors at M_L and M_R.' I shall take up this point below.

OI says that the λ-probability – the 'hidden' or common cause probability – of the outcome of a measurement at M_L is independent of the outcome of any measurement at M_R. That is, λ screens off M_L-outcomes from M_R-outcomes, and similarly for M_R-outcomes with respect to M_L-outcomes.

PI says that the λ-probability of the outcome of any measurement at M_L depends only on what is measured at M_L, and not on the observable (the 'parameter setting') measured at M_R (and similarly for M_R with respect to M_L).

If both these conditions are satisfied, the systems S_L and S_R are independent, given λ, in the sense (OI) that the probability of the outcome of any measurement on one system is independent of the outcome of any measurement on the other system, and also in the sense (PI) that the probability of the outcome of any measurement on one system is independent of what observable is measured on the other system. So λ is the 'whole story' about the systems S_L and S_R. The specification of λ characterizes S_L and S_R as fully as possible with respect to possible measurements on the systems. Any remaining distribution in M_L-measurement outcomes for a particular value of λ can only be due to independent random factors at M_L and M_R.

To see that conditional statistical independence as formulated by equation (2.3) follows from the two conditions OI and PI for the observables A and B, note that:

$$p_\lambda^{AB}(a\&b) \equiv p_\lambda^{AB}(a|b)p_\lambda^{AB}(b)$$

from the definition of conditional probability. By OI:

$$p_\lambda^{AB}(a\&b) = p_\lambda^{AB}(a)p_\lambda^{AB}(b)$$

and by PI:

$$p_\lambda^{AB}(a\&b) = p_\lambda^{A}(a)p_\lambda^{B}(b) \tag{2.8}$$

The superscripts are redundant for the probabilities in (2.8), so (2.3) follows:

$$p_\lambda(a\&b) = p_\lambda(a)p_\lambda(b)$$

To derive PI from conditional statistical independence, note that conditional statistical independence as expressed by (2.8) holds for b_+ and b_-:

$$p_\lambda^{AB}(a\&b_+) = p_\lambda^A(a)p_\lambda^B(b_+)$$
$$p_\lambda^{AB}(a\&b_-) = p_\lambda^A(a)p_\lambda^B(b_-)$$

adding these two equations yields (2.7a):

$$p_\lambda^{AB}(a) = p_\lambda^A(a)$$

since $p_\lambda^B(b_+) + p_\lambda^B(b_-) = 1$. (Note that the probability $p_\lambda^{AB}(a)$ is derived as a marginal probability here, with respect to a B-measurement on the system S_R, while $p_\lambda^A(a)$ has the significance of the λ-probability of the outcome a for an A-measurement on S_L when no measurement is performed on S_R.) Obviously, a similar derivation holds for (2.7b):

$$p_\lambda^{AB}(b) = p_\lambda^B(b)$$

and for other observables.

To derive OI from conditional statistical independence, we exploit the derivation of PI from conditional statistical independence and then show that OI follows from conditional statistical independence and PI. So assume (2.8):

$$p_\lambda^{AB}(a\&b) = p_\lambda^A(a)p_\lambda^B(b)$$

and also (2.7a) and (2.7b):

$$p_\lambda^{AB}(a) = p_\lambda^A(a)$$

$$p_\lambda^{AB}(b) = p_\lambda^B(b)$$

from which it follows that:

$$p_\lambda^{AB}(a\&b) = p_\lambda^{AB}(a)p_\lambda^{AB}(b) \tag{2.9}$$

Condition (2.9), not be to confused with conditional statistical independence as expressed by (2.8):

$$p_\lambda^{AB}(a\&b) = p_\lambda^A(a)p_\lambda^B(b),$$

is definitionally equivalent to OI as formulated above, by the definition of conditional probability:

$$\begin{aligned} p_\lambda^{AB}(a\&b) &\equiv p_\lambda^{AB}(a)p_\lambda^{AB}(b|a) \\ &\equiv p_\lambda^{AB}(b)p_\lambda^{AB}(a|b) \end{aligned}$$

Substituting for $p_\lambda^{AB}(a\&b)$ in (2.9) yields (2.6b):

$$p_\lambda^{AB}(b|a) = p_\lambda^{AB}(b)$$

and (2.6a):

$$p_\lambda^{AB}(a|b) = p_\lambda^{AB}(a)$$

(Similarly, the definition of OI in terms of the joint probability follows from these two equations.)

Van Fraassen (1982) calls this definitional equivalent of OI 'causality.' Howard refers to the condition as 'factorizability' (1989, p. 231), but note that conditional statistical independence is usually referred to as 'factorizability' in the literature.

The derivation of OI and PI from conditional statistical independence here follows Howard's analysis (1989, p. 231). While the unpacking of conditional statistical independence as two independent conditions capturing different aspects of separability and locality is generally attributed to Jarrett (1984), all the essential points are already in van Fraassen's anti-realist argument (1982), that there are quantum phenomena involving separated systems that exhibit statistical correlations for which there can be no common cause.

A common position in the literature appears to be that the failure of conditional statistical independence should be understood as a failure of outcome independence rather than parameter independence. Otherwise, it is argued, the possibility of influencing the statistics of measurement outcomes on a system by manipulating a measurement parameter under our control on a distant system – for a given λ-value or suitably restricted range of λ-values – would allow the sort of superluminal signalling between spatio-temporally separate events that conflicts with special relativity. No such instantaneous signalling is possible from a violation of outcome independence alone, simply because the relevant measurement outcomes are only constrained stochastically and are not under our control (even for a given λ-value or suitably restricted range of λ-values). As Shimony (1984a, b, 1986) puts it, dropping OI and keeping PI allows a 'peaceful coexistence' between quantum mechanics and relativity: we avoid action at a distance (the failure of PI) but have to live with a sort of 'passion at a distance' (the failure of OI) exhibited by certain quantum phenomena. See also Jarrett (1984) and Howard (1985, 1989).

This view has been challenged by several authors. Maudlin (1994, especially chapter 4) discusses the issue in a very clear and thorough analysis of Bell's inequality and relativity. Jones and Clifton (1993) show that, under certain circumstances, violations of completeness can lead to superluminal signalling, no less than violations of locality.

There is a difference between Jarrett's original formulation of his completeness and locality conditions (in which he explicitly takes account of hidden states of the measuring instruments, as well as the common cause λ and the observables measured by the instruments), and the conditions OI and PI. Jones and Clifton note that the probability of a particular measurement outcome at M_L (for certain values of λ and parameter settings of M_L and M_R, and certain hidden states of M_R) might depend (in part) on the values of hidden variables parametrizing hidden states of the instrument M_L (and similarly for M_R). It might also be the case that the probability of a particular measurement outcome at M_L (for certain values of λ and certain parameter settings of M_L and M_R, and certain hidden states of M_L), conditional on a certain measurement outcome at M_R, is independent of variations in the hidden variables of M_R that leave this measurement outcome at M_R unchanged (and similarly for M_R-outcomes with respect to the hidden variables of M_L). They refer to the former condition as 'measurement contextualism' and the latter as 'constrained locality,' to distinguish it from Jarrett's locality condition, which requires measurement outcomes at one measuring instrument to be probabilistically independent of both the observable measured by the distant instrument and the hidden states of the distant instrument. It would then follow – assuming measurement contextualism and constrained locality – that manipulating the hidden variables of M_R could affect certain measurement outcomes at M_R probabilistically, and hence – if completeness fails – could influence the probabilities of certain measurement outcomes at M_L (and similarly for the hidden variables of M_L and measurement outcomes at M_R). Constrained locality ensures that manipulating the hidden states of the instrument M_R as a stochastic 'trigger' for certain outcomes at M_R does not have a causal effect on the outcome at M_L, by some route other than the causal link between outcomes at M_R and outcomes at M_L characterized by the failure of completeness, that cancels its effect via the outcome at M_R. In effect, a failure of completeness under these circumstances entails a failure of locality (because measurement outcomes at one measuring instrument will depend probabilistically on variations in the hidden variables of the distant instrument), and hence the possibility of superluminal signalling.

In a deterministic universe, where all the λ-probabilities are 0 or 1, Jarrett's completeness condition, and hence OI, is automatically satisfied (because conditionalizing a probability that is 0 or 1 leaves the probability unchanged, or equivalently, if $p_\lambda^{AB}(a)$ is equal to 0 or 1 and $p_\lambda^{AB}(b)$ is equal to 0 or 1, then $p_\lambda^{AB}(a\&b) = p_\lambda^{AB}(a)p_\lambda^{AB}(b)$). One might think that fixing a parameter λ, with respect to which completeness or OI holds for the statistics of measurement outcomes on separated systems in an indeterministic universe, means that we have characterized the systems as completely as possible (hence Jarrett's term 'completeness'), and that the condition can therefore be distinguished from Jarrett locality or PI as doing the work of a stochastic version of Einstein's separability condition. From the Jones and Clifton argument, it seems rather that these two conditions together capture different aspects of separability and locality associated with the notion of a common cause for the statistical correlations of separated systems in an indeterministic universe. It is only with respect to the common cause (if such a cause exists) that we are entitled to regard the two systems as separable, in the sense that the outcome of any measurement on each system separately depends solely on the stochastic 'being-thus' of the system (through the common cause), the observable measured, and local hidden variables of the measuring instrument – there are no other causally relevant factors. In other words, given the common cause, each system manifests its properties separately to the measuring instruments M_L and M_R, without reference to the other system.

Whether or not the experiments that have actually been performed do rule out common causes for the statistical correlations of certain quantum phenomena – whether the experimental counts can be taken as measuring the probabilities in the Clauser–Horne inequalities – depends, of course, on certain assumptions about how the measuring instruments function. While most physicists take the Aspect *et al.*

experiments (1981, 1982a, b) as ruling out 'local hidden variables' or common causes, there is a significant minority view that regards this question as still open. See Ballentine (1987) for a review of experiments designed to test various versions of Bell's inequality, and Home and Selleri (1991) for a detailed critical analysis of what the different experimental results do and do not show.

3

The Kochen and Specker 'no go' theorem

> We wish to entertain the heretical view that the results of a measurement are not brought into being by the act of measurement itself. This heresy takes the state vector to describe an ensemble of systems and maintains that in each individual member of that ensemble every observable does indeed have a definite value, which the measurement merely reveals when carried out on that particular individual system. . . . To this kind of talk the well-trained quantum mechanician says 'Rubbish!' and gets back to serious business. But is it possible to offer a better rejoinder? Is it possible to demonstrate not only that the innocent view is at odds with the prevailing orthodoxy, but that it is, in fact, directly refuted by the quantum-mechanical formalism itself, without any appeal to an interpretation of that formalism?
>
> *Mermin (1993, pp. 804–5)*

3.1 The 'colouring' problem

Gleason's theorem (1957) shows that the set of quantum states is complete, in the sense that all possible probability measures μ definable on the lattice $\mathscr{L}$ of quantum propositions, represented by Hilbert space projection operators, are generated by the density operators of pure or mixed states by the Born rule.[59] This result might seem to foreclose the possibility of hidden variables underlying the quantum statistics. In fact, what Gleason's theorem excludes is a class of hidden variable reconstructions of the quantum statistics satisfying a certain constraint.

The standard Kolmogorov formulation of probability theory defines probability as a measure on a σ-field of subsets of a set, equivalently as a measure on a Boolean algebra of events. A generalized probability measure μ on a non-Boolean lattice $\mathscr{L}$ is required to satisfy the usual Kolmogorov conditions for a probability measure on each Boolean sublattice of $\mathscr{L}$. It follows from these conditions that if a and b are 'compatible' propositions belonging to the same Boolean sublattice of $\mathscr{L}$, then

[59] See the appendix, section A.2.

$$\mu(a \vee b) = \mu(a) + \mu(b) - \mu(a \wedge b)$$

If μ is a dispersion-free measure, h, assigning the probability 0 or 1 to each proposition in $\mathscr{L}$, then

$$h(a \wedge b) = h(a)h(b)$$

(because if either or both of the propositions a, b are assigned probability 0, the Boolean meet or conjunction $a \wedge b$ – the lattice infimum – is assigned probability 0, and if both a and b are assigned probability 1, $a \wedge b$ is assigned probability 1). So

$$h(a \vee b) = h(a) + h(b) - h(a)h(b)$$

Also

$$h(a^{\perp}) = 1 - h(a)$$

A dispersion-free state therefore defines what I called a 'Boolean homomorphism' on $\mathscr{L}$ in section 1.3 – a map $h: \mathscr{L} \rightarrow \{0, 1\}$ onto the 2-element Boolean algebra $\{0, 1\}$, satisfying the conditions:

$$\left.\begin{aligned} h(a \vee b) &= h(a) \vee h(b) \\ h(a \wedge b) &= h(a) \wedge h(b) \\ h(a^{\perp}) &= h(a)^{\perp} \end{aligned}\right\} \qquad (3.1)$$

on each Boolean sublattice of $\mathscr{L}$, since

$$\begin{aligned} &h(a) \vee h(b) = h(a) + h(b) - h(a)h(b) \\ &h(a) \wedge h(b) = h(a)h(b) \\ &h(a^{\perp}) = 1 - h(a) \end{aligned}$$

We can think of the map h as defining a bivalent truth-value assignement to the quantum propositions, which is classical on each Boolean sublattice of $\mathscr{L}$ (taking 1 as 'true,' 0 as 'false,' the lattice infimum or Boolean meet '$\wedge$' as conjunction, the lattice supremum or Boolean join '$\vee$' as disjunction, and the lattice or Boolean orthocomplement '$^{\perp}$' as negation). Then $a \vee b$ is true if and only if at least one of the propositions a or b is true, $a \wedge b$ is true if and only if both a and b are true, and $a^{\perp}$ is true if and only if a is false. Equivalently, h defines a valuation, v, assigning values to all quantum observables, where the values assigned to compatible observables are required to 'mesh' in a certain way (since any observable can be represented as a linear combination of projection operators or propositions with eigenvalues as coefficients, by the spectral representation theorem[60]). Gleason's theorem excludes hidden variable reconstructions of the quantum statistics in which the hidden variables parametrize dispersion-free probability measures represented by Boolean homomorphisms on $\mathscr{L}$ – no such dispersion-free measures are generated by any of the states in the set of pure and mixed

[60] See the appendix, section A.1.

quantum states, and these states exhaust the probability measures that can be generated on $\mathcal{L}$.

The Kochen and Specker theorem establishes this corollary of Gleason's theorem directly, for a relatively small finite sublattice of the lattice of quantum propositions $\mathcal{L}_3$ in $\mathcal{H}_3$. As Kochen and Specker see it (1967, p. 70), this avoids a possible objection 'that in fact it is not physically meaningful to assume that there are a continuous number of quantum mechanical propositions.' In his 1966 review article, Bell proved a version of the corollary to Gleason's theorem, exploiting a continuum of quantum propositions. As Mermin (1993) shows, Bell's proof can be reformulated using only a finite number of propositions or observables.

Kochen and Specker put the hidden variable problem this way: Suppose you have a theory with a set of observables $A, B, C, \ldots$ and a set of states $\{W\}$ that assign probabilities to ranges of values of these observables, as in quantum mechanics. Can you introduce a space X of 'hidden' states such that the observables $A, B, C, \ldots$ are represented as real-valued functions $f_A, f_B, f_C, \ldots$ on X:

$$f_A: X \to \mathbb{R}, \text{ etc.,}$$

where $\mathbb{R}$ here represents the set of real numbers, so that the probability assigned by a state W to a range of values E of the observable A is representable as the measure of the set of hidden states for which A takes a value in the range E, and similarly for ranges of values of observables $B, C, \ldots$:

$$\text{prob}_W(a \in E) = \rho_W(f_A^{-1}(E)), \text{ etc.,} \tag{3.2a}$$

where ρ is the distribution over hidden states associated with the state W? (So, for example, in the case of quantum mechanics, $\text{prob}_W(a \in E)$ would be the quantum mechanical probability assigned by the quantum state W to the range of values E of the observable A.) Equivalently, in terms of expectation values:

$$\text{Exp}_W(A) = \int_X f_A(x)\mathrm{d}\rho_W(x) \tag{3.2b}$$

As Kochen and Specker point out, without some further constraint any statistical theory can be given a hidden variable reconstruction in this sense. For example, suppose there are only two observables, A and B. Introduce hidden variables x_A and x_B, each ranging over the real line $\mathbb{R}$, for A and B. Let the value of x_A determine the value of A and the value of x_B determine the value of B. That is, suppose there are prediction functions $\phi_A(x_A)$ and $\phi_B(x_B)$ that associate values of the hidden variables x_A and x_B with values of the observables A and B, respectively. Define the space X as the Cartesian plane parametrized by x_A and x_B. The value of the function f_A on X representing the observable A is defined at the point (x_A, x_B) as the value assigned to A by x_A; that is, $f_A(x_A, x_B) = \phi_A(x_A)$, for all values of x_B. Similarly, $f_B(x_A, x_B) = \phi_B(x_B)$, for all values of x_A.

Define the probability measure on X corresponding to the state W as the product measure

$$\rho_W = \rho_{WA} \cdot \rho_{WB},$$

where

$$\rho_{WA}(\phi_A^{-1}(E)) = \text{prob}_W(a \in E)$$
$$\rho_{WB}(\phi_B^{-1}(F)) = \text{prob}_W(b \in F)$$

Then

$$f_A^{-1}(E) = \phi_A^{-1}(E) \times \mathbb{R}$$
$$f_B^{-1}(F) = \mathbb{R} \times \phi_B^{-1}(F)$$

where '$\times$' represents the Cartesian product (that is, $U \times V$ is the set of all ordered pairs of elements from the sets U and V, respectively), and so

$$\begin{aligned} \rho_W(f_A^{-1}(E)) &= \rho_{WA}(\phi_A^{-1}(E)) \cdot \rho_{WB}(\mathbb{R}) \\ &= \rho_{WA}(\phi_A^{-1}(E)) \cdot 1 \\ &= \text{prob}_W(a \in E) \end{aligned}$$

and similarly

$$\begin{aligned} \rho_W(f_B^{-1}(F)) &= \rho_{WB}(\mathbb{R}) \cdot \rho_{WB}(\phi_B^{-1}(F)) \\ &= 1 \cdot \rho_{WB}(\phi_B^{-1}(F)) \\ &= \text{prob}_W(b \in F) \end{aligned}$$

In words: the set of points ('hidden states') in X for which A takes a value in the set E is the set of sequences (x_A, x_B), where x_A is restricted to the set $\phi_A^{-1}(E)$ yielding an A-value in the range E, and x_B ranges over the real line $\mathbb{R}$ and so contributes unit probability to the product; and similarly for B lying in the range F.

Evidently, this trivial construction can be extended to any number of observables and, as Kochen and Specker show, even to a continuous infinity of observables. The crucial feature of this construction is the association of each observable with an *independent* random variable on the measure space X. Presumably, the observables of the original theory satisfy certain algebraic or functional relationships. Some observables might be functions of other observables – for example, an observable might be the square of another observable, or the sum of two other observables. There is no requirement that any of these relationships is preserved in this hidden variable reconstruction of the statistics, and it is just the lack of any such constraint that makes the construction work for *any* theory. Since hidden variables in this sense can always be introduced to reproduce the statistics of any theory, the mere possibility of the construction tells us nothing about the character of the statistics of the theory in question, and is of no interest.

The problem of hidden variables for quantum mechanics only becomes interesting, then, if we impose constraints on the hidden variables. We can then hope to see quantum mechanics as incompatible with hidden variables satisfying various constraints, and in this way learn something about the peculiar nature of the quantum statistics.

Kochen and Specker require that the values assigned by hidden variables or hidden states to a set of *mutually compatible* observables[61] of a quantum system, represented by pairwise commuting self-adjoint operators, should preserve the functional relations holding among these observables. As Mermin formulates the requirement, if a set of such observables $A, B, C, \ldots$ satisfies a functional identity of the form

$$f(A, B, C, \ldots) = 0$$

then the values $v(A), v(B), v(C), \ldots$ assigned to these observables by the hidden variables of a quantum system should satisfy the same functional identity:

$$f(v(A), v(B), v(C), \ldots) = 0 \tag{3.3}$$

(For example, if the constraint requires that $A = B^2$, or $A - B^2 = 0$, then the value assigned to the observable A should be the square of the value assigned to the observable B.)

The earlier von Neumann proof in effect imposed a functional relationship constraint for any set of observables, commuting or noncommuting. Specifically, von Neumann required that if

$$C = A + B$$

then

$$\text{Exp}(C) = \text{Exp}(A) + \text{Exp}(B)$$

for all states, including dispersion-free states, so that

$$v(C) = v(A) + v(B)$$

whether or not A and B commute. (The functional identity in this case is, of course, $C - A - B = 0$.) This is the requirement that Bell and Mermin characterize as 'silly' (Mermin, 1993, p. 805, cited in section 2.1) because it fails to hold in general for the eigenvalues of noncommuting observables, even in a 2-dimensional Hilbert space, where a hidden variable reconstruction of the quantum statistics in the Kochen and Specker sense can be shown to be possible. (For example, the eigenvalues of σ_x and σ_y are ± 1, but the eigenvalues of $\sigma_x + \sigma_y$ are $\pm\sqrt{2}$.) Von Neumann justified the requirement on the basis that it holds for expectation values in quantum mechanics. But there is no *a priori* reason why this relation could not be recovered for expectation values computed for hidden variable distributions associated with quantum states, while failing for more general distributions over the hidden variables, including

[61] In Kochen and Specker's terminology, mutually 'commeasurable' observables.

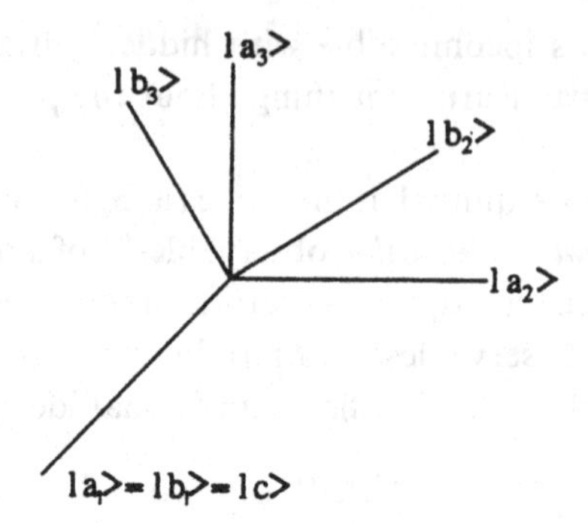

Figure 3.1. Eigenvectors of two noncommuting observables A and B in $\mathcal{H}_3$, with a common eigenvector $|a_1\rangle = |b_1\rangle = |c\rangle$.

dispersion-free hidden variable distributions (concentrated at points in the hidden variable space X).[62]

The motivation for Bell's theorem arose, partly, from Bell's (1966) critique of the functional relationship constraint on value assignments to compatible observables in his own proof of the corollary to Gleason's theorem as a 'no go' hidden variables theorem. Consider two noncommuting observables, A with eigenvectors $|a_1\rangle$, $|a_2\rangle$, $|a_3\rangle$, and B with eigenvectors $|b_1\rangle$, $|b_2\rangle$, $|b_3\rangle$, where $|a_1\rangle = |b_1\rangle = |c\rangle$ is a common eigenvector, and $|b_2\rangle$, $|b_3\rangle$ are obtained by rotating $|a_2\rangle$, $|a_3\rangle$ through an acute angle in the plane orthogonal to $|c\rangle$, as shown in figure 3.1.

Since

$$\begin{aligned} I &= P_c + P_{a_2} + P_{a_3} \\ &= P_c + P_{b_2} + P_{b_3} \end{aligned}$$

the constraint requires that

$$\begin{aligned} v(I) &= 1 \\ &= v(P_c) + v(P_{a_2}) + v(P_{a_3}) \\ &= v(P_c) + v(P_{b_2}) + v(P_{b_3}) \end{aligned}$$

and so

$$v(P_{a_2}) + v(P_{a_3}) = v(P_{b_2}) + v(P_{b_3}) = 1 - v(P_c)$$

If $v(P_c) = 1$, $v(P_{a_2}) = v(P_{a_3}) = 0$ and $v(P_{b_2}) = v(P_{b_3}) = 0$. That is, if the propositions represented by any pair of projection operators onto orthogonal rays in a plane in $\mathcal{H}_3$ are assigned the value 0 ('false') by the hidden variables, then any proposition represented by a projection operator onto a ray in the plane must be assigned the value

[62] Hidden variable models of a spin-$\frac{1}{2}$ system have been constructed by Kochen and Specker, and by Bell. See Bell (1966), Kochen and Specker (1967), Mermin (1993). Note that these models do not yield the quantum statistics for sequences of measurements, or for conditional probabilities, unless the hidden variables are randomized between measurements. See Bub (1973, 1976).

0 (since the rotation from $|a_2\rangle, |a_3\rangle$ to $|b_2\rangle, |b_3\rangle$ is through an arbitrary acute angle and so covers all rays in the plane). Thus, a constraint on value assignments to *incompatible* propositions represented by noncommuting projection operators follows from the homomorphism constraint, or the constraint on value assignments to *compatible* observables. Alternatively, if A is assigned the value a_1 corresponding to the common eigenvector $|a_1\rangle = |c\rangle$, then the functional relationship constraint requires that the noncommuting observable B must be assigned the eigenvalue b_1 corresponding to the eigenvector $|b_1\rangle = |c\rangle$. Bell's observation is that measurement of P_c in the context of P_{a_2}, P_{a_3} (that is, via an A-measurement) cannot be required, *a priori*, to yield the same value as a measurement of P_c in the context of P_{b_2}, P_{b_3} (that is, via a B-measurement) (Bell, 1966, p. 450):

It was tacitly assumed that measurement of an observable must yield the same value independently of what other measurements may be made simultaneously. . . . These different possibilities require different experimental arrangements; there is no *a priori* reason to believe that the results . . . should be the same. The result of an observation may reasonably depend not only on the state of the system (including hidden variables) but also on the complete disposition of the apparatus.

Kochen and Specker formulate the noncontextuality requirement a little differently from Mermin. The observable $g(A)$ is defined for every observable A and Borel function $g: \mathbb{R} \rightarrow \mathbb{R}$ by the condition:

$$\mu_{g(A)\psi}(S) = \mu_{A\psi}(g^{-1}(S)), \text{ for all quantum states } \psi \text{ and Borel sets } S,$$

where μ is a probability measure on the real line, assuming that $X = Y$ if $\mu_{X\psi} = \mu_{Y\psi}$ for all ψ. Kochen and Specker associate with each observable A a function f_A mapping values of the hidden variable onto values of A. They require that:

$$f_{g(A)} = g(f_A) \tag{3.4}$$

for every Borel function g and observable A. (In Mermin's notation, this would be the requirement that:

$$v(g(A)) = g(v(A)),$$

where $v(X)$ is the value assigned to an observable X for specific values of the hidden variables.) With sums and products defined for mutually compatible observables only, the observables of a quantum mechanical system form a partial algebra, and the idempotent observables form a partial Boolean algebra.[63] Kochen and Specker show that a necessary condition for the assignment of values to all the observables of a quantum mechanical system, in such a way as to satisfy the functional relationship

[63] A partial Boolean algebra can be characterized as the union of a family of Boolean algebras, 'pasted together' so that the maximum and minimum elements of all the Boolean algebras are identified (and in general certain other elements as well), and so that, for every n-tuple of pairwise compatible elements (it suffices to consider triples), there exists a Boolean algebra in the family containing these elements. Two elements are compatible if and only if they belong to a common Boolean algebra.

constraint, is the existence of an embedding of the partial algebra of observables into a commutative algebra; equivalently, the embedding of the partial Boolean algebra of idempotent observables into a Boolean algebra.[64] A necessary and sufficient condition for the existence of an embedding of a partial Boolean algebra into a Boolean algebra is that, for every pair of distinct elements p, q in the partial Boolean algebra, there exists a homomorphism h onto the 2-element Boolean algebra $\{0, 1\}$ such that $h(p) \neq h(q)$.

As Kochen and Specker show, there are no 2-valued homomorphisms on the partial Boolean algebra of projection operators on a Hilbert space of three or more dimensions (much less any embeddings). This is the corollary to Gleason's theorem proved by Bell in the 1966 review article. For the homomorphism condition implies that for every orthogonal n-tuple of 1-dimensional projection operators or corresponding rays in $\mathscr{H}_n$, one projection operator or ray is mapped onto 1 ('true') and the remaining $n - 1$ projection operators or rays are mapped onto 0 ('false'). This is shown to be impossible for a particular set of 43 orthogonal triples of rays that can be constructed from 117 appropriately chosen rays in $\mathscr{H}_3$: any assignment of 1's and 0's to this set of orthogonal triples satisfying the homomorphism condition involves a contradiction.

Hidden variables theories that satisfy the Kochen and Specker constraint are termed 'noncontextual.' While Bell argued in the 1966 review article that the general requirement of noncontextuality cannot be justified on physical grounds, he pointed out that a special case of noncontextuality is physically plausible. An observable A can be compatible with an observable B and also with an observable C, while B and C are incompatible (represented by noncommuting operators). In this case A and B will be representable as functions of an observable X, while A and C will be representable as functions of an observable Y, incompatible with X. If A denotes an observable of a system S, and B and C denote incompatible observables of a system S', space-like separated from S, then the functional relationship constraint, that the value of A as a function of X should be the same as the value of A as a function of Y, becomes a locality condition. Bell (1966) argued that the general functional relationship constraint cannot be justified physically, but this weaker locality condition can. Bell's theorem (1964)[65] shows that the locality condition cannot be satisfied in general in a Hilbert space of four or more dimensions. There are sets of observables for which value assignments satisfying the locality condition are inconsistent with the quantum statistics.[66] From this perspective, it appears that Bell's theorem is a stronger version of the Kochen and Specker theorem: any hidden variable theory satisfying the Kochen and Specker constraint will necessarily satisfy the locality condition, but not conversely. Putting it differently, the Kochen and Specker theorem shows that the value assigned to an observable A in a hidden variable theory will in general have to depend on what other observables are measured together with A. That is, the value of A will depend on the

[64] An embedding of a partial Boolean $\mathscr{L}$ into a Boolean algebra $\mathscr{B}$ is a homomorphism $h: \mathscr{L} \rightarrow \mathscr{B}$ that is 1–1 into $\mathscr{B}$.

[65] Recall that the 1964 paper was written after the 1966 review article!

[66] More recent versions of Bell's theorem, such as the Greenberger–Horne–Zeilinger theorem (1989), do not require statistical arguments, as we saw in section 2.2, or minimize the statistics needed (for example, Hardy, 1992).

complete experimental arrangement in which A is measured, or on the measurement context. Bell's theorem establishes that the value assigned to A must depend on the complete experimental arrangement, even when two alternative arrangements differ only in a region space-like separated from the region in which A is measured.

The Kochen and Specker theorem in $\mathscr{H}_3$ can be regarded as the solution to a colouring problem on the surface of the unit sphere. An assignment of values to all the observables in $\mathscr{H}_3$ that preserves the functional relations holding among mutually compatible observables (represented by commuting self-adjoint operators) is possible only if truth values – 1's and 0's – can be assigned to all the quantum propositions (idempotent observables, represented by projection operators), in such a way as to define a Boolean homomorphism on the lattice of propositions, $\mathscr{L}$. Equivalently, regarding $\mathscr{L}$ as a partial Boolean algebra, in which lattice infima and suprema are defined for mutually compatible elements only, the map is a 2-valued homomorphism on the partial Boolean algebra. This requires, as we have seen, that every orthogonal triple of rays is assigned one 1 and two 0's (which corresponds to selecting one of the three possible eigenvalues for every maximal observable). If two orthogonal triples of rays share a common ray, then an assignment that selects this common ray (that is, a map that assigns this ray a 1) must select this ray for both orthogonal triples. The Kochen and Specker theorem shows that no such assignment of 1's and 0's is possible to all the rays in $\mathscr{H}_3$.

Associating the colour green with the truth value 1 and red with the truth value 0 (that is, green for 'yes' or 'true,' and red for 'no' or 'false'), this means that we cannot colour the surface of the unit sphere in $\mathscr{H}_3$ with two colours, green and red, such that for every orthogonal triple of points on the sphere (associated with an orthogonal triple of rays that cut the sphere at these points), one and only one point is coloured green and the remaining two points are coloured red. The impossibility of such a colouring – a result that holds equally for the surface of a sphere in ordinary (real) Euclidean space – reflects a topological property of the surface of a sphere. So a hidden variable reconstruction of the quantum statistics, satisfying the functional relationship constraint, is impossible because you can't colour a ball with red and green dots in such a way that every orthogonal triple of dots includes one green dot and two red dots!

Kochen and Specker demonstrated the impossibility of this colouring on a particular set of 117 points on the surface of the unit sphere. For every colouring of the 117 points as green or red, Kochen and Specker showed, in effect, that there would exist at least one orthogonal triple of points violating the colouring constraint.

A spin-1 system provides a physical realization illustrating this impossible colouring. The three 0-eigenvectors of the spin component observables S_x, S_y, S_z in three orthogonal directions form an orthogonal triple.[67] The 1-eigenplanes of S_x^2, S_y^2, S_z^2 intersect at right angles in the rays spanned by the vectors $|S_x = 0\rangle$, $|S_y = 0\rangle$, $|S_z = 0\rangle$,

[67] See the appendix, section A.4. Note that the directions x, y, z here refer to directions of the spin components in real space. They should not be confused with directions in Hilbert space. Each spin component, S_x say, has three eigenvectors, $|S_x = -1\rangle$, $|S_x = 0\rangle$, $|S_x = +1\rangle$, and these define three orthogonal directions in Hilbert space, that is, three orthogonal points on the unit sphere in $\mathscr{H}_3$.

and so their projection operators commute. These projection operators are just S_x^2, S_y^2, S_z^2, respectively. That is:

$$S_x^2 = 1 \cdot P_{x=0}^{\perp} + 0 \cdot P_{x=0}, \text{ etc.}$$

in the spectral representation, where $P_{x=0} = |S_x = 0\rangle\langle S_x = 0|$. As shown in the appendix, section A.4, the observable

$$\begin{aligned} H &= aS_x^2 + bS_y^2 + cS_z^2 \\ &= (b + c)P_{x=0} + (a + c)P_{y=0} + (a + b)P_{z=0} \end{aligned}$$

with distinct values for a, b, c, is a maximal observable with eigenvalues $b + c$, $a + c$, $a + b$. The mutually compatible nonmaximal observables S_x^2, S_y^2, S_z^2 are all functions of H, and so they can all be measured simultaneously by measuring H. When H takes the value $b + c$, S_x^2 takes the value 0; when H takes the value $a + c$, S_y^2 takes the value 0; and when H takes the value $a + b$, S_z^2 takes the value 0. Kochen and Specker (1967) outline a procedure for measuring H.

Consider a particular set of 117 directions in real space and the 117 spin component observables associated with these directions. Each squared spin component observable is associated with a 0-eigenray and a 1-eigenplane in $\mathscr{H}_3$. That is, the 117 directions in real space correspond to 117 0-eigenrays that cut the unit sphere in $\mathscr{H}_3$ at 117 points (and 117 planes orthogonal to these 117 rays). Every assignment of 1's and 0's to these 117 rays or points on the unit sphere (or every colouring of these points as green or red) amounts to a set of predictions as to which of the 117 spin component observables have eigenvalue 0 and which have eigenvalue nonzero (± 1). The Kochen and Specker proof now shows (if the 117 directions are chosen appropriately) that for every such prediction there are three orthogonal directions x, y, z among the 117 directions in real space for which the prediction violates the requirement that exactly one of the three components has the eigenvalue 0. In effect, no assignment of values to the nonmaximal squared spin component observables for the 117 directions, and the maximal H-observables associated with these squared spin components, can be consistent with the requirement that the value assignment should preserve the functional relations holding among mutually compatible observables. The 117 directions are chosen in such a way that a given squared spin component observable is a function of multiple noncommuting H-observables (that is, a given 0-eigenray of some spin component will belong to multiple orthogonal triples of 0-eigenrays).

As Kochen and Specker observed, the impossibility of colouring these 117 points, subject to the colouring constraint, means that there is a classical tautology that corresponds to the null space or null projection operator in $\mathscr{H}_3$ – a quantum mechanical contradiction.

To see this, consider the proposition.

$$P_i \dot{\vee} P_j \dot{\vee} P_k$$

where '$\dot{\vee}$' represents exclusive disjunction. That is, the proposition is true if and only if exactly one of P_i, P_j, or P_k is true. Conjoin this proposition with propositions of the same form, where i, j, k range over indices corresponding to all the orthogonal triples of directions in the 117 directions of the Kochen and Specker theorem. That is, construct the proposition:

$$\bigwedge(P_i \dot{\vee} P_j \dot{\vee} P_k)$$

where the symbol '$\bigwedge$' represents conjunction over the orthogonal triples. It turns out that there are a total of 43 orthogonal triples, so the proposition is a conjunction of 43 disjunctions. By the theorem, this proposition is logically false. For the proposition to be true, each of the 43 disjunctions $P_i \dot{\vee} P_j \dot{\vee} P_k$ would have to be true. But each such disjunction is true if and only if exactly one of P_i, P_j, or P_k is true, and the remaining two propositions are false, and this we know is impossible by the Kochen and Specker theorem. It follows that the negation of this proposition:

$$Q = \neg \bigwedge(P_i \dot{\vee} P_j \dot{\vee} P_k)$$

is a classical tautology. If we identify the three propositions P_i, P_j, P_k in each exclusive disjunction with three orthogonal 1-dimensional projection operators onto an orthogonal triple of rays in $\mathscr{H}_3$, and interpret exclusive disjunction as the lattice supremum of orthogonal (and hence commuting) projection operators, then each disjunction corresponds to the projection operator:

$$P_i \vee P_j \vee P_k = P_i + P_j + P_k = I$$

(where the symbol '$\vee$' here represents the lattice supremum). Interpreting conjunction as the lattice infimum for commuting projection operators, the conjunction of 43 disjunctions corresponds to the unit operator:

$$\bigwedge(P_i \vee P_j \vee P_k) = I \wedge I \wedge \ldots \wedge I = I \cdot I \cdot \ldots \cdot I = I$$

Taking negation as the orthogonal complement, the negation of this conjunction corresponds to the null operator, which represents a quantum proposition that is never true:

$$(\bigwedge(P_i \vee P_j \vee P_k))^{\perp} = I^{\perp} = 0$$

If we think of each orthogonal triple of projection operators P_i, P_j, P_k in this proposition as the projection operators in the spectral measure of a maximal observable, then what the proposition says is this: 'It is not the case that: the observable A has one of the values a_1, or a_2, or a_3, exclusively (where these are the three possible eigenvalues of A), and the observable B has one of the values b_1, or b_2, or b_3, exclusively (where these are the three possible eigenvalues of B), and . . . ,' where A, B, . . . are maximal observables corresponding to the orthogonal triples of rays in the 117 rays of the Kochen and Specker theorem. What we know from the theorem is that there is no selection function that will select a value for A, and a value for B, and . . . . Hence, by the theorem, this proposition is a classical tautology. But if we identify the proposition: 'The observable A has the value a_i' with the 1-dimensional projection operator P_{a_i} in $\mathscr{H}_3$, and interpret 'or' as lattice supremum for *commuting* projection operators (that is, projection operators corresponding to compatible quantum propositions belonging to a Boolean sublattice of $\mathscr{L}$), 'and' as lattice infimum for *commuting* projection operators, and 'not' as lattice orthocomplement, then the proposition: 'The observable A has one of the values a_1, or a_2, or a_3, exclusively' is identified with the unit operator:

$$P_{a_1} \vee P_{a_2} \vee P_{a_3} = P_{a_1} + P_{a_2} + P_{a_3} = I$$

and similarly for the observable B, etc., and the negation of the corresponding conjunction of exclusive disjunctions is identified with the null operator.

Notice that in the construction of this proposition all conjunctions (infima) and disjunctions (suprema) involve compatible propositions represented by commuting projection operators that belong to the same Boolean sublattice of $\mathscr{L}$. Any classical tautology that can be interpreted quantum mechanically in this way (so as to respect compatibility constraints in forming conjunctions and disjunctions), and that corresponds to a contingent quantum proposition (a proposition that can be false for some quantum states), or to a quantum contradiction (a proposition that is false for all quantum states) will be associated with a collection of Hilbert space subspaces that cannot be coloured by two colours, according to colouring constraints that reflect the conditions of a Boolean homomorphism on $\mathscr{L}$.

A Boolean homomorphism on $\mathscr{L}$ is a classical truth-value assignment on each Boolean sublattice of $\mathscr{L}$. So the impossibility of colouring a collection of Hilbert space subspaces in this sense corresponds to the impossibility of assigning truth values – 1's and 0's, greens and reds – to the associated propositions in such a way that *compatible* propositions (represented by subspaces that are the ranges of commuting projection operators) are interpreted classically – that is, such that lattice infima and suprema of propositions that belong to the same Boolean sublattice of $\mathscr{L}$ are interpreted as classical conjunction and disjunction, respectively, and the lattice orthocomplement is interpreted as classical negation. (So conjunction is interpreted as Boolean meet, disjunction as Boolean join, and negation as Boolean complement.) The existence of such uncolourable collections of subspaces is a consequence of the existence of collections of observables represented by self-adjoint Hilbert space operators that cannot be assigned values simultaneously, if the value assignment is required to preserve the functional relationships holding among the observables.

3.2 Schütte's tautology

In a letter to Specker dated April 22, 1965, the logician Kurt Schütte[68] constructed a classical tautology in eleven propositional variables that corresponds to a contingent proposition in $\mathscr{H}_3$, if the propositional variables are interpreted as projection operators onto certain rays in $\mathscr{H}_3$. In this proposed interpretation, all conjunctions and disjunctions are interpreted as infima and suprema, respectively, of compatible

[68] Bell and Machover (1977, p. 48) credit Schütte with discovering the tree method (or tableau method) as a formal system for first-order logic in 1956, independently of Beth and Hintikka who are usually credited with originating the method around the same time.

quantum propositions represented by commuting projection operators that belong to the same Boolean sublattice of $\mathscr{L}_3$.

With the author's permission, I reproduce the letter here:[69]

PHILOSOPHISCHES SEMINAR
DER UNIVERSITÄT KIEL
Prof. Dr. Kurt Schütte

23 Kiel, den 22 April 1965
Olshausenstrasse 40–46

Herrn
Prof. Dr.
Ernst P. Specker
Steinbruechelstrasse 18
Zürich 7/53 / Schweiz
Lieber Herr Specker,

die Formel, die ich Ihnen angegeben hatte, lässt sich noch etwas vereinfachen. Man Kommt mit 11 Aussagenvariablen aus.

(1) $d_1 \to \neg b_2$
(2) $d_1 \to \neg b_3$
(3) $d_2 \to a_2 \vee b_2$
(4) $d_2 \to \neg b_3$
(5) $d_3 \to \neg b_2$
(6) $d_3 \to (a_1 \vee a_2 \to b_3)$
(7) $d_4 \to a_2 \vee b_2$
(8) $d_4 \to (a_1 \vee a_2 \to b_3)$
(9) $(a_2 \vee c_1) \vee (b_3 \vee d_1)$
(10) $(a_2 \vee c_2) \vee (a_1 \vee b_1 \to d_1)$
(11) $c_2 \to b_3 \vee d_2$
(12) $c_1 \to b_1 \vee d_2$
(13) $(a_2 \vee c_1) \vee [(a_1 \vee a_2 \to b_3) \to d_3]$
(14) $(a_2 \vee c_2) \to (b_1 \vee d_3)$
(15) $c_2 \to [(a_1 \vee a_2 \to b_3) \to d_4]$
(16) $c_1 \to (a_1 \vee b_1 \to d_4)$
(17) $(a_1 \to a_2) \vee b_1$

F sei die Formel: (1) $\wedge \ldots \wedge$ (16) $\to$ (17)
F ist klassisch allgemeingültig, denn (17) folgt klassisch aus (1) – (16).
F ist nicht gültig im E^3, wie folgende Belegung zeigt:

$a_1 = (1, 0, 0)$ $b_1 = (0, 1, 1)$ $c_1 = (1, 0, 2)$ $d_1 = (-1, 1, 1)$
$a_2 = (0, 1, 0)$ $b_2 = (1, 0, 1)$ $c_2 = (2, 0, 1)$ $d_2 = (1, -1, 1)$
$b_3 = (1, 1, 0)$ $d_3 = (1, 1, -1)$
$d_4 = (1, 1, 1)$

Hierbei ergibt sich E^3 für jede der Formeln (1) – (16) und (17) $= \overline{(1,\ 0,\ 0)}$, also $F = \overline{(1,\ 0,\ 0)}$.

Herzliche Grüsse
Ihr
Kurt Schütte

I shall show that Schütte's tautology yields an uncolourable set of 33 rays in $\mathscr{H}_3$. So implicit in Schütte's tautology is a 33-ray proof of the Kochen and Specker theorem in $\mathscr{H}_3$. The 33 rays are closely related to the 31 rays of the Conway and Kochen proof,[70]

[69] Schütte's letter is discussed briefly in a doctoral dissertation by E. Clavadetscher-Seeberger (1983, pp. 83–4), a student of Specker's, and by Karl Svozil (1995). See also Svozil and Tkadlec (1996). I am indebted to Rob Clifton and Karl Svozil for a copy of the letter.

[70] The proof is unpublished. See Peres (1993, p. 197).

the world record for the smallest uncolourable set of rays in $\mathscr{H}_3$. Surprisingly, Kochen and Specker chose not to exploit Schütte's observation in their 1967 'no go' theorem, but opted instead for a less economical 117-ray proof. It took almost ten years before Jost (1976) reduced the number of rays to 109.

To facilitate the interpretation of Schütte's propositions in terms of Hilbert space subspaces, it will be convenient to express the conditionals $X \rightarrow Y$ as disjunctions $\neg X \vee Y$. Then Schütte's tautology F is a proposition of the following form:

$$[\neg(1) \vee \neg(2) \vee \ldots \vee \neg(16)] \vee (17) \qquad \text{(F)}$$

Here (1), . . . , (17) denote seventeen propositions constructed as logical combinations of eleven propositional variables, which I label with capital letters to conform to my notation for propositions:

A_1	B_1	C_1	D_1
A_2	B_2	C_2	D_2
	B_3		D_3
			D_4

The seventeen propositions can be expressed in terms of disjunctions and negations as follows:

(1) $\neg D_1 \vee \neg B_2$
(2) $\neg D_1 \vee \neg B_3$
(3) $\neg D_2 \vee (A_2 \vee B_2)$
(4) $\neg D_2 \vee \neg B_3$
(5) $\neg D_3 \vee \neg B_2$
(6) $\neg D_3 \vee [\neg(A_1 \vee A_2) \vee B_3]$
(7) $\neg D_4 \vee (A_2 \vee B_2)$
(8) $\neg D_4 \vee [\neg(A_1 \vee A_2) \vee B_3]$
(9) $(A_2 \vee C_1) \vee (B_3 \vee D_1)$
(10) $(A_2 \vee C_2) \vee [\neg(A_1 \vee B_1) \vee D_1]$
(11) $\neg C_2 \vee (B_3 \vee D_2)$
(12) $\neg C_1 \vee (B_1 \vee D_2)$
(13) $(A_2 \vee C_1) \vee \{\neg[\neg(A_1 \vee A_2) \vee B_3] \vee D_3\}$
(14) $(A_2 \vee C_2) \vee (B_1 \vee D_3)$
(15) $\neg C_2 \vee \{\neg[\neg(A_1 \vee A_2) \vee B_3] \vee D_4\}$
(16) $\neg C_1 \vee [\neg(A_1 \vee B_1) \vee D_4]$
(17) $(\neg A_1 \vee A_2) \vee B_1$

A straightforward way to see that Schütte's formula F is a classical tautology (a proposition that is true for all possible assignments of truth values to the eleven propositional variables) is to derive a contradiction from the supposition that F is false.

Since F is disjunction, F can be false only if each disjunct, $\neg(1), \ldots, \neg(16)$, (17), is false; that is, if (1), . . . , (16) are all true, and (17) is false. Similarly, the disjunction (17) can be false only if A_1 is true, A_2 is false, and B_1 is false. It now follows fairly easily (but tediously) that if A_1 is true, A_2 false, and B_1 false, then (1), . . . , (16) cannot all be true without generating a contradiction. For a proof, see section 3.4.

Now interpret the eleven propositional variables as quantum propositions represented by projection operators onto a particular set of 1-dimensional subspaces or rays in $\mathscr{H}_3$, as follows (where a ray is indicated by a vector spanning the ray):

$$A_1: \begin{pmatrix}1\\0\\0\end{pmatrix} \quad B_1: \begin{pmatrix}0\\1\\1\end{pmatrix} \quad C_1: \begin{pmatrix}1\\0\\2\end{pmatrix} \quad D_1: \begin{pmatrix}-1\\1\\1\end{pmatrix}$$

$$A_2: \begin{pmatrix}0\\1\\0\end{pmatrix} \quad B_2: \begin{pmatrix}1\\0\\1\end{pmatrix} \quad C_2: \begin{pmatrix}2\\0\\1\end{pmatrix} \quad D_2: \begin{pmatrix}1\\-1\\1\end{pmatrix}$$

$$B_3: \begin{pmatrix}1\\1\\0\end{pmatrix} \quad D_3: \begin{pmatrix}1\\1\\-1\end{pmatrix}$$

$$D_4: \begin{pmatrix}1\\1\\1\end{pmatrix}$$

Under the above interpretation, each of the propositions (1), . . . , (16) represents the quantum proposition corresponding to the unit operator, or the whole Hilbert space $\mathscr{H}_3$.

Consider, for example, the proposition (1):

$$\neg D_1 \vee \neg B_2$$

Notice that D_1 and B_2 are interpreted as orthogonal and hence compatible quantum propositions corresponding to the rays $\begin{pmatrix}-1\\1\\1\end{pmatrix}$ and $\begin{pmatrix}1\\0\\1\end{pmatrix}$, respectively.[71] It follows that $\neg D_1$ is represented by the plane orthogonal to the ray $\begin{pmatrix}-1\\1\\1\end{pmatrix}$ and $\neg B_2$ is represented by the plane orthogonal to $\begin{pmatrix}1\\0\\1\end{pmatrix}$. The two planes, which I shall denote here by $\begin{pmatrix}-1\\1\\1\end{pmatrix}^{\perp}$ and $\begin{pmatrix}1\\0\\1\end{pmatrix}^{\perp}$, respectively, intersect at right angles and are hence compatible (represented by commuting projection operators). So the disjunction $\neg D_1 \vee \neg B_2$ corresponds to the supremum of the two planes, which is just the span of the planes, $\mathscr{H}_3$.

As another example, take the proposition (6):

$$\neg D_3 \vee [\neg(A_1 \vee A_2) \vee B_3]$$

The propositions A_1 and A_2 are interpreted as orthogonal, and hence compatible, quantum propositions represented by the rays $\begin{pmatrix}1\\0\\0\end{pmatrix}$ and $\begin{pmatrix}0\\1\\0\end{pmatrix}$, respectively. The span of these two 1-dimensional subspaces is the plane orthogonal to $\begin{pmatrix}0\\0\\1\end{pmatrix}$. Hence, the proposition $\neg(A_1 \vee A_2)$ corresponds to the ray $\begin{pmatrix}0\\0\\1\end{pmatrix}$. The proposition B_3 corresponds to the ray $\begin{pmatrix}1\\1\\0\end{pmatrix}$, which is orthogonal to $\begin{pmatrix}0\\0\\1\end{pmatrix}$. So the propositions $\neg(A_1 \vee A_2)$ and B_3 correspond to compatible 1-dimensional subspaces and their disjunction corresponds

[71] For brevity in the following, I shall refer, e.g., to 'the ray (or 1-dimensional subspace) $\begin{pmatrix}1\\0\\0\end{pmatrix}$' rather than 'the ray (or 1-dimensional subspace) spanned by the vector $\begin{pmatrix}1\\0\\0\end{pmatrix}$.'

to the plane spanned by these subspaces. I shall denote this plane by:

$$\begin{pmatrix}0\\0\\1\end{pmatrix} \vee \begin{pmatrix}1\\1\\0\end{pmatrix}$$

The proposition $\neg D_3$ corresponds to the plane $\begin{pmatrix}1\\1\\-1\end{pmatrix}^{\perp}$. The two planes, $\begin{pmatrix}0\\0\\1\end{pmatrix} \vee \begin{pmatrix}1\\1\\0\end{pmatrix}$ and $\begin{pmatrix}1\\1\\-1\end{pmatrix}^{\perp}$ intersect at right angles in the ray $\begin{pmatrix}1\\1\\2\end{pmatrix}$. (This can be checked by writing down the matrices of the projection operators onto the planes. See section 3.4.) So, again, the propositions $\neg D_3$ and $\neg(A_1 \vee A_2) \vee B_3$ correspond to compatible subspaces, and their disjunction corresponds to the span of these subspaces, which is just $\mathscr{H}_3$.

Since each of the propositions (1), . . . , (16) is interpreted as the quantum proposition corresponding to $\mathscr{H}_3$, the negations, $\neg$(1), . . . , $\neg$(16), correspond to the null space, and their disjunction – a disjunction of compatible propositions – corresponds to the null space (the span of the null space with itself, iterated sixteen times).

The proposition (17):

$$(\neg A_1 \vee A_2) \vee B_1$$

however, corresponds to the plane orthogonal to $\begin{pmatrix}1\\0\\0\end{pmatrix}$. Since A_2 is interpreted as the ray $\begin{pmatrix}0\\1\\0\end{pmatrix}$, which is orthogonal to $\begin{pmatrix}1\\0\\0\end{pmatrix}$ and so lies in the plane $\begin{pmatrix}1\\0\\0\end{pmatrix}^{\perp}$ representing $\neg A_1$, the propositions $\neg A_1$ and A_2 correspond to compatible Hilbert space subspaces. Their disjunction $\neg A_1 \vee A_2$ corresponds to the span of $\begin{pmatrix}0\\1\\0\end{pmatrix}$ and $\begin{pmatrix}1\\0\\0\end{pmatrix}^{\perp}$, which is just $\begin{pmatrix}1\\0\\0\end{pmatrix}^{\perp}$. The proposition B_1 is interpreted as the ray $\begin{pmatrix}0\\1\\1\end{pmatrix}$, which also lies in the plane $\begin{pmatrix}1\\0\\0\end{pmatrix}^{\perp}$ (since it is orthogonal to $\begin{pmatrix}1\\0\\0\end{pmatrix}$), and so the disjunction of $\neg A_1 \vee A_2$ with B_2 corresponds to the span of $\begin{pmatrix}1\\0\\0\end{pmatrix}^{\perp}$ and $\begin{pmatrix}0\\1\\1\end{pmatrix}$, which again yields the plane $\begin{pmatrix}1\\0\\0\end{pmatrix}^{\perp}$.

It follows that the proposition F

$$[\neg(1) \vee \neg(2) \vee \ldots \vee \neg(16)] \vee (17)$$

corresponds to the span of the null space and the plane $\begin{pmatrix}1\\0\\0\end{pmatrix}^{\perp}$, since $\neg(1) \vee \neg(2) \vee \ldots \vee \neg(16)$ corresponds to the null space, and the null space and the plane $\begin{pmatrix}1\\0\\0\end{pmatrix}^{\perp}$ are compatible. This is just the plane $\begin{pmatrix}1\\0\\0\end{pmatrix}^{\perp}$. So F is a classical tautology corresponding to a contingent quantum mechanical proposition: for the quantum state represented by the ray $\begin{pmatrix}1\\0\\0\end{pmatrix}$, the proposition represented by the plane $\begin{pmatrix}1\\0\\0\end{pmatrix}^{\perp}$ is false (or, more generally, for any quantum state not lying in the plane $\begin{pmatrix}1\\0\\0\end{pmatrix}^{\perp}$, the proposition represented by this plane has a nonzero probability of being false).

Each of Schütte's seventeen propositions is constructed as a disjunction of component propositions corresponding to two planes in $\mathscr{H}_3$ that intersect at right angles. In some cases (for example, (1)), a plane is generated as the plane orthogonal to one of the eleven originally defined rays corresponding to the eleven propositional variables. In other cases, a plane is generated as the span of two of these rays (for example, (3)), or as the span of an original ray with a ray that is orthogonal to the plane generated as the span of two other original rays (for example, (6)), or as the span of an original ray with a ray that is orthogonal to the plane generated as the span of a second original ray with a ray that is orthogonal to the span of two other original rays (for example, (13)). What we have, in effect, is a *state-dependent proof of the impossibility of colouring the eleven rays and the finite set of planes generated in this way by these rays*; that is, the impossibility of assigning truth values to the quantum propositions associated with these 1- and 2-dimensional subspaces of $\mathscr{H}_3$ in such a way as to satisfy the homomorphism constraint, if the ray $\begin{pmatrix}1\\0\\0\end{pmatrix}$ is assigned the truth value 1 and the plane $\begin{pmatrix}1\\0\\0\end{pmatrix}^\perp$ representing F is assigned the truth value 0. It follows that the eleven rays generate a partial Boolean algebra of subspaces of $\mathscr{H}_3$ containing the eleven rays and associated planes that cannot be embedded into a Boolean algebra, because there are no 2-valued homomorphisms on this partial Boolean subalgebra.

To generate a state-independent proof, notice that we can construct a similar classical tautology, F', that corresponds to the plane $\begin{pmatrix}0\\1\\0\end{pmatrix}^\perp$ by rotating the eleven rays through a 90° angle about the ray $\begin{pmatrix}0\\0\\1\end{pmatrix}$ and interpreting the propositional variables as the quantum propositions represented by the projection operators onto the rotated rays.[72]

The rotation matrix

$$\begin{pmatrix}0 & -1 & 0\\1 & 0 & 0\\0 & 0 & 1\end{pmatrix}$$

transforms the eleven rays as follows:

$$\begin{pmatrix}1\\0\\0\end{pmatrix}\to\begin{pmatrix}0\\1\\0\end{pmatrix}\quad\begin{pmatrix}0\\1\\1\end{pmatrix}\to\begin{pmatrix}-1\\0\\1\end{pmatrix}\quad\begin{pmatrix}1\\0\\2\end{pmatrix}\to\begin{pmatrix}0\\1\\2\end{pmatrix}\quad\begin{pmatrix}-1\\1\\1\end{pmatrix}\to\begin{pmatrix}-1\\-1\\1\end{pmatrix}$$

$$\begin{pmatrix}0\\1\\0\end{pmatrix}\to\begin{pmatrix}-1\\0\\0\end{pmatrix}\quad\begin{pmatrix}1\\0\\1\end{pmatrix}\to\begin{pmatrix}0\\1\\1\end{pmatrix}\quad\begin{pmatrix}2\\0\\1\end{pmatrix}\to\begin{pmatrix}0\\2\\1\end{pmatrix}\quad\begin{pmatrix}1\\-1\\1\end{pmatrix}\to\begin{pmatrix}1\\1\\1\end{pmatrix}$$

$$\begin{pmatrix}1\\1\\0\end{pmatrix}\to\begin{pmatrix}-1\\1\\0\end{pmatrix}\quad\begin{pmatrix}1\\1\\-1\end{pmatrix}\to\begin{pmatrix}-1\\1\\-1\end{pmatrix}$$

$$\begin{pmatrix}1\\1\\1\end{pmatrix}\to\begin{pmatrix}-1\\1\\1\end{pmatrix}$$

[72] I owe this observation to Rob Clifton.

The vector $\begin{pmatrix}-1\\0\\0\end{pmatrix}$ represents the same ray as the vector $\begin{pmatrix}1\\0\\0\end{pmatrix}$ – they simply point in opposite directions. Similarly, $\begin{pmatrix}-1\\-1\\1\end{pmatrix}$ represents the same ray as $\begin{pmatrix}1\\1\\-1\end{pmatrix}$, and $\begin{pmatrix}-1\\1\\-1\end{pmatrix}$ represents the same ray as $\begin{pmatrix}1\\-1\\1\end{pmatrix}$.

The rays $\begin{pmatrix}0\\1\\2\end{pmatrix}$ and $\begin{pmatrix}0\\2\\1\end{pmatrix}$ are 'new' rays, not generated from the original set of eleven rays. All the other rotated rays are either original rays, or can be generated from the original rays. The ray $\begin{pmatrix}-1\\0\\1\end{pmatrix}$ can be generated as the ray orthogonal to the span of $\begin{pmatrix}0\\1\\0\end{pmatrix}$ and $\begin{pmatrix}0\\0\\1\end{pmatrix}$; that is, $\begin{pmatrix}-1\\0\\1\end{pmatrix} = \left[\begin{pmatrix}0\\1\\0\end{pmatrix} \vee \begin{pmatrix}0\\0\\1\end{pmatrix}\right]^{\perp}$. Similarly, the ray $\begin{pmatrix}-1\\1\\0\end{pmatrix}$ can be generated as the ray $\left[\begin{pmatrix}0\\0\\1\end{pmatrix} \vee \begin{pmatrix}1\\1\\0\end{pmatrix}\right]^{\perp}$, and $\begin{pmatrix}0\\0\\1\end{pmatrix}$ can be generated as the ray $\left[\begin{pmatrix}1\\0\\0\end{pmatrix} \vee \begin{pmatrix}0\\1\\0\end{pmatrix}\right]^{\perp}$.

If we interpret the propositional variables as quantum propositions associated with the rotated rays, then:

$$(17) \rightarrow (17)'$$

where (17)′ corresponds to the plane:

$$\left[\begin{pmatrix}0\\1\\0\end{pmatrix}^{\perp} \vee \begin{pmatrix}1\\0\\0\end{pmatrix}\right] \vee \begin{pmatrix}-1\\0\\1\end{pmatrix} = \begin{pmatrix}0\\1\\0\end{pmatrix}^{\perp}$$

(Since both $\begin{pmatrix}1\\0\\0\end{pmatrix}$ and $\begin{pmatrix}-1\\0\\1\end{pmatrix}$ are orthogonal to $\begin{pmatrix}0\\1\\0\end{pmatrix}$ and hence lie in the plane $\begin{pmatrix}0\\1\\0\end{pmatrix}^{\perp}$, $\begin{pmatrix}0\\1\\0\end{pmatrix}^{\perp} \vee \begin{pmatrix}1\\0\\0\end{pmatrix} = \begin{pmatrix}0\\1\\0\end{pmatrix}^{\perp}$, and $\begin{pmatrix}0\\1\\0\end{pmatrix}^{\perp} \vee \begin{pmatrix}-1\\0\\1\end{pmatrix} = \begin{pmatrix}0\\1\\0\end{pmatrix}^{\perp}$.) Each of the propositions (1)′, . . . , (16)′ still corresponds to the whole space $\mathscr{H}_3$, so the tautology F' corresponds to the plane $\begin{pmatrix}0\\1\\0\end{pmatrix}^{\perp}$.

Similarly, rotating 90° about $\begin{pmatrix}0\\1\\0\end{pmatrix}$ with the rotation matrix:

$$\begin{pmatrix}0 & 0 & -1\\0 & 1 & 0\\1 & 0 & 0\end{pmatrix}$$

yields the transformation:

$$\begin{pmatrix}1\\0\\0\end{pmatrix} \rightarrow \begin{pmatrix}0\\0\\1\end{pmatrix} \quad \begin{pmatrix}0\\1\\1\end{pmatrix} \rightarrow \begin{pmatrix}-1\\1\\0\end{pmatrix} \quad \begin{pmatrix}1\\0\\2\end{pmatrix} \rightarrow \begin{pmatrix}-2\\0\\1\end{pmatrix} \quad \begin{pmatrix}-1\\1\\1\end{pmatrix} \rightarrow \begin{pmatrix}-1\\1\\-1\end{pmatrix}$$

$$\begin{pmatrix}0\\1\\0\end{pmatrix} \rightarrow \begin{pmatrix}0\\1\\0\end{pmatrix} \quad \begin{pmatrix}1\\0\\1\end{pmatrix} \rightarrow \begin{pmatrix}-1\\0\\1\end{pmatrix} \quad \begin{pmatrix}2\\0\\1\end{pmatrix} \rightarrow \begin{pmatrix}-1\\0\\2\end{pmatrix} \quad \begin{pmatrix}1\\-1\\1\end{pmatrix} \rightarrow \begin{pmatrix}-1\\-1\\1\end{pmatrix}$$

$$\begin{pmatrix}1\\1\\0\end{pmatrix} \rightarrow \begin{pmatrix}0\\1\\1\end{pmatrix} \qquad \begin{pmatrix}1\\1\\-1\end{pmatrix} \rightarrow \begin{pmatrix}1\\1\\1\end{pmatrix}$$

$$\begin{pmatrix}1\\1\\1\end{pmatrix} \rightarrow \begin{pmatrix}-1\\1\\1\end{pmatrix}$$

In this case, no 'new' rays are generated by the rotation. (The ray $\begin{pmatrix}-1\\0\\2\end{pmatrix}$ can be generated as $\left[\begin{pmatrix}0\\1\\0\end{pmatrix}\vee\begin{pmatrix}2\\0\\1\end{pmatrix}\right]^{\perp}$, and the ray $\begin{pmatrix}-2\\0\\1\end{pmatrix}$ can be generated as $\left[\begin{pmatrix}0\\1\\0\end{pmatrix}\vee\begin{pmatrix}1\\0\\2\end{pmatrix}\right]^{\perp}$.) The transformation:

$$(17)\to(17)''$$

yields the proposition (17)″, which corresponds to the plane:

$$\left[\begin{pmatrix}0\\0\\1\end{pmatrix}^{\perp}\vee\begin{pmatrix}0\\1\\0\end{pmatrix}\right]\vee\begin{pmatrix}-1\\1\\0\end{pmatrix}=\begin{pmatrix}0\\0\\1\end{pmatrix}^{\perp}$$

Each of the propositions (1)″, . . . , (16)″ corresponds to the whole space $\mathscr{H}_3$, so the tautology F'' corresponds to the plane $\begin{pmatrix}0\\0\\1\end{pmatrix}^{\perp}$.

Since (17) corresponds to the plane $\begin{pmatrix}1\\0\\0\end{pmatrix}^{\perp}$, (17)′ corresponds to the plane $\begin{pmatrix}0\\1\\0\end{pmatrix}^{\perp}$, and (17)″ corresponds to the plane $\begin{pmatrix}0\\0\\1\end{pmatrix}^{\perp}$, and these planes are mutually compatible, we can interpret the conjunction:

$$F\wedge F'\wedge F''$$

as the proposition corresponding to the lattice infimum of the compatible planes $\begin{pmatrix}1\\0\\0\end{pmatrix}^{\perp}$, $\begin{pmatrix}0\\1\\0\end{pmatrix}^{\perp}$, and $\begin{pmatrix}0\\0\\1\end{pmatrix}^{\perp}$, which is just the intersection of the planes. This yields the proposition corresponding to the null space, or the null projection operator, the quantum proposition that is false for all quantum states.

Now, since F, F', and F'' are classical tautologies, the conjunction $F\wedge F'\wedge F''$ is a classical tautology. So we have a classical tautology that corresponds to a quantum mechanical contradiction, just as in the Kochen and Specker proof. In this form, Schütte's construction yields a *state-independent proof of the Kochen and Specker theorem: the partial Boolean algebra generated by the thirteen rays consisting of Schütte's eleven rays and the two 'new' rays derived from the two rotations*[73] cannot be coloured consistently with the colouring constraint of the homomorphism condition, and hence is not embeddable into a Boolean algebra.

This version of the Kochen and Specker 'no go' theorem involves the impossibility of colouring a certain set of *rays and planes* in $\mathscr{H}_3$. To exploit Schütte's tautology to construct a colouring theorem for a set of *rays* in $\mathscr{H}_3$, notice that each of the seventeen

[73] It might appear that the rotation about $\begin{pmatrix}0\\1\\0\end{pmatrix}$ is redundant, since the classical tautology $F\wedge F'$, which corresponds to the ray $\begin{pmatrix}0\\0\\1\end{pmatrix}$, already yields a state-independent Kochen and Specker proof: a set of rays and planes that can't be coloured, whatever colour is assigned to $\begin{pmatrix}0\\0\\1\end{pmatrix}$. This is of course correct, but the relevant point is that the tautology $F\wedge F'\wedge F''$, which corresponds to the null space, yields the *same* set of rays and planes as $F\wedge F'$.

propositions corresponds to one or more orthogonal triples of rays, via the planes that intersect at right angles.

Consider, for example, the proposition (1):

$$\neg D_1 \vee \neg B_2$$

the two planes involved here are $\begin{pmatrix}-1\\1\\1\end{pmatrix}^{\perp}$ and $\begin{pmatrix}1\\0\\1\end{pmatrix}^{\perp}$. They intersect at right angles in the ray $\begin{pmatrix}1\\2\\-1\end{pmatrix}$ and so define an orthogonal triple of rays:

$$\begin{pmatrix}1\\2\\-1\end{pmatrix} \quad \begin{pmatrix}1\\0\\1\end{pmatrix} \quad \begin{pmatrix}1\\-1\\-1\end{pmatrix}$$

where $\begin{pmatrix}1\\0\\1\end{pmatrix}$ is orthogonal to $\begin{pmatrix}1\\2\\-1\end{pmatrix}$ in the plane $\begin{pmatrix}-1\\1\\1\end{pmatrix}^{\perp}$, and $\begin{pmatrix}1\\-1\\-1\end{pmatrix}$ is orthogonal to $\begin{pmatrix}1\\2\\-1\end{pmatrix}$ in the plane $\begin{pmatrix}1\\0\\1\end{pmatrix}^{\perp}$.

Some of the propositions yield more than one orthogonal triple. For example, the proposition (3):

$$\neg D_2 \vee (A_2 \vee B_2)$$

yields the orthogonal triple:

$$\begin{pmatrix}1\\2\\1\end{pmatrix} \quad \begin{pmatrix}1\\0\\-1\end{pmatrix} \quad \begin{pmatrix}1\\-1\\1\end{pmatrix}$$

from the intersection of the two planes corresponding to the propositions $\neg D_2$ and $A_2 \vee B_2$. The propositions A_2 and B_2 correspond to orthogonal rays $\begin{pmatrix}0\\1\\0\end{pmatrix}$ and $\begin{pmatrix}1\\0\\1\end{pmatrix}$, and so generate a second orthogonal triple:

$$\begin{pmatrix}0\\1\\0\end{pmatrix} \quad \begin{pmatrix}1\\0\\1\end{pmatrix} \quad \begin{pmatrix}1\\0\\-1\end{pmatrix}$$

(For details, see section 3.4.)

It turns out that Schütte's tautology F, via the component propositions (1), ..., (17), yields 26 orthogonal triples of rays involving a total of 37 distinct rays. Rotating through 90° about $\begin{pmatrix}0\\0\\1\end{pmatrix}$ yields an additional 10 new orthogonal triples involving 12 new rays. (The 90° rotation about $\begin{pmatrix}0\\1\\0\end{pmatrix}$, of course, yields no new rays.)

Tables 3.1 and 3.2 list the 37 rays and the 26 orthogonal triples of rays generated by F. (For convenience, I label each ray as a triple of integers, corresponding to the – possibly unnormalized – components of the associated vector.)

Since these 37 rays generate a collection of rays and planes in $\mathscr{H}_3$ that cannot be coloured if the ray $\begin{pmatrix}1\\0\\0\end{pmatrix}$ orthogonal to the plane $\begin{pmatrix}1\\0\\0\end{pmatrix}^{\perp}$ representing the proposition F is

Table 3.1. *Rays generated by F*

(1)	0	0	1	(14)	1	0	2	(27)	2	1	−1
(2)	0	1	0	(15)	1	0	−2	(28)	2	−1	1
(3)	1	0	0	(16)	2	0	1	(29)	2	−1	−1
(4)	0	1	1	(17)	2	0	−1	(30)	1	5	2
(5)	0	1	−1	(18)	1	1	2	(31)	1	5	−2
(6)	1	0	1	(19)	1	1	−2	(32)	1	−5	2
(7)	1	0	−1	(20)	1	−1	2	(33)	1	−5	−2
(8)	1	1	0	(21)	1	−1	−2	(34)	2	5	1
(9)	1	−1	0	(22)	1	2	1	(35)	2	5	−1
(10)	1	1	1	(23)	1	2	−1	(36)	2	−5	1
(11)	1	−1	1	(24)	1	−2	1	(37)	2	−5	−1
(12)	1	1	−1	(25)	1	−2	−1				
(13)	1	−1	−1	(26)	2	1	1				

Table 3.2. *Orthogonal triples generated by F*

(1)	0	0	1	0	1	0	1	0	0
(2)	0	0	1	1	1	0	1	−1	0
(3)	0	1	0	1	0	1	1	0	−1
(4)	0	1	0	1	0	2	2	0	−1
(5)	0	1	0	1	0	−2	2	0	1
(6)	1	0	0	0	1	1	0	1	−1
(7)	0	1	1	1	1	−1	2	−1	1
(8)	0	1	1	1	−1	1	2	1	−1
(9)	0	1	−1	1	1	1	2	−1	−1
(10)	0	1	−1	1	−1	−1	2	1	1
(11)	1	0	1	1	1	−1	1	−2	−1
(12)	1	0	1	1	−1	−1	1	2	−1
(13)	1	0	−1	1	1	1	1	−2	1
(14)	1	0	−1	1	−1	1	1	2	1
(15)	1	1	0	1	−1	1	1	−1	−2
(16)	1	1	0	1	−1	−1	1	−1	2
(17)	1	−1	0	1	1	1	1	1	−2
(18)	1	−1	0	1	1	−1	1	1	2
(19)	1	0	2	2	1	−1	2	−5	−1
(20)	1	0	2	2	−1	−1	2	5	−1
(21)	1	0	−2	2	1	1	2	−5	1
(22)	1	0	−2	2	−1	1	2	5	1
(23)	2	0	1	1	1	−2	1	−5	−2
(24)	2	0	1	1	−1	−2	1	5	−2
(25)	2	0	−1	1	1	2	1	−5	2
(26)	2	0	−1	1	−1	2	1	5	2

assigned the colour green (1 or 'true'), we know that the rays cannot be coloured green and red in such a way that one and only one ray in each orthogonal triple is coloured green, and the remaining two rays are coloured red.

Now, notice that each of the eight rays with a '5' as one component, the rays labelled:

$$\begin{array}{rrrcrrr} 1 & 5 & 2 & \quad & 2 & 5 & 1 \\ 1 & 5 & -2 & & 2 & 5 & -1 \\ 1 & -5 & 2 & & 2 & -5 & 1 \\ 1 & -5 & -2 & & 2 & -5 & -1 \end{array}$$

each occur in only one orthogonal triple. Also, the four rays:

$$\begin{array}{rrr} 1 & 2 & 1 \\ 1 & 2 & -1 \\ 1 & -2 & 1 \\ 1 & -2 & -1 \end{array}$$

each occur in only one orthogonal triple. So, whatever colour is assigned to these twelve rays will not constrain the colouring of any of the other rays. (The propositions corresponding to these rays can be assigned any truth values, without constraining the truth values assigned to the other propositions.) Removing these rays from the set of 37 rays leaves a set of 25 rays that cannot be coloured, if the ray $\begin{pmatrix}1\\0\\0\end{pmatrix}$ is assigned the colour green.

The 25 rays yield a state-dependent proof of the Kochen and Specker theorem. For a state-independent proof, we add the 12 new rays and the 10 new orthogonal triples, derived by rotation, to the original set of 37 rays and 26 orthogonal triples (tables 3.3 and 3.4).

This yields a total of 49 rays and 36 orthogonal triples. Now the only rays that occur in only one orthogonal triple are the 16 rays with a '5' as component. (The four rays:

$$\begin{array}{rrr} 1 & 2 & 1 \\ 1 & 2 & -1 \\ 1 & -2 & 1 \\ 1 & -2 & -1 \end{array}$$

now occur in some of the new orthogonal triples.) Removing these 16 rays from the 49 rays yields the following set of 33 rays that cannot be coloured in $\mathscr{H}_3$, shown in table 3.5.

The relationship of this set of 33 rays to the 31-ray Conway and Kochen proof can be seen on the cube shown in figure 3.2.

The relevant rays are represented by the black, white, and grey dots (which indicate the points where the rays, originating at the center of the cube, cut the surface of the cube). The Conway and Kochen proof uses the 31 rays defined by the black and grey dots. (Note that some of the dots indicate the same ray. For example, the dots $-2\ -2\ 2$ and $2\ 2\ -2$ lie on the same ray, which is represented in the list by the coordinates $1\ 1\ -1$.) The 33-ray proof derived from Schütte's tautology uses the rays represented by

Table 3.3. *New rays generated by F′*

0	1	2
0	1	−2
0	2	1
0	2	−1
5	1	2
5	1	−2
5	−1	2
5	−1	−2
5	2	1
5	2	−1
5	−2	1
5	−2	−1

Table 3.4. *New orthogonal triples generated by F′*

1	0	0	0	1	2	0	2	−1
1	0	0	0	1	−2	0	2	1
0	1	2	1	−2	1	5	2	−1
0	1	2	1	2	−1	5	−2	1
0	1	−2	1	−2	−1	5	2	1
0	1	−2	1	2	1	5	−2	−1
0	2	1	1	1	−2	5	−1	2
0	2	1	1	−1	2	5	1	−2
0	2	−1	1	1	2	5	−1	−2
0	2	−1	1	−1	−2	5	1	2

the black dots and the six rays represented by the white dots:

$$\begin{array}{rrr} 1 & 0 & 2 \\ 1 & 1 & 2 \\ 1 & -1 & 2 \\ 2 & 0 & -1 \\ 2 & 1 & -1 \\ 2 & -1 & -1 \end{array}$$

but does not use the four rays represented by the grey dots:

$$\begin{array}{rrr} 1 & 2 & 0 \\ -1 & 2 & 0 \\ 2 & 1 & 0 \\ 2 & -1 & 0 \end{array}$$

The 33 rays derived from Schütte's tautology are different from the 33 rays in Peres's (1991)[74] proof of the Kochen and Specker theorem. Peres's 33 rays are related to the

[74] Reprinted in Peres (1993, pp. 196–200).

Table 3.5. 33 *uncolourable rays in* $\mathcal{H}_3$

(1)	0	0	1	(18)	1	0	2
(2)	0	1	0	(19)	1	0	−2
(3)	1	0	0	(20)	2	0	1
(4)	0	1	1	(21)	2	0	−1
(5)	0	1	−1	(22)	1	1	2
(6)	1	0	1	(23)	1	1	−2
(7)	1	0	−1	(24)	1	−1	2
(8)	1	1	0	(25)	1	−1	−2
(9)	1	−1	0	(26)	1	2	1
(10)	1	1	1	(27)	1	2	−1
(11)	1	−1	1	(28)	1	−2	1
(12)	1	1	−1	(29)	1	−2	−1
(13)	1	−1	−1	(30)	2	1	1
(14)	0	1	2	(31)	2	1	−1
(15)	0	1	−2	(32)	2	−1	1
(16)	0	2	1	(33)	2	−1	−1
(17)	0	2	−1				

rays in figure 3.2 as follows. Discard the four rays associated with the corners of the cube in figure 3.2:

$$\begin{matrix} 2 & 2 & 2 \\ 2 & 2 & -2 \\ -2 & 2 & 2 \\ -2 & 2 & -2 \end{matrix}$$

and replace each remaining occurrence of '2' in the coordinates of the black, white, and grey rays by '$\sqrt{2}$.' This yields $31 + 6 - 4 = 33$ rays, with different orthogonality relations to those in figure 3.2. The Peres rays are represented on the cube shown in figure 3.3.

Penrose (1994a) has pointed out that Peres's 33 rays are defined by a polyhedron that can be constructed by superimposing a cube with two other cubes, each obtained by rotating the first cube 90° about one of two perpendicular lines connecting its center to the midpoints of two of its edges, and that this polyhedron is exhibited in Escher's engraving 'Waterfall' (on top of the left tower, see figure 3.4). The 33 rays connect the common center of the three interpenetrating cubes to the vertices and to the midpoints of the faces and edges.[75]

So Escher's 'impossible world' can be interpreted as an imaginative depiction of

[75] Both Mermin (1993, p. 809) and Peres (1993, p. 212) mention Penrose's curious observation. See also Zimba and Penrose (1993). Escher's 'Waterfall,' and 'Ascending and Descending' were inspired by 'impossible' objects designed, respectively, by Roger Penrose and his father, the geneticist Lionel Penrose, and reported in a paper published in the *British Journal of Psychology* in 1958. The Penroses sent a copy of the paper to Escher. See John Horgan (1993).

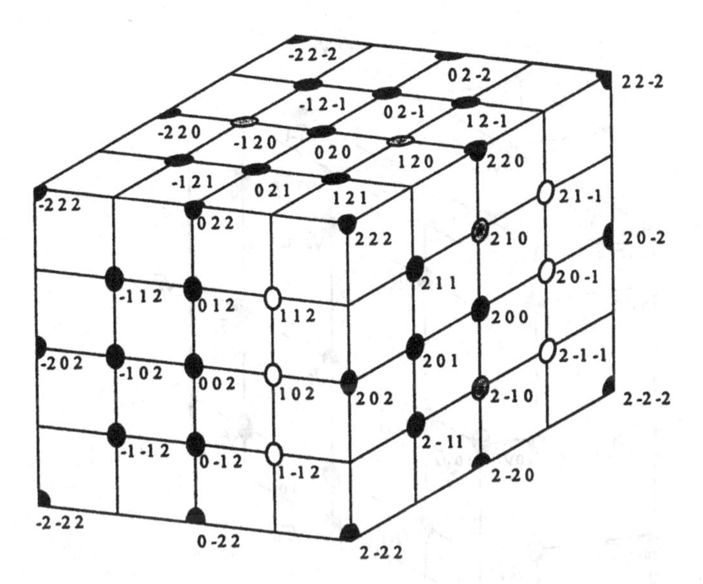

Figure 3.2. Cube representing the 31 rays of the Conway and Kochen proof and the 33 rays derived from Schütte's tautology. From a diagram in Peres (1993, p. 114).

different Boolean perspectives 'pasted together' in such a way as to form a structure that cannot be embedded into a single Boolean framework, with an encoding of the Hilbert space directions that generate a proof of non-embeddability (or, equivalently, define an interpretation of the propositional variables in a classical tautology under which the tautology turns out to be false). In this sense, 'Waterfall' is a precise representation of the way in which a quantum world is nonclassical!

3.3 Four-dimensional uncolourable configurations

The argument presented at the end of section 2.2 in terms of two spin-$\frac{1}{2}$ particles (analogous to the Greenberger–Horne–Zeilinger three-particle counter-argument to EPR) that didn't work as a rebuttal to EPR can be extended to a 9-operator Kochen and Specker theorem in $\mathscr{H}_4$ (see Peres, 1991; Peres, 1993, p. 189; Mermin, 1990c, 1993).

Recall the argument: Since the observables $\sigma_x^1, \sigma_x^2, \sigma_x^1\sigma_x^2$ and $\sigma_y^1, \sigma_y^2, \sigma_y^1\sigma_y^2$ of two spin-$\frac{1}{2}$ particles form (complete) commuting sets, the eigenvalues of these observables must satisfy the conditions:

$$v(\sigma_x^1\sigma_x^2) = v(\sigma_x^1)v(\sigma_x^2)$$
$$v(\sigma_y^1\sigma_y^2) = v(\sigma_y^1)v(\sigma_y^2)$$

The singlet state $|\psi\rangle$ is a simultaneous eigenstate of $\sigma_x^1\sigma_x^2$ and $\sigma_y^1\sigma_y^2$ with eigenvalue -1, so in the singlet state:

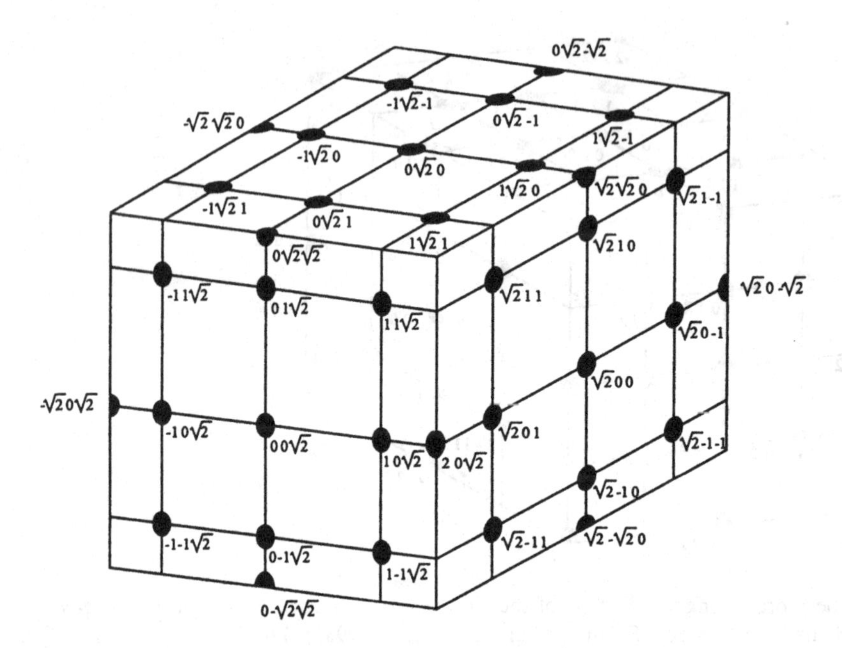

Figure 3.3. Cube representing the 33 rays of the Peres proof. From a diagram in Peres (1993, p. 198).

$$v(\sigma_x^1\sigma_x^2) = v(\sigma_y^1\sigma_y^2) = -1$$

If each of the observables σ_x^1, σ_x^2, σ_y^1, σ_y^2, $\sigma_x^1\sigma_x^2$, and $\sigma_y^1\sigma_y^2$ has a determinate value that is one of its eigenvalues, then these determinate values must satisfy the condition:

$$v(\sigma_x^1\sigma_x^2)v(\sigma_y^1\sigma_y^2) = v(\sigma_x^1)v(\sigma_x^2)v(\sigma_y^1)v(\sigma_y^2) = 1$$

But since the operators $\sigma_x^1\sigma_y^2$ and $\sigma_y^1\sigma_x^2$ commute, and $(\sigma_x^1\sigma_y^2)\cdot(\sigma_y^1\sigma_x^2) = \sigma_z^1\sigma_z^2$, and $|\psi\rangle$ is also an eigenstate of $\sigma_z^1\sigma_z^2$ with eigenvalue -1, it follows that if each of the observables $\sigma_x^1\sigma_y^2$ and $\sigma_y^1\sigma_x^2$ has a determinate value that is one of its eigenvalues, then these determinate values must satisfy the condition:

$$v(\sigma_x^1\sigma_y^2)v(\sigma_y^1\sigma_x^2) = v(\sigma_x^1)v(\sigma_y^2)v(\sigma_y^1)v(\sigma_x^2) = -1$$

The two conditions:

$$v(\sigma_x^1)v(\sigma_x^2)v(\sigma_y^1)v(\sigma_y^2) = 1$$
$$v(\sigma_x^1)v(\sigma_y^2)v(\sigma_y^1)v(\sigma_x^2) = -1$$

are inconsistent, so the eight observables σ_x^1, σ_x^2, σ_y^1, σ_y^2, $\sigma_x^1\sigma_x^2$, $\sigma_y^1\sigma_y^2$, $\sigma_x^1\sigma_y^2$, $\sigma_y^1\sigma_x^2$, cannot all have determinate values simultaneously for the two-particle system in the singlet state $|\psi\rangle$, if these values are eigenvalues.

Figure 3.4. 'Waterfall.' M.C. Escher, Lithograph, 1961, © 1996 M. C. Escher/ Cordon Art – Baarn – Holland.

This is a state-dependent Kochen–Specker theorem in $\mathscr{H}_4$.[76] Adding the observable $\sigma_z^1\sigma_z^2$ yields a 9-observable state-independent theorem. Exploiting only the functional relationship constraint on the assignment of values to the operators, we have for any state in $\mathscr{H}_4$ (since the operators $\sigma_x^1\sigma_x^2$ and $\sigma_y^1\sigma_y^2$ commute, and $(\sigma_x^1\sigma_x^2)\cdot(\sigma_y^1\sigma_y^2) = -(\sigma_z^1\sigma_z^2)$):

$$
\begin{gathered}
v(\sigma_x^1\sigma_x^2)v(\sigma_x^1)v(\sigma_x^2) = 1 \\
v(\sigma_y^1\sigma_y^2)v(\sigma_y^1)v(\sigma_y^2) = 1 \\
v(\sigma_x^1\sigma_y^2)v(\sigma_x^1)v(\sigma_y^2) = 1 \\
v(\sigma_y^1\sigma_x^2)v(\sigma_y^1)v(\sigma_x^2) = 1 \\
v(\sigma_x^1\sigma_y^2)v(\sigma_y^1\sigma_x^2)v(\sigma_z^1\sigma_z^2) = 1 \\
v(\sigma_x^1\sigma_x^2)v(\sigma_y^1\sigma_y^2)v(\sigma_z^1\sigma_z^2) = -1
\end{gathered}
$$

These equations are inconsistent. Each observable occurs exactly twice on the left side of the equations and takes the value ± 1, so the product of the left sides of the equations yields the number 1, but the product of the right sides yields -1.

The observables in each of the six equations form a complete commuting set and define a basis of four mutually orthogonal rays. None of the basis sets have any rays in common, so there are 24 rays in all. The 24 rays generate an uncolourable Kochen–Specker configuration in $\mathscr{H}_4$ (shown in table 3.6) where each ray is denoted by a quadruple of numbers representing the coordinates of a (possibly unnormalized) vector in the ray.

We know this configuration is uncolourable, because if it were colourable, values could be assigned simultaneously to the nine observables σ_x^1, σ_x^2, σ_y^1, σ_y^2, $\sigma_x^1\sigma_x^2$, $\sigma_y^1\sigma_y^2$, $\sigma_z^1\sigma_z^2$, $\sigma_x^1\sigma_y^2$, $\sigma_y^1\sigma_x^2$ consistently with the functional relationship constraint, and this is impossible. Peres (1993, pp. 200–1) gives a direct proof for the 24 rays.

Kernaghan (1994) has shown how to reduce this 24-ray configuration to a 20-ray uncolourable configuration by excluding the rays in boldface. I reproduce below Kernaghan's 20-ray 'no go' theorem in $\mathscr{H}_4$, and an unpublished proof by Clifton[77] further reducing Kernaghan's configuration to 15 uncolourable rays. Clifton's proof exploits the colouring constraints implied by the homomorphism condition for orthogonal pairs and triples of rays spanning 2-dimensional and 3-dimensional subspaces of $\mathscr{H}_4$, as well as the constraints for orthogonal quadruples of rays spanning $\mathscr{H}_4$.

Kernaghan constructs the eleven basis sets or orthogonal quadruples of rays in $\mathscr{H}_4$ shown in table 3.7 out of the twenty rays.

If each of these 20 rays is assigned a 1 (green) or a 0 (red), and the assignment satisfies the homomorphism condition, then the values assigned to the four rays in each of the eleven basis sets must sum to 1. Since each ray appears an even number of times (either twice or four times) in the eleven basis sets, this yields an inconsistent set of eleven

[76] Actually, the contradiction can be derived for the six observables σ_x^1, σ_x^2, σ_y^1, σ_y^2, $\sigma_x^1\sigma_y^2$, $\sigma_y^1\sigma_x^2$, if we observe that the pairs of observables (σ_x^1, σ_x^2), (σ_y^1, σ_y^2), $(\sigma_x^1\sigma_y^2, \sigma_y^1{\sigma_x}^2)$ must yield opposite measured values in the singlet state, and we assume that the hypothetical determinate values of these six observables are the values revealed by measurement.

[77] Private communication.

Table 3.6. *24 uncolourable rays in $\mathscr{H}_4$*

1	0	0	0	0	1	0	0	0	0	1	0	0	0	0	1
−1	1	1	1	1	−1	1	1	1	1	−1	1	1	1	1	−1
0	1	1	0	0	1	−1	0	1	0	0	−1	**1**	**0**	**0**	**1**
0	0	1	1	0	0	1	−1	1	−1	0	0	**1**	**1**	**0**	**0**
1	0	1	0	0	1	0	1	0	1	0	−1	**1**	**0**	**−1**	**0**
1	1	1	1	1	1	−1	−1	1	−1	−1	1	**1**	**−1**	**1**	**−1**

Table 3.7. *Orthogonal quadruples formed by Kernaghan's 20 uncolourable rays*

(1)	1	0	0	0	0	1	0	0	0	0	1	0	0	0	0	1
(2)	1	0	0	0	0	1	0	0	0	0	1	1	0	0	1	−1
(3)	1	0	0	0	0	0	1	0	0	1	0	1	0	1	0	−1
(4)	1	0	0	0	0	0	0	1	0	1	1	0	0	1	−1	0
(5)	−1	1	1	1	1	−1	1	1	1	1	−1	1	1	1	1	−1
(6)	−1	1	1	1	1	1	−1	1	1	0	1	0	0	1	0	−1
(7)	1	−1	1	1	1	1	−1	1	0	1	1	0	1	0	0	−1
(8)	1	1	−1	1	1	1	1	−1	0	0	1	1	1	−1	0	0
(9)	0	1	−1	0	1	0	0	−1	1	1	1	1	1	−1	−1	1
(10)	0	0	1	−1	1	−1	0	0	1	1	1	1	1	1	−1	−1
(11)	1	0	1	0	0	1	0	1	1	1	−1	−1	1	−1	−1	1

equations, where the left sides of the equations sum to an even number, while the right sides sum to an odd number (eleven):

$$v(1,0,0,0) + v(0,1,0,0) + v(0,0,1,0) + v(0,0,0,1) = 1$$
$$v(1,0,0,0) + v(0,1,0,0) + v(0,0,1,1) + v(0,0,1,-1) = 1$$
$$v(1,0,0,0) + v(0,0,1,0) + v(0,1,0,1) + v(0,1,0,-1) = 1$$
$$v(1,0,0,0) + v(0,0,0,1) + v(0,1,1,0) + v(0,1,-1,0) = 1$$
$$v(-1,1,1,1) + v(1,-1,1,1) + v(1,1,-1,1) + v(1,1,1,-1) = 1$$
$$v(-1,1,1,1) + v(1,1,-1,1) + v(1,0,1,0) + v(0,1,0,-1) = 1$$
$$v(1,-1,1,1) + v(1,1,-1,1) + v(0,1,1,0) + v(1,0,0,-1) = 1$$
$$v(1,1,-1,1) + v(1,1,1,-1) + v(0,0,1,1) + v(1,-1,0,0) = 1$$
$$v(0,1,-1,0) + v(1,0,0,-1) + v(1,1,1,1) + v(1,-1,-1,1) = 1$$
$$v(0,0,1,-1) + v(1,-1,0,0) + v(1,1,1,1) + v(1,1,-1,-1) = 1$$
$$v(1,0,1,0) + v(0,1,0,1) + v(1,1,-1,-1) + v(1,-1,-1,1) = 1$$

Kernaghan's proof exploits the representation of $\mathscr{H}_4$ as the span of different orthogonal quadruples of rays. Clifton's 15-ray argument dispenses with the rays:

$$\begin{array}{cccc} 0 & 1 & 0 & 0 \\ 1 & 0 & 0 & 0 \\ 1 & 1 & 1 & 1 \\ 1 & 1 & -1 & 1 \\ 1 & -1 & 1 & 1 \end{array}$$

by considering the representation of certain 2-dimensional and 3-dimensional subspaces of $\mathscr{H}_4$ as spans of different orthogonal pairs and triples of rays, respectively.

In the following six identities, (a,b,c,d) represents the projection operator onto the corresponding ray:

(1) $(0,0,1,0) + (0,0,0,1) = (0,0,1,1) + (0,0,1,-1)$
(2) $(-1,1,1,1) + (1,1,1,-1) = (0,1,1,0) + (1,0,0,-1)$
(3) $(0,1,0,-1) + (0,0,1,0) + (0,1,0,1) = (0,1,1,0) + (0,1,-1,0) + (0,0,0,1)$
(4) $(-1,1,1,1) + (1,0,1,0) + (0,1,0,-1) = (0,0,1,1) + (1,-1,0,0) + (1,1,1,-1)$
(5) $(0,1,-1,0) + (1,0,0,-1) + (1,-1,-1,1) = (0,0,1,-1) + (1-1,0,0) + (1,1,-1-1)$
(6) $(1,0,1,0) + (0,1,0,1) + (1,1,-1,-1) + (1,-1,-1,1) = I$

The first identity expresses two distinct ways of representing the plane $[(1,0,0,0) \vee (0,1,0,0)]^{\perp}$ as the span of orthogonal rays – two different ways of representing the projection operator onto this plane as a sum of projection operators onto a pair of orthogonal rays spanning the plane. The pair of orthogonal rays associated with the right side of equation (1) is rotated 45° with respect to the pair of orthogonal rays associated with the left side. Similarly, the second identity expresses two distinct ways of representing the plane $[(1,0,0,1) \vee (0,1,-1,0)]^{\perp}$ as the span of orthogonal rays. Equations (3), (4), and (5) express distinct ways of representing 3-dimensional subspaces of $\mathscr{H}_4$ as the spans of orthogonal triples of rays. For example, equation (3) corresponds to the 3-dimensional subspace $(1,0,0,0)^{\perp}$. Finally, equation (6) represents $\mathscr{H}_4$ as the span of four mutually orthogonal rays.

A value assignment to the 15 rays satisfying the homomorphism condition must satisfy the six identities, because the value assigned to a 2-dimensional subspace or a 3-dimensional subspace must be the sum of the values assigned to any mutually orthogonal set of rays that span the subspace. (Equivalently, this follows from the functional relationship condition: the value assigned to a sum of mutually orthogonal and hence compatible projection operators must be the sum of the values assigned to the projection operators.) So we get six equations for the value assignments to the 1-dimensional projection operators, and it is easily seen that these equations are inconsistent.

Since each ray or 1-dimensional projection operator occurs exactly twice in the six equations, when we add the equations for the value assignments we get the valuation for each of the 15 projection operators (1 or 0) occurring either twice on the left side, or twice on the right side, or once on the left side and once on the right side, plus an additional 1 on the right side from the value assigned to the unit operator I in equation (6). Since the paired values that occur on the left and right sides cancel out (these are the values assigned to the projection operators $(0,0,0,1)$, $(1,1,1,-1)$, $(1,0,0,-1)$, $(0,1,-1,0)$, $(1,1,-1,-1)$), we are left with the inconsistent requirement that $2 \times$ (sum of values assigned to certain projection operators) $= 2 \times$ (sum of values assigned to certain other projection operators) $+ 1$. That is, again an even number is equated with an odd number.

Clifton's colouring problem is defined by more restrictive colouring constraints than the problem considered by Peres and Kernaghan, which requires only that every orthogonal quadruple of rays is assigned one 1 ('green') and three 0's ('red'). Kernaghan's 20-ray configuration in $\mathcal{H}_4$ is a *critical* configuration, in the sense that every subconfiguration is colourable (with respect to the colouring constraint for orthogonal quadruples of rays). Removing any single ray from the 20-ray configuration yields a colourable configuration, and any subconfiguration of a colourable configuration is colourable (because there are fewer orthogonality relations to constrain the colouring). There are 95 other 20-ray critical uncolourable subconfigurations in Peres's 24-ray configuration, and 16 critical uncolourable 18-ray subconfigurations discovered by Cabello, Estebaranz, and García-Alcaine (1996a).[78] A 28-ray uncolourable configuration in $\mathcal{H}_4$, abstracted by Zimba and Penrose (1993) from a 40-ray uncolourable configuration (with essentially complex rays) obtained by Penrose (1994a), is critical.[79] In $\mathcal{H}_3$, the 31-ray Conway and Kochen configuration, Peres's 33-ray configuration,

[78] The 18-ray configurations each yield a 10-ray state-dependent Kochen–Specker theorem in $\mathcal{H}_4$. Kernaghan and Peres (1995) have found a 36-ray uncolourable configuration in $\mathcal{H}_8$ that yields a 13-ray state-dependent Kochen–Specker theorem in $\mathcal{H}_8$. This seems to be a record in terms of the ratio d/n, where d is the dimension of the Hilbert space and n the number of rays (a reasonable measure of the size of the set of rays, since the maximum number of mutually orthogonal rays, and hence the colouring constraint, depends on the dimension of the space).

[79] Penrose's 1994 paper was circulated as a manuscript in 1992. Peres (1993, p. 201, and 1991, p. L177) claims that removing any single ray from his 24-ray uncolourable configuration in $\mathcal{H}_4$ yields a colourable configuration, but this is incorrect (and the error is acknowledged in the paperback edition of 1993). From Kernaghan's construction, it is clear that removing any ray that does not belong to Kernaghan's 20-ray subconfiguration will leave the remaining 23-ray configuration uncolourable.

and the 33-ray configuration derived from Schütte's tautology are all critical. This can be checked by Peres's (1993, chapter 7) FORTRAN program for the colourability of a finite set of rays in $\mathscr{H}_3$. The generalization of the program to $\mathscr{H}_4$ shows that the 18-ray subconfigurations are the smallest uncolourable subconfigurations in Peres's 24-ray uncolourable configuration.[80] How to construct the smallest critical uncolourable configurations in an n-dimensional Hilbert space (or, for that matter, in $\mathscr{H}_3$ or $\mathscr{H}_4$) is an open problem.

Uncolourable configurations can be associated with classical tautologies that correspond to quantum contradictions, under appropriate identifications of the propositional variables with subspaces of a Hilbert space (where conjunction and disjunction are interpreted as the span and intersection, respectively, of the subspaces representing the propositions, negation is interpreted as the orthocomplement, and compatibility constraints are respected). So there must be a classical tautology associated with the uncolourable configurations in $\mathscr{H}_4$ derived from the 9-observable Peres–Mermin state-independent Kochen and Specker theorem. In fact, the relevant tautology is:

$$[(P_1 \leftrightarrow P_2) \leftrightarrow (Q_1 \leftrightarrow \neg Q_2)] \leftrightarrow [(P_1 \leftrightarrow Q_2) \leftrightarrow (Q_1 \leftrightarrow \neg P_2)],$$

where '$\leftrightarrow$' denotes the biconditional 'if and only if.' For a proof, see the following section.

3.4 Proofs and constructions

I. *Proof that Schütte's formula F is a classical tautology*
Schütte's formula F is the proposition:

$$[\neg(1) \vee \neg(2) \vee \ldots \vee \neg(16)] \vee (17)$$

where the propositions (1), (2), . . . , (17) are constructed as follows:

(1) $\neg D_1 \vee \neg B_2$
(2) $\neg D_1 \vee \neg B_3$
(3) $\neg D_2 \vee (A_2 \vee B_2)$
(4) $\neg D_2 \vee \neg B_3$
(5) $\neg D_3 \vee \neg B_2$
(6) $\neg D_3 \vee [\neg(A_1 \vee A_2) \vee B_3]$
(7) $\neg D_4 \vee (A_2 \vee B_2)$
(8) $\neg D_4 \vee [\neg(A_1 \vee A_2) \vee B_3]$
(9) $(A_2 \vee C_1) \vee (B_3 \vee D_1)$
(10) $(A_2 \vee C_2) \vee [\neg(A_1 \vee B_1) \vee D_1]$
(11) $\neg C_2 \vee (B_3 \vee D_2)$
(12) $\neg C_1 \vee (B_1 \vee D_2)$
(13) $(A_2 \vee C_1) \vee \{\neg[\neg(A_1 \vee A_2) \vee B_3] \vee D_3\}$
(14) $(A_2 \vee C_2) \vee (B_1 \vee D_3)$
(15) $\neg C_2 \vee \{\neg[\neg(A_1 \vee A_2) \vee B_3] \vee D_4\}$
(16) $\neg C_1 \vee [\neg(A_1 \vee B_1) \vee D_4]$
(17) $(\neg A_1 \vee A_2) \vee B_1$

To show that F is a classical tautology, I shall derive a contradiction from the

[80] Cabello, Estebaranz, and García-Alcaine (1996a).

supposition that F is false – that (1), (2), . . . , (16) are all true, and (17) is false. If (17) is false, then A_1 is true, A_2 false, and B_1 false.

Consider two cases: B_3 true, and B_3 false.

Case 1 (A_1 true, A_2 false, B_1 false, B_3 true)

From (2) – that is, if (2) is true – since $\neg B_3$ is false, $\neg D_1$ must be true, and so D_1 must be false. From (10), since A_1 is true, $A_1 \vee B_1$ is true, and so $\neg(A_1 \vee B_1)$ is false. Since A_2 is false, D_1 is false, and $\neg(A_1 \vee B_1)$ is false, C_2 must be true. From (15), since B_3 is true, $\neg(A_1 \vee A_2) \vee B_3$ is true, and so $\neg[\neg(A_1 \vee A_2) \vee B_3]$ is false. Since $\neg C_2$ is false and $\neg[\neg(A_1 \vee A_2) \vee B_3]$ is false, D_4 must be true. From (4), since $\neg B_3$ is false, $\neg D_2$ must be true, so D_2 must be false. From (12), since B_1 is false and D_2 is false, $\neg C_1$ must be true, and so C_1 must be false. From (13), since C_1 is false, A_2 is false, and $\neg[\neg(A_1 \vee A_2) \vee B_3]$ is false, D_3 *must be true.* From (7), since $\neg D_4$ is false and A_2 is false, B_2 must be true. But now from (5), if $\neg B_2$ is false, $\neg D_3$ must be true, and so D_3 *must be false.* Hence, a contradiction follows if B_3 is true.

Case 2 (A_1 true, A_2 false, B_1 false, B_3 false*)*

From (8), D_4 must be false if A_1 is true and B_3 is false. Hence, from (16), C_1 must be false. Since C_1 is false A_2 is false, and B_3 is false, D_1 must be true for (9) to be true. Since D_1 is true, B_2 must be false for (1) to be true. Hence, from (3), D_2 must be false. From (11), since D_2 is false and B_3 is false, C_2 *must be false.* From (6), since B_3 is false and $\neg(A_1 \vee A_2)$ is false, D_3 must be false. But now from (14), since A_2 is false, B_1 is false, and D_3 is false, C_2 *must be true.* Hence, a contradiction follows if B_3 is false.

Since B_3 must be either true or false, the proposition F cannot be false for any assignment of truth values to the eleven propositional variables. So F is a classical tautology.

Readers familiar with tree proofs will notice that Case 1 can be set out in the form of a closed tree for the sentences:

$$\neg(17), B_3, (2), (10), (15), (4), (12), (13), (7), (5)$$

If the rules for disjunctions and negations of disjunctions are applied to the sentences in this order, then each sentence yields only one open branch, until the two branches derived from (5), which both close. Similarly, Case 2 can be set out in the form of a closed tree for the sentences:

$$\neg(17), \neg B_3, (8), (16), (9), (1), (3), (11), (6), (14)$$

The two closed trees demonstrate that $\neg(17)$, (2), (10), (15), (4), (12), (13), (7), (5) are unsatisfiable with B_3 and $\neg(17)$, (8), (16), (9), (1), (3), (11), (6), (14) are unsatisfiable with $\neg B_3$. Since F can be expressed as:

$$[(1) \wedge (2) \wedge \ldots \wedge (16)] \rightarrow (17),$$

where '$\wedge$' denotes conjunction and '$\rightarrow$' denotes the conditional (Schütte's original formulation), it follows that the sentences (1), (2), ..., (16), $\neg$(17) are unsatisfiable, which means that $\neg\{[(1)\wedge(2)\wedge\ldots\wedge(16)]\rightarrow(17)\}$ is unsatisfiable, and so F is a classical tautology.

II. *Constructing the Schütte planes and rays*

To see how to construct the intersecting orthogonal planes defined by the propositions (1), (2), ..., (17) in Schütte's tautology F, and the orthogonal triples of rays defined by these intersecting planes, consider, for example, proposition (6):

$$\neg D_3 \vee [\neg(A_1 \vee A_2) \vee B_3]$$

Since A_1 and A_2 correspond to the orthogonal rays or 1-dimensional subspaces spanned by the unit vectors $\begin{pmatrix}1\\0\\0\end{pmatrix}$ and $\begin{pmatrix}0\\1\\0\end{pmatrix}$, respectively, their disjunction $A_1 \vee A_2$ corresponds to the span of these 1-dimensional subspaces, which I represent for brevity as:

$$\begin{pmatrix}1\\0\\0\end{pmatrix} \vee \begin{pmatrix}0\\1\\0\end{pmatrix}$$

This is the plane orthogonal to $\begin{pmatrix}0\\0\\1\end{pmatrix}$. Hence $\neg(A_1 \vee A_2)$ corresponds to the 1-dimensional subspace spanned by the unit vector $\begin{pmatrix}1\\0\\0\end{pmatrix}$. The proposition $\neg B_3$ corresponds to the 1-dimensional subspace spanned by the vector $\begin{pmatrix}0\\0\\1\end{pmatrix}$. So $\neg(A_1 \vee A_2) \vee B_3$ corresponds to the plane spanned by $\begin{pmatrix}1\\0\\0\end{pmatrix}$ and $\begin{pmatrix}0\\0\\1\end{pmatrix}$, which I represent as:

$$\begin{pmatrix}0\\0\\1\end{pmatrix} \vee \begin{pmatrix}1\\1\\0\end{pmatrix}$$

The matrices of the projection operators onto the 1-dimensional subspaces spanned by $\begin{pmatrix}1\\0\\0\end{pmatrix}$ and $\begin{pmatrix}0\\0\\1\end{pmatrix}$ are derived as the matrix products of the associated normalized row and column vectors (recall the Dirac notation $|\psi\rangle\langle\psi|$ for the projection operator onto the 1-dimensional subspace spanned by the unit vector $|\psi\rangle$):

$$(0,0,1)\cdot\begin{pmatrix}0\\0\\1\end{pmatrix} = \begin{pmatrix}0&0&0\\0&0&0\\0&0&1\end{pmatrix}$$

$$\tfrac{1}{\sqrt{2}}(1,1,0)\cdot\tfrac{1}{\sqrt{2}}\begin{pmatrix}1\\1\\0\end{pmatrix} = \tfrac{1}{2}\begin{pmatrix}1&1&0\\1&1&0\\0&0&0\end{pmatrix}$$

(where the '$\cdot$' here represents the matrix product). The matrix of the projection operator onto the plane spanned by these orthogonal 1-dimensional subspaces is therefore:

$$\begin{pmatrix}0&0&0\\0&0&0\\0&0&1\end{pmatrix}+\tfrac{1}{2}\begin{pmatrix}1&1&0\\1&1&0\\0&0&0\end{pmatrix}=\tfrac{1}{2}\begin{pmatrix}1&1&0\\1&1&0\\0&0&2\end{pmatrix}$$

The proposition $\neg D_3$ corresponds to the plane $\begin{pmatrix}1\\1\\-1\end{pmatrix}^{\perp}$. The matrix representing this plane is:

$$\begin{pmatrix}1&0&0\\0&1&0\\0&0&1\end{pmatrix}-\tfrac{1}{\sqrt{3}}(1,1,-1)\cdot\tfrac{1}{\sqrt{3}}\begin{pmatrix}1\\1\\-1\end{pmatrix}$$

$$=\begin{pmatrix}1&0&0\\0&1&0\\0&0&1\end{pmatrix}-\tfrac{1}{3}\begin{pmatrix}1&1&-1\\1&1&-1\\-1&1&1\end{pmatrix}$$

$$=\tfrac{1}{3}\begin{pmatrix}2&-1&1\\-1&2&1\\1&1&2\end{pmatrix}$$

It is easily checked that the matrices representing these two planes commute and that their product is the matrix:

$$\tfrac{1}{6}\begin{pmatrix}1&1&2\\1&1&2\\2&2&4\end{pmatrix}$$

which is the matrix of the projection operator onto the 1-dimensional subspace spanned by the unit vector $\frac{1}{\sqrt{6}}\begin{pmatrix}1\\1\\2\end{pmatrix}$.

The proposition (6) generates three orthogonal triples of rays. Firstly, an orthogonal triple is generated by the intersection of the above two planes:

$$\tfrac{1}{\sqrt{6}}\begin{pmatrix}1\\1\\2\end{pmatrix},\ \tfrac{1}{\sqrt{2}}\begin{pmatrix}1\\-1\\0\end{pmatrix},\ \tfrac{1}{\sqrt{3}}\begin{pmatrix}1\\1\\-1\end{pmatrix}$$

Here $\frac{1}{\sqrt{2}}\begin{pmatrix}1\\-1\\0\end{pmatrix}$ represents the ray orthogonal to $\frac{1}{\sqrt{6}}\begin{pmatrix}1\\1\\2\end{pmatrix}$ in the plane corresponding to $\neg(A_1 \vee A_2) \vee B_3$:

$$\begin{pmatrix}1&0&0\\0&1&0\\0&0&1\end{pmatrix}-\tfrac{1}{2}\begin{pmatrix}1&1&0\\1&1&0\\0&0&2\end{pmatrix}=\tfrac{1}{2}\begin{pmatrix}1&-1&0\\-1&1&0\\0&0&0\end{pmatrix}$$

(which is the projection operator onto the 1-dimensional subspace spanned by the unit vector $\frac{1}{\sqrt{2}}\begin{pmatrix}1\\-1\\0\end{pmatrix}$), and $\frac{1}{\sqrt{3}}\begin{pmatrix}1\\1\\-1\end{pmatrix}$ represents the ray orthogonal to $\frac{1}{\sqrt{6}}\begin{pmatrix}1\\1\\2\end{pmatrix}$ in the plane corresponding to $\neg D_3$:

$$\begin{pmatrix}1&0&0\\0&1&0\\0&0&1\end{pmatrix}-\tfrac{1}{2}\begin{pmatrix}2&-1&1\\-1&2&1\\1&1&2\end{pmatrix}=\tfrac{1}{2}\begin{pmatrix}1&1&-1\\1&1&-1\\-1&-1&1\end{pmatrix}$$

(which is the projection operator onto the 1-dimensional subspace spanned by the unit vector $\frac{1}{\sqrt{3}}\begin{pmatrix}1\\1\\-1\end{pmatrix}$).

A second orthogonal triple is generated from the pair of orthogonal rays $\begin{pmatrix}1\\0\\0\end{pmatrix}, \begin{pmatrix}0\\1\\0\end{pmatrix}$ corresponding to the propositions A_1 and A_2 in the disjunction $A_1 \vee A_2$:

$$\begin{pmatrix}1\\0\\0\end{pmatrix}, \begin{pmatrix}0\\1\\0\end{pmatrix}, \begin{pmatrix}0\\0\\1\end{pmatrix}$$

Finally, a third orthogonal triple is generated from the pair of orthogonal rays $\begin{pmatrix}0\\0\\1\end{pmatrix}$ corresponding to $\neg(A_1 \vee A_2)$, and $\frac{1}{\sqrt{2}}\begin{pmatrix}1\\1\\0\end{pmatrix}$ corresponding to B_3:

$$\begin{pmatrix}0\\0\\1\end{pmatrix}, \frac{1}{\sqrt{2}}\begin{pmatrix}1\\1\\0\end{pmatrix}, \frac{1}{\sqrt{2}}\begin{pmatrix}1\\-1\\0\end{pmatrix}$$

Proceeding in this way, the propositions (1), (2), ..., (16) generate the orthogonal triples of rays given in table 3.8 (where, for brevity, I denote each ray by a triple of integers representing the – possibly unnormalized – components of a vector spanning the ray, and I ignore repeat generations):[81]

Table 3.9 lists the orthogonality relations among the members of the 33-ray uncolourable set. (The column on the right lists all the rays that are orthogonal to the ray on the left.)

III. *Direct proof that Schütte's 25 rays are uncolourable*

The proof that F is a tautology can be extended to a direct 'brute force' state-independent proof that the 33 rays cannot be coloured consistently, or to a direct state-dependent proof that the 25 rays (the 33 rays excluding 012, 01 – 2, 021, 02 – 1, 121, 12 – 1, 1 – 21, 1 – 2 – 1) cannot be coloured consistently with the assignment green (1 or 'true') to the ray 100. I reproduce here a proof for the 25 rays (which could also be set out in the form of a tree proof, as in I above).

Consider two cases, 110 green, and 110 red. In the following, the numbers refer to the number of the orthogonal triple in the list of 26 orthogonal triples generated by F in table 3.2.

Case 1 (100 green, 110 green)

From (1), since 100 is green, 010 is red. From (6), since 100 is green, 011 is red and 01 – 1 is red. From (16), since 110 is green, 1 – 1 – 1 is red. From (10), since 1 – 1 – 1 is red and 01 – 1 is red, 211 is green. So from (21), 10 – 2 is red. From (5), since 10 – 2 is red and 010 is red, 201 is green. So from (23), 11 – 2 is red. From (2), since 110 is green,

[81] No new orthogonal triples are generated by (17).

1 − 10 is red. From (17), since 11 − 2 is red and 1 − 10 is red, 111 is green. From (15), since 110 is green 1 − 11 is red. From (8), since 011 is red and 1 − 11 is red, 21 − 1 is green. So from (19), 102 is red. From (4), since 102 is red and 010 is red, 20 − 1 is green. So from (25), 112 is red. From (18), since 112 is red and 1 − 10 is red, *11 − 1 is green.* From (13), since 111 is green (from (17), above), 10 − 1 is red. So from (3), since 10 − 1 is red and 010 is red (from (1), above), 101 is green. But now, from (11), *11 − 1 is red.* Hence, a contradiction follows if 110 is green.

Case 2 (100 *green,* 110 *red*)

From (1), since 100 is green, 001 is red and 010 is red. From (2), since 001 is red and 110 is red, 1 − 10 is green. So from (17), 111 is red. From (6), since 100 is green, 01 − 1 is red and 011 is red. From (9), since 111 is red and 01 − 1 is red, 2 − 1 − 1 is green. So from (20), 102 is red. From (4), since 010 is red and 102 is red, 20 − 1 is green. So from (26), 1 − 12 is red. From (16), since 1 − 12 is red and 110 is red, 1 − 1 − 1 is green. So from (12), 101 is red. From (3), since 101 is red and 010 is red, 10 − 1 is green. So from (14), 1 − 11 is red. From (15), since 110 is red and 1 − 11 is red, 1 − 1 − 2 is green. So from (24), *201 is red.* From (18), since 1 − 10 is green (from (2), above), 11 − 1 is red. From (7), since 011 is red (from (6), above) and 11 − 1 is red, 2 − 11 is green. So from (22), 10 − 2 is red. But now from (5), since 010 is red (from (1), above) and 10 − 2 is red, *201 is green.* Hence, a contraction follows if 110 is red.

Since 110 is either green or red, the 25 rays cannot be coloured consistently with the constraint that one ray in each orthogonal triple is assigned the colour green, and the remaining two rays are assigned the colour red.

IV. *A classical tautology that corresponds to a quantum contradiction in* $\mathscr{H}_4$
Consider the six identities in Mermin's proof of the 9-observable Peres–Mermin state-independent Kochen–Specker theorem in $\mathscr{H}_4$:

$$v(\sigma_x^1\sigma_x^2)v(\sigma_x^1)v(\sigma_x^2) = 1$$
$$v(\sigma_y^1\sigma_y^2)v(\sigma_y^1)v(\sigma_y^2) = 1$$
$$v(\sigma_x^1\sigma_y^2)v(\sigma_x^1)v(\sigma_y^2) = 1$$
$$v(\sigma_y^1\sigma_x^2)v(\sigma_y^1)v(\sigma_x^2) = 1$$
$$v(\sigma_x^1\sigma_y^2)v(\sigma_y^1\sigma_x^2)v(\sigma_z^1\sigma_z^2) = 1$$
$$v(\sigma_x^1\sigma_x^2)v(\sigma_y^1\sigma_y^2)v(\sigma_z^1\sigma_z^2) = -1$$

The first identity constrains the values of σ_x^1 and σ_x^2 to have the same sign when $v(\sigma_x^1\sigma_x^2) = 1$ and different signs when $v(\sigma_x^1\sigma_x^2) = -1$. Similarly, the second identity constrains the value of σ_y^1 and σ_y^2 to have the same sign when $v(\sigma_y^1\sigma_y^2) = 1$ and different signs when $v(\sigma_y^1\sigma_y^2) = -1$. The sixth identity constrains the values of $\sigma_x^1\sigma_x^2$ and $\sigma_y^1\sigma_y^2$ to have different signs when $v(\sigma_z^1\sigma_z^2) = 1$ and the same sign when $v(\sigma_z^1\sigma_z^2) = -1$. These three identities together therefore require that (i) when $v(\sigma_z^1\sigma_z^2) = 1$, the spins σ_x^1, σ_x^2 have the same sign if and only if the spins σ_y^1, σ_y^2 have different signs, and (ii) when

Table 3.8. *Orthogonal triples of rays generated by propositions (1) to (16)*

(1)	1	0	1	1	−1	−1	1	2	−1
(2)	1	1	0	1	−1	−1	1	−1	2
(3)	1	0	−1	1	−1	1	1	2	1
	0	1	0	1	0	1	1	0	−1
(4)	1	1	0	1	−1	1	1	−1	−2
(5)	1	0	1	1	1	−1	1	−2	−1
(6)	1	−1	0	1	1	−1	1	1	2
	0	0	1	0	1	0	1	0	0
(7)	0	0	1	1	1	0	1	−1	0
	1	0	−1	1	1	1	1	−2	1
(8)	1	−1	0	1	1	1	1	1	−2
(9)	2	0	−1	1	−1	2	1	5	2
(10)	1	0	−2	2	1	1	2	−5	1
	0	1	−1	1	−1	−1	2	1	1
(11)	2	0	1	1	−1	−2	1	5	−2
(12)	1	0	2	2	1	−1	2	−5	−1
	0	1	1	1	−1	1	2	1	−1
(13)	2	0	−1	1	1	2	1	−5	2
	0	1	0	1	0	2	2	0	−1
(14)	1	0	−2	2	−1	1	2	5	1
	0	1	0	1	0	−2	2	0	1
	0	1	1	1	1	−1	2	−1	1
(15)	2	0	1	1	1	−2	1	−5	−2
(16)	1	0	2	2	−1	−1	2	5	−1
	1	0	0	0	1	1	0	1	−1
	0	1	−1	1	1	1	2	−1	−1

Table 3.9. *Orthogonality relations for Schütte's 33-ray uncolourable set*

Ray	Orthogonal rays							
0 0 1	0 1 0	1 0 0	1 1 0	1 −1 −0				
0 1 0	0 0 1	1 0 0	1 0 1	1 0 −1	1 0 2	1 0 −2	2 0 1	2 0 −1
1 0 0	0 0 1	0 1 0	0 1 1	0 1 −1	0 1 2	0 1 −2	0 2 1	0 2 −1
0 1 1	1 0 0	0 1 −1	1 1 −1	1 −1 1	2 1 −1	2 −1 1		
0 1 −1	1 0 0	0 1 1	1 1 1	1 −1 −1	2 1 1	2 −1 −1		
1 0 1	0 1 0	1 0 −1	1 1 −1	1 −1 −1	1 2 −1	1 −2 −1		
1 0 −1	0 1 0	1 0 1	1 1 1	1 −1 1	1 2 1	1 −2 1		
1 1 0	0 0 1	1 −1 0	1 −1 −1	1 −1 2	1 −1 1	1 −1 −2		
1 −1 0	0 0 1	1 1 0	1 1 1	1 1 −1	1 1 2	1 1 −2		
1 1 1	0 1 −1	1 0 −1	1 −1 0	1 1 −2	1 −2 1	2 −1 −1		
1 −1 1	0 1 1	1 0 −1	1 1 0	1 −1 −2	1 2 1	2 1 −1		
1 1 −1	0 1 1	1 0 1	1 −1 0	1 1 2	1 −2 −1	2 −1 1		
1 −1 −1	0 1 −1	1 0 1	1 1 0	1 −1 2	1 2 −1	2 1 1		
0 1 2	1 0 0	0 2 −1	1 2 −1	1 −2 1				
0 1 −2	1 0 0	0 2 1	1 −2 −1	1 −2 −1				
0 2 1	1 0 0	0 1 −2	1 1 −2	1 −1 2				
0 2 −1	1 0 0	0 1 2	1 1 2	1 −1 −2				
1 0 2	0 1 0	2 0 −1	2 1 −1	2 −1 −1				
1 0 −2	0 1 0	2 0 1	2 1 1	2 −1 1				
2 0 1	0 1 0	1 0 −2	1 1 −2	1 −1 −2				
2 0 −1	0 1 0	1 0 2	1 1 2	1 −1 2				
1 1 2	1 −1 0	1 1 −1	0 2 −1	2 0 −1				
1 1 −2	1 −1 0	1 1 1	0 2 1	2 0 1				
1 −1 2	1 1 0	1 −1 −1	0 2 1	2 0 −1				
1 −1 −2	1 1 0	1 −1 1	0 2 −1	2 0 1				
1 2 1	1 0 −1	1 −1 1	0 1 −2					
1 2 −1	1 0 1	1 −1 −1	0 1 2					
1 −2 1	1 0 −1	1 1 1	0 1 2					
1 −2 −1	1 0 1	1 1 −1	0 1 2					
2 1 1	0 1 −1	1 −1 −1	1 0 −2					
2 1 −1	0 1 1	1 −1 1	1 0 2					
2 −1 1	0 1 1	1 1 −1	1 0 −2					
2 −1 −1	0 1 −1	1 1 1	1 0 2					

$v(\sigma_z^1\sigma_z^2) = -1$, the spins σ_x^1, σ_x^2 have the same sign if and only if the spins σ_y^1, σ_y^2 have the same sign.

Similarly, the third, fourth, and fifth identities require that (iii) when $v(\sigma_z^1\sigma_z^2) = 1$, σ_x^1 and σ_y^2 have the same sign if and only if σ_y^1 and σ_x^1 have the same sign, and (iv) when $v(\sigma_z^1\sigma_z^2) = -1$, σ_x^1 and σ_y^2 have the same sign if and only if σ_y^1 and σ_x^2 have different signs.

Let P_s^{xx} denote the proposition 'the values of σ_x^1 and σ_x^2 have the same sign,' and P_d^{xx} denote the proposition 'the values of σ_x^1 and σ_x^2 have different signs,' and similarly for P_s^{yy} and P_d^{yy}. Also, let P_s^{xy} denote the proposition 'the values of σ_x^1 and σ_y^2 have the same sign,' and P_d^{xy} denote the proposition 'the values of σ_x^1 and σ_y^2 have different signs,' and similarly for P_s^{yx}, P_d^{yx}. Since either $v(\sigma_z^1\sigma_z^2) = 1$ or $v(\sigma_z^1\sigma_z^2) = -1$, we can sum up the (inconsistent) constraints expressed by the six identities in the following proposition C (where '$\leftrightarrow$' denotes the biconditional 'if and only if'):

$$[(P_s^{xx} \leftrightarrow P_d^{yy}) \wedge (P_s^{xy} \leftrightarrow P_s^{yx})] \vee [(P_s^{xx} \leftrightarrow P_s^{yy}) \wedge (P_s^{xy} \leftrightarrow P_d^{yx})] \qquad (C)$$

Now, $P \leftrightarrow Q$ is true if and only if P and Q are either both true or both false, so $P \leftrightarrow Q$ is logically equivalent to $(P \wedge Q) \vee (\neg P \wedge \neg Q)$. Since P_s^{yy} is true if and only if P_d^{yy} is false, $P_s^{xx} \leftrightarrow P_s^{yy}$ is logically equivalent to $P_s^{xx} \leftrightarrow \neg P_d^{yy}$, which is logically equivalent to $\neg(P_s^{xx} \leftrightarrow P_d^{yy})$. Similarly, since P_d^{yx} is true if and only if P_s^{yx} is false, $P_s^{xy} \leftrightarrow P_d^{yx}$ is logically equivalent to $P_s^{xy} \leftrightarrow \neg P_s^{yx}$, which is logically equivalent to $\neg(P_s^{xy} \leftrightarrow P_s^{yx})$. Substituting equivalent propositions in the second disjunct of C, it follows that C is logically equivalent to:

$$(P_s^{xx} \leftrightarrow P_d^{yy}) \leftrightarrow (P_s^{xy} \leftrightarrow P_s^{yx})$$

Since C represents a classical contradiction, the negation

$$\neg[(P_s^{xx} \leftrightarrow P_d^{yy}) \leftrightarrow (P_s^{xy} \leftrightarrow P_s^{yx})]$$

represents a classical tautology. This proposition is logically equivalent to:

$$(P_s^{xx} \leftrightarrow P_d^{yy}) \leftrightarrow \neg(P_s^{xy} \leftrightarrow P_s^{yx})$$

or

$$(P_s^{xx} \leftrightarrow P_d^{yy}) \leftrightarrow (P_s^{xy} \leftrightarrow \neg P_s^{yx})$$

or

$$(P_s^{xx} \leftrightarrow P_d^{yy}) \leftrightarrow (P_s^{xy} \leftrightarrow P_d^{yx}) \qquad (T)$$

To exhibit the logical form of T as a classical tautology explicitly, let P_{x+}^1, P_{x-}^1 denote the propositions 'the value of σ_x^1 is $+1$' and 'the value of σ_x^1 is -1,' respectively, and let P_{y+}^1, P_{y-}^1 denote the propositions 'the value of σ_y^1 is $+1$' and 'the value of σ_y^1 is -1,' respectively, and similarly for P_{x+}^2, P_{x-}^2; P_{y+}^2, P_{y-}^2. Then T can be expressed as:

$$[(P_{x+}^1 \leftrightarrow P_{x+}^2) \leftrightarrow (P_{y+}^1 \leftrightarrow P_{y-}^2)] \leftrightarrow [(P_{x+}^1 \leftrightarrow P_{y+}^2) \leftrightarrow (P_{y+}^1 \leftrightarrow P_{x-}^2)]$$

or, writing P_1 for P_{x+}^1, P_2 for P_{x+}^2, Q_1 for P_{y+}^1, and Q_2 for P_{y+}^2, as:

Table 3.10. *Truth value assignments that make T_1 true*

P_1	P_2	Q_1	Q_2
1	1	1	0
1	1	0	1
0	0	1	0
0	0	0	1
1	0	1	1
0	1	1	1
1	0	0	0
0	1	0	0

$$[(P_1 \leftrightarrow P_2) \leftrightarrow (Q_1 \leftrightarrow \neg Q_2] \leftrightarrow [(P_1 \leftrightarrow Q_2) \leftrightarrow (Q_1 \leftrightarrow \neg P_2)],$$

which I abbreviate as:

$$T_1 \leftrightarrow T_2$$

To see that this proposition is a classical tautology, note that T_1 is true if and only if P_1 and P_2 have the same truth value and Q_1 and Q_2 have different truth values, or P_1 and P_2 have different truth values and Q_1 and Q_2 have the same truth value. So T_1 is true for the eight combinations of truth values given in table 3.10 (denoting 'true' by 1, and 'false' by 0), and in no other cases:

The proposition T_2 is true if and only if P_1 and Q_2 have the same truth value and Q_1 and P_2 have different truth values, or P_1 and Q_2 have different truth values and Q_1 and P_2 have the same truth values. So T_2 is true for the same eight combinations of truth values as T_1, and in no other cases. Since T_1 is true if and only if T_2 is true, $T_1 \leftrightarrow T_2$ is a classical tautology.

The proposition $T_1 \leftrightarrow T_2$ corresponds to the null space in the Hilbert space $\mathscr{H}(S_1) \otimes \mathscr{H}(S_2)$ of a pair of spin-$\frac{1}{2}$ particles S_1 and S_2 (and hence to a quantum contradiction), under an appropriate interpretation of the propositional variables P_1, P_2, Q_1, Q_2 as quantum propositions represented by projection operators onto subspaces of $\mathscr{H}(S_1) \otimes \mathscr{H}(S_2)$. To see this, interpret the propositions P^1_{x+}, P^1_{x-} introduced above as quantum propositions represented by the projection operators onto the 2-dimensional eigenspaces corresponding to the eigenvalues $+1$ and -1, respectively, of the spin operator σ^1_x, and P^1_{y+}, P^1_{y-} as quantum propositions represented by the projection operators onto the 2-dimensional eigenspaces corresponding to the eigenvalues $+1$ and -1 of σ^1_y, respectively, and similarly for P^2_{x+}, P^2_{x-}; P^2_{y+}, P^2_{y-}. That is:

$$P^1_{x+} = |\sigma^1_x = +1\rangle\langle\sigma^1_x = +1| \otimes I^2,\ P^1_{x-} = |\sigma^1_x = -1\rangle\langle\sigma^1_x = -1| \otimes I^2, \text{ etc.}$$
$$P^2_{x+} = I^1 \otimes |\sigma^2_x = +1\rangle\langle\sigma^2_x = +1|,\ P^2_{x-} = I^1 \otimes |\sigma^2_x = -1\rangle\langle\sigma^2_x = -1|, \text{ etc.}$$

Identifying P_1, P_2, Q_1, Q_2 with $P^1_{x+}, P^2_{x+}, P^1_{y+}, P^2_{y+}$, respectively, the proposition T_1, which is logically equivalent to:

$$\{[(P_1 \wedge P_2) \vee (\neg P_1 \wedge \neg P_2)] \wedge [(Q_1 \wedge \neg Q_2) \vee (\neg Q_1 \wedge Q_2)]\}$$
$$\vee \{[(P_1 \wedge \neg P_2) \vee (\neg P_1 \wedge P_2)] \wedge [(Q_1 \wedge Q_2) \vee (\neg Q_1 \wedge \neg Q_2)]\},$$

is then interpreted as the quantum proposition T_1':

$$(P^1_{x+}P^2_{x+} + P^1_{x-}P^2_{x-})(P^1_{y+}P^2_{y-} + P^1_{y-}P^2_{y+}) + (P^1_{x+}P^2_{x-} + P^1_{x-}P^2_{x+})(P^1_{y+}P^2_{y+} + P^1_{y-}P^2_{y-})$$

and the proposition T_2, which is logically equivalent to:

$$\{[(P_1 \wedge Q_2) \vee (\neg P_1 \wedge \neg Q_2)] \wedge [(Q_1 \wedge \neg P_2) \vee (\neg Q_1 \wedge P_2)]\}$$
$$\vee \{[(P_1 \wedge \neg Q_2) \vee (\neg P_1 \wedge Q_2)] \wedge [(Q_1 \wedge P_2) \vee (\neg Q_1 \wedge \neg P_2)]\},$$

is interpreted as the quantum proposition T_2':

$$(P^1_{x+}P^2_{y+} + P^1_{x-}P^2_{y-})(P^1_{y+}P^2_{x-} + P^1_{y-}P^2_{x+}) + (P^1_{x+}P^2_{y-} + P^1_{x-}P^2_{y+})(P^1_{y+}P^2_{x+} + P^1_{y-}P^2_{x-})$$

The term

$$(P^1_{x+}P^2_{x+} + P^1_{x-}P^2_{x-})(P^1_{y+}P^2_{y-} + P^1_{y-}P^2_{y+})$$

in T_1' represents the ray that is the intersection of the plane represented by the projection operator:

$$(P^1_{x+}P^2_{x+} + P^1_{x-}P^2_{x-})$$

associated with + + or − − for the eigenvalues of σ^1_x, σ^2_x, respectively (that is, 'same sign'), and the plane represented by the projection operator:

$$(P^1_{y+}P^2_{y-} + P^1_{y-}P^2_{y+})$$

associated with + − or − + for the eigenvalues of σ^1_y, σ^2_y, respectively (that is, 'different sign'). These two planes intersect at right angles, and so the projection operator onto their intersection is just the product of the (commuting) projection operators onto the planes. (Note that $P^1_{x+}P^2_{x+}$, $P^1_{x-}P^2_{x-}$ and $P^1_{y+}P^2_{y-}$, $P^1_{y-}P^2_{y+}$ are projection operators onto two orthogonal pairs of rays that span these planes, so the projection operators onto the spans of these orthogonal rays are just the sums of the projection operators onto the rays. Similar remarks apply to the pairs P^1_{x+}, P^2_{x-}; P^1_{x-}, P^2_{x+}; P^1_{y+}, P^2_{y+}; P^1_{y-}, P^2_{y-}, which are all pairs of projection operators onto planes that intersect at right angles.)

Similarly, the term:

$$(P^1_{x+}P^2_{x-} + P^1_{x-}P^2_{x+})(P^1_{y+}P^2_{y+} + P^1_{y-}P^2_{y-})$$

in T_1' represents the ray that is the intersection of the plane represented by the projection operator:

$$P^1_{x+}P^2_{x-} + P^1_{x-}P^2_{x+}$$

associated with + − or − + for the eigenvalues of σ_x^1, σ_x^2, respectively (that is, 'different sign'), and the plane represented by the projection operator:

$$P_{y+}^1 P_{y+}^2 + P_{y-}^1 P_{y-}^2$$

associated with + + or − − for the eigenvalues of σ_y^1, σ_y^2, respectively (that is, 'same sign').

The quantum proposition T_1' can therefore be expressed as:

$$P_s^{xx} P_d^{yy} + P_d^{xx} P_s^{yy}$$

where P_s^{xx} here denotes the projection operator onto the plane associated with the eigenvalues + + or − − for σ_x^1, σ_x^2, respectively, and P_d^{xx} denotes the projection operator onto the plane associated with the eigenvalues + − or − + for σ_x^1, σ_x^2, respectively, and similarly for P_s^{yy} and P_d^{yy}.

Similarly, the quantum proposition T_2' can be expressed as:

$$P_s^{xy} P_d^{yx} + P_d^{xy} P_s^{yx}$$

Now

$$P_s^{xx} P_d^{yy} + P_d^{xx} P_s^{yy} = P_s^{zz}$$

and

$$P_s^{xy} P_d^{yx} + P_d^{xy} P_s^{yx} = P_d^{zz}$$

It follows that the classical tautology T is interpreted as the quantum proposition:

$$P_s^{zz} \leftrightarrow P_d^{zz} = (P_s^{zz} \wedge P_d^{zz}) \vee (P_d^{zz} \wedge P_s^{zz})$$

which is the zero operator onto the null space.

To see that $P_s^{xx} P_d^{yy} + P_d^{xx} P_s^{yy} = P_s^{zz}$, recall that in the spectral representation:

$$\sigma_x = P_{x+} - P_{x-}$$

$$I = P_{x+} + P_{x-}$$

so

$$P_{x+} = \tfrac{1}{2}(I + \sigma_x)$$
$$P_{x-} = \tfrac{1}{2}(I - \sigma_x)$$

and similarly for σ_y, σ_z.

It follows that:

$$P_s^{xx} \equiv P_{x+}^1 P_{x+}^2 + P_{x-}^1 P_{x-}^2$$
$$= \tfrac{1}{2}(I + \sigma_x^1 \sigma_x^2)$$

and

$$P_d^{xx} \equiv P_{x+}^1 P_{x-}^2 + P_{x-}^1 P_{x+}^2$$

$$= \tfrac{1}{2}(I - \sigma_x^1\sigma_x^2)$$

and similarly:

$$P_s^{xy} = \tfrac{1}{2}(I + \sigma_x^1\sigma_y^2)$$
$$P_d^{xy} = \tfrac{1}{2}(I - \sigma_x^1\sigma_y^2), \text{ etc.}$$

Now

$$\begin{aligned} P_s^{xx}P_d^{yy} &= \tfrac{1}{2}(I + \sigma_x^1\sigma_x^2)\cdot\tfrac{1}{2}(I - \sigma_y^1\sigma_y^2) \\ &= \tfrac{1}{4}(I + \sigma_x^1\sigma_x^2 - \sigma_y^1\sigma_y^2 + \sigma_z^1\sigma_z^2) \\ &= P_d^{yy}P_s^{xx} \end{aligned}$$

That is, the projection operators P_s^{xx} and P_d^{yy} commute, and since their commutator is nonzero, they must represent two planes that intersect at right angles. Similarly, P_d^{xx}, P_s^{yy} commute and represent two planes that intersect at right angles in the ray:

$$P_d^{xx}P_s^{yy} = \tfrac{1}{4}(I - \sigma_x^1\sigma_x^2 + \sigma_y^1\sigma_y^2 + \sigma_z^1\sigma_z^2)$$

So:

$$P_s^{xx}P_d^{yy} + P_d^{xx}P_s^{yy} = \tfrac{1}{4}(I + \sigma_z^1\sigma_z^2) = P_s^{zz}$$

That is, $P_s^{xx}P_d^{yy}$ and $P_d^{xx}P_s^{yy}$ are the projection operators onto two orthogonal rays (as can easily be checked) that span the plane P_s^{zz}.

Similarly, P_s^{xy} and P_d^{yx} commute, and P_d^{xy} and P_s^{yx} commute, and

$$P_s^{xy}P_d^{yx} = \tfrac{1}{4}(I + \sigma_x^1\sigma_y^2 - \sigma_y^1\sigma_x^2 - \sigma_z^1\sigma_z^2)$$
$$P_d^{xy}P_s^{yx} = \tfrac{1}{4}(I - \sigma_x^1\sigma_y^2 + \sigma_y^1\sigma_x^2 - \sigma_z^1\sigma_z^2)$$

These two rays are orthogonal, and so the projection operator onto their span is:

$$P_s^{xy}P_d^{yx} + P_d^{xy}P_s^{yx} = \tfrac{1}{2}(I - \sigma_z^1\sigma_z^2) = P_d^{zz}$$

4

The problem of interpretation[82]

> It seems to me that the concept of probability is terribly mishandled these days. Probability surely has as its substance a statement as to whether something *is* or *is not* the case – an uncertain statesment, to be sure. But nevertheless it has meaning only if one is indeed convinced that the something in question quite definitely either *is* or *is not* the case. A probabilistic assertion presupposes the full reality of its subject.
>
> *Schrödinger, from a letter to Einstein dated 18 November, 1950. See Przibram (1967, p. 37)*

4.1 The problem defined

Suppose that quantum mechanics had been invented as a noncommutative generalization of classical mechanics by a particularly enterprising mathematics student in the early 1800s, before small clouds began appearing on the horizon of classical mechanics, and long before it became expedient to consider subjective perception and the intellectual inner life of the individual in discussions of measurement in physics.[83] Imagine the student defending this construction as a viable mechanical theory to a somewhat skeptical but very learned mathematician.[84]

The student argues that in this noncommutative mechanics the analogue of a classical state (represented by a point in phase space) is represented by a ray (a 1-dimensional subspace) in Hilbert space. As a dynamical state, the ray evolves in time via unitary transformations. As a property state – a maximal specification of the properties of the system at a particular time – the ray can be interpreted as assigning

[82] As indicated in the preface, the proof of the theorem in section 4.3 and much of the analysis in sections 4.1 and 4.2 are taken from a paper co-authored with Rob Clifton (Bub and Clifton, 1996). The theorem supersedes my earlier results along these lines (Bub, 1992a, b, 1993a, 1994a, b, 1995a, c, 1996).

[83] The references are to Kelvin's (1901) remark about to two 'small clouds' (the luminiferous ether and the Maxwell–Boltzmann energy equipartition theorem) in the otherwise clear sky of nineteenth century physics, which presaged the development of relativity and quantum mechanics, and to von Neumann's discussion of measurement, quoted from von Neumann (1955, p. 418) in section 1.4.

[84] In a comment on a 1925 letter from Jordan, Heisenberg refers to 'the learned Göttingen mathematicians' who 'talk so much about Hermitian matrices' and remarks that he does not even know what a matrix is. Quoted from Jammer (1966, p. 207, footnote 37).

probabilities to properties associated with ranges of values of the dynamical variables. Properties assigned probabilities 1 or 0 by a ray (represented by subspaces that contain the ray or are orthogonal to the ray) are taken as determinate in the state represented by the ray (that is, as either obtaining or not obtaining for the system, so that the corresponding propositions are either true or false), and the dynamical variables associated with these properties are likewise taken as determinate (in the sense of having determinate values corresponding to the determinate properties). Properties assigned probabilities that are between 0 and 1 are indeterminate in the state, and the corresponding observables are indeterminate.

'What on earth do you mean?' says the mathematician, who has been working on a formulation of probability theory that will later be re-discovered by Kolmogorov. 'These numbers between 0 and 1 that you call probabilities generated by one of your "states" can't be represented on a probability space in terms of a measure over different possible maximal sets of properties that obtain for the system, different property states in your sense. You can't even assign values to all dynamical variables simultaneously without violating the functional relations between these variables.' (Here the mathematician – who is *very* learned – quickly derives a result that anticipates the Kochen and Specker theorem.)

'Well,' responds the student, 'I think of a probability assigned to a proposition asserting that the value of a dynamical variable lies in a certain range as the probability of *finding the proposition to be true on measurement*, or the probability of *finding the value of the dynamical variable in the range on measurement*. The new noncommutative mechanics can be applied to physical systems so as to provide probabilities for the results of measurements by some agent or device external to the systems. In fact, dynamical variables should really be thought of as observables that have no determinate value until they are measured, unless the ray representing the state of the system lies in one of the eigenspaces of the operator representing the observable, in which case the measurement result is certain. We have to accept that measurement or the related process of subjective perception is a new entity relative to the physical environment and is not reducible to the latter. We must always divide the world into two parts, the one being the observed system, the other the observer. In the former, we can follow up all physical processes (in principle at least) arbitrarily precisely. In the latter, this is meaningless. Indeed, experience only makes statements of this type: an observer has made a certain (subjective) description; and never any like this: a physical quantity has a certain value.'[85]

'I think you've been spending too much time with physicists – or possibly shopkeepers and engineers,' replies the mathematician coldly.[86] 'This notion of

[85] The reader will recognize these sentiments as von Neumann's (1955, p. 418 and p. 420), quoted in section 1.4. Of course, von Neumann was a learned mathematician *par excellence*, but he largely abdicates this rôle in his discussion of measurement in quantum mechanics in favour of that of a (rather uncritical) metaphysician.

[86] Here the mathematician echoes Einstein's remark in a letter to Schrödinger, quoted in section 1.1 (Przibram, 1967, p. 39), and (later) Bell's suggestion that a fundamental theory of motion should be formulated in terms of 'beables,' quoted in section 1.3.

"measurement" is completely undefined dynamically, as is the notion of an observable having no determinate value at one time and coming to have a determinate value at some other time as the outcome of a "measurement." And it's not at all clear that an *internal* account of measurement is possible in this theory, along the lines of the standard account in a theory of mechanics, where a measurement is represented, in principle, as an interaction between two mechanical systems. If this new theory is going to be acceptable as a theory of motion, what you want is an interpretation in terms of what one might call, abusing language, "*be*ables" rather than "*observ*ables." The first question you should try and answer is: What are the maximal subsets of determinate properties of a system, or the maximal subsets of propositions to which truth values can be assigned in the usual way – call these the *determinate* subsets – such that the probabilities defined by a state for the properties in a determinate subset can be represented in terms of a probability measure over the different possible maximal sets of co-obtaining properties in the subset, or the different possible truth value assignments to the propositions in the subset? In other words, you need to introduce a probability space over a range of possibilities in some way to make sense of the probabilities defined by your states, where property states constituted by selected subsets of these possibilities may be realized while the complementary subsets are unrealized. The property states you introduce defined by your 0,1-probability assignments won't work at all, because they make an internal account of measurement impossible.' (Here the mathematician elaborates an amusing and instructive example involving a mouse in a closed box.) 'You also have to show how these property states evolve in time consistently with the unitary dynamics for the rays, and how to represent measurement interactions internally in the theory. If you come up with something, let me know.'

It is hard to imagine the learned mathematician being satisfied with von Neumann's justification of the projection postulate as a solution to the measurement problem, or even the introduction of a non-unitary 'collapse' dynamics, because this would surely seem like a blatantly *ad hoc* move to avoid the problem. (I defer a discussion of decoherence and related moves to chapter 8.)

The student's notion of a property state is the orthodox Dirac–von Neumann notion in quantum mechanics. As I pointed out in section 1.3, both Dirac (1958, pp. 46, 47) and von Neumann (1955, p. 253) explicitly propose that an observable has a determinate (definite, sharp) value for a system in a given quantum state if and only if the state is an eigenstate of the observable.[87] This orthodox interpretation principle selects a particular set of observables that have determinate values in a given quantum state; equivalently, a particular set of idempotent observables or propositions, represented by projection operators, that have determinate truth values. If the quantum state is represented by a ray or 1-dimensional projection operator e spanned by the unit vector $|e\rangle$, these are the propositions p such that $e \leq p$ or $e \leq p^{\perp}$ (where the relation '$\leq$' denotes

[87] Von Neumann formulates this principle for properties represented by projection operators (see von Neumann, 1955, principle β on p. 253), but this is of course equivalent to the formulation in terms of general observables.

subspace inclusion and $p^{\perp}$ denotes the subspace orthogonal to p, or the corresponding relations for projection operators).

In section 1.4, I showed that the orthodox interpretation leads to the measurement problem, which Dirac and von Neumann resolve formally by invoking quantum jumps or a projection postulate that characterizes the 'collapse' or projection of the quantum state of a system onto an eigenstate of the measured observable. Dynamical 'collapse' interpretations of quantum mechanics (Bohm and Bub, 1966a; Ghirardi, Rimini, and Weber, 1986; Ghirardi, Pearle, and Rimini, 1990) keep the orthodox interpretation of the quantum state and modify the unitary, Schrödinger dynamics of the theory to achieve the required state evolution for both measurement and nonmeasurement interactions.

'No collapse' interpretations (apart from Everett's interpretation, which I discuss in Chapter 8) avoid the measurement problem by selecting other sets of observables as determinate for a system in a given quantum state. For example, certain versions of the 'modal' interpretation[88] (Kochen, 1985; Dieks, 1988, 1989a, 1994a,b) exploit the biorthogonal decomposition theorem to select a 'preferred' set of determinate observables for a system S as a subsystem of a composite system $S + S^*$ in a state $e \in \mathscr{H} \otimes \mathscr{H}^*$. Bohm's 'causal' interpretation (Bohm, 1952a; Bohm and Hiley, 1993) selects position in configuration space as a preferred always-determinate observable for any quantum state, and certain other observables can be taken as determinate at a given time together with this preferred always-determinate observable, depending on the state at that time. Bohr's complementarity interpretation (1934, 1939, 1948, 1961, 1963) selects as determinate an observable associated with an individual quantum phenomenon manifested in a measurement interaction involving a specific classically describable experimental arrangement, and certain other observables, regarded as measured in the interaction, can be taken as determinate together with this observable and the quantum state. Chapters 6 and 7 deal with these interpretations and their relation to the orthodox (Dirac–von Neumann) interpretation.

There are restrictions on what sets of observables can be taken as simultaneously determinate without contradiction, if the attribution of determinate values to observables is required to satisfy certain constraints. The 'no go' theorems for 'hidden variables' underlying the quantum statistics discussed in chapters 2 and 3 provide a series of such results. So, in fact, the options for a 'no collapse' interpretation of quantum mechanics in the sense considered here are rather limited.

In the previous chapter, I reviewed the work of several authors who considered the problem of constructing the smallest critical configurations of rays in a Hilbert space of a given dimension that generate a Kochen–Specker contradiction: 'uncolourable' configurations with the smallest possible number of rays, such that every subconfiguration is colourable. For $\mathscr{H}_3$, the original (1967) 117-ray configuration of Kochen and Specker was followed by Jost's (1976) 109-ray configuration, Peres's (1991) 33-ray

[88] The idea of a modal interpretation of quantum mechanics was first introduced by van Fraassen (1979, 1981, 1991).

configuration, a 33-ray configuration of essentially complex rays obtained by Penrose (1994a), and the 31-ray Conway and Kochen configuration (and, of course, the 33-ray configuration derived from Schütte's tautology).[89] There are similar results for $\mathscr{H}_4$: Penrose's (1994a) 40-ray configuration, the 28-ray configuration of Zimba and Penrose (1993), Peres's (1991) 24-ray configuration, Kernaghan's (1994) 20-ray configuration, and the 18-ray configuration of Cabello, Estebaranz, and García-Alcaine (1996a). For $\mathscr{H}_8$, there is the 36-ray configuration of Kernaghan and Peres (1995), and Cabello, Estebaranz, and García-Alcaine (1996c) have found a 29-ray configuration in $\mathscr{H}_5$, a 31-ray configuration in $\mathscr{H}_6$, and a 34-ray configuration in $\mathscr{H}_7$. Some of these configurations (Penrose's 40-ray configuration and Peres's 24-ray configuration in $\mathscr{H}_4$, for example) were later found to have uncolourable subconfigurations.

The question of how small we can make the set of observables and still generate a Kochen–Specker contradiction is important in revealing structural features of Hilbert space, but of no immediate significance for a 'no collapse' interpretation of quantum mechanics. The relevant question really concerns the converse issue. To provide a 'no collapse' interpretation of quantum mechanics, we want to know how *large* we can take the set of determinate observables *without* generating a Kochen–Specker contradiction. That is, we want a characterization of the maximal sets of observables that can be taken as having determinate (but perhaps unknown) values for a given quantum state, subject to the functional relationship constraint, or the maximal sets of propositions that can be taken as having determinate truth values, where a truth-value assignment is defined by a 2-valued homomorphism. (As it turns out, even the orthodox interpretation selects such a maximal set. See section 5.1.)

More precisely, the propositions or yes–no experiments pertaining to a quantum mechanical system form a lattice $\mathscr{L}$ isomorphic to the lattice of subspaces or corresponding projection operators of a Hilbert space. We know that we cannot assign truth values to all the propositions in $\mathscr{L}$ in such a way as to satisfy the functional relationship constraint, or even the weaker locality condition, for all observables generated as spectral measures over these propositions. That is, we cannot take all the propositions in $\mathscr{L}$ as determinately true or false if truth values are assigned subject to these constraints. So the probabilities defined by the quantum state cannot be interpreted epistemically and represented as measures over the different possible truth value assignments to all the propositions in $\mathscr{L}$. But we also know that any single observable can be taken as determinate for any quantum state (since the propositions associated with an observable generate a Boolean algebra), so we may suppose that fixing a quantum state represented by a ray e in $\mathscr{H}$ and an arbitrary 'preferred' observable R as determinate places restrictions on what propositions can be taken as determinate for e in addition to R-propositions.

The natural question for a 'no collapse' interpretation of quantum mechanics would

[89] Independently of Jost, Peres and Ron (1988) have found a 109-ray uncolourable configuration in $\mathscr{H}_3$. Clifton (1993) has formulated an 8-ray Kochen and Specker argument in $\mathscr{H}_3$, but the proof requires quantum statistics to derive a contradiction.

then appear to be: What are the *maximal* sublattices $\mathscr{D}(e,R)$ of $\mathscr{L}$ to which truth values can be assigned, where each assignment of truth values is defined by a 2-valued homomorphism on $\mathscr{D}(e,R)$, and the probabilities defined by e for mutually compatible sets of propositions in $\mathscr{D}(e,R)$ can be represented as measures over the different possible truth value assignments to $\mathscr{D}(e,R)$? We should require that $\mathscr{D}(e,R)$ is invariant under lattice automorphisms that preserve e and R (so that $\mathscr{D}(e,R)$ is genuinely selected by e and R, and the lattice structure of $\mathscr{H}$), and that $\mathscr{D}(e,R)$ is unaffected if the quantum system S is regarded as a subsystem of a composite system $S + S'$, where S is not 'entangled' with S'. I shall refer to these sublattices as the 'determinate' sublattices of $\mathscr{L}$.

In a classical world, where the properties of systems 'fit together' in a Boolean lattice $\mathscr{B}$, all properties in $\mathscr{B}$ are simultaneously determinate, so there is one fixed determinate sublattice of properties defined by $\mathscr{B}$ itself, which specifies the collection of possible properties at all times. The actual property state at time t, given by the actual values of the dynamical variables at t, is defined by some particular 2-valued homomorphism on $\mathscr{B}$, corresponding to a particular atom in $\mathscr{B}$ (a 1-point subset in the classical phase space). Dynamical change is tracked by the evolution of atoms in $\mathscr{B}$, so dynamical change applies directly to what is actual in a classical world, and the dynamical state coincides with the property state.

In a quantum world, where the properties of systems 'fit together' in a non-Boolean lattice $\mathscr{L}$, not embeddable into a Boolean lattice, the properties that are simultaneously determinate at any time t form a proper sublattice of $\mathscr{L}$. These are the maximal collections of properties that are possible at time t, in the sense that the actual collection of properties at t is defined by a 2-valued homomorphism that selects a maximal subcollection of properties *in the determinate sublattice at* t. These 2-valued homomorphisms on the determinate sublattices define the property states of quantum mechanics. Dynamical change in a quantum world is tracked by the evolution of atoms in $\mathscr{L}$, just as dynamical change in a classical world is tracked by the evolution of atoms in the Boolean lattice $\mathscr{B}$. So dynamical states do not coincide with property states. The atoms of $\mathscr{L}$ represent dynamical states that evolve unitarily in time according to Schrödinger's equation of motion. The difference between a classical world and a quantum world is that the classical dynamics directly constrains the temporal evolution of what is *actual*, through the dynamical evolution of the property state, while the quantum dynamics directly constrains the temporal evolution of what is *possible* (and what is *probable*), through the dynamical evolution of the determinate sublattice defined by the quantum state and the preferred observable.

The idea here is that what is possible in a quantum world, given a dynamical quantum state represented by an atom $e \in \mathscr{L}$, is a certain sublattice $\mathscr{D}(e,R)$ of properties in $\mathscr{L}$, with probabilities – in the standard, Kolmogorov sense – defined over alternative subsets of properties in this sublattice. It is these possibilities and probabilities that evolve dynamically in a quantum world. What is actual at any time is some maximal subset of properties in $\mathscr{D}(e,R)$, with respect to a range of such alternative subsets in $\mathscr{D}(e,R)$, so that the probabilities defined by e on $\mathscr{D}(e,R)$ can be interpreted in the

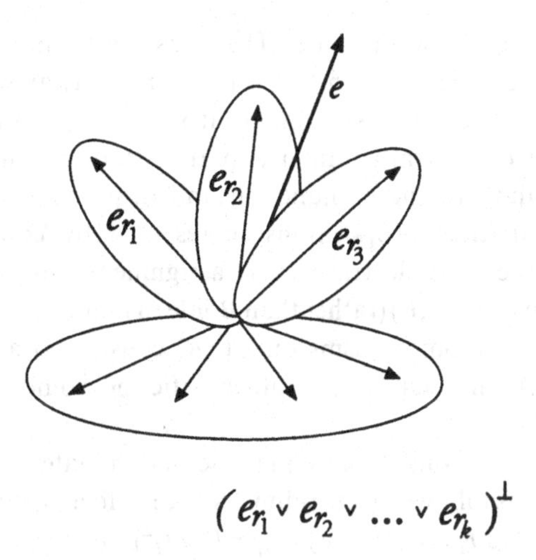

Figure 4.1. Generators of the lattice $\mathcal{L}_{e_{r_1} e_{r_2} \ldots e_{r_k}}$: the k projections e_{r_i} and all the rays in the subspace $(e_{r_1} \vee e_{r_2} \ldots \vee e_{r_k})^{\perp}$.

standard way as measures over the range of alternatives. In section 5.2 I consider how the quantum dynamics indirectly constrains the temporal evolution of actual properties in a quantum world.

The problem of characterizing the maximal sets of observables or propositions that can be taken as (simultaneously) determinate, without generating a Kochen–Specker contradiction, is not well defined *unless* we impose constraints on the sets – there are clearly many different infinite sets of propositions that can be assigned determinate truth values without contradiction in this sense. The problem is only interesting relative to the requirement that a maximal set of determinate propositions is an extension of some physically significant algebraic structure of determinate propositions. Since the dynamical variables in classical mechanics are all simultaneously determinate, and any single observable in quantum mechanics can be taken as determinate, the proposal here is to consider what part of the non-Boolean lattice of quantum propositions can be added to the Boolean algebra of propositions defined by the spectral measure of a particular quantum mechanical observable R, for a given quantum state e, before this extended structure becomes too 'large' to support sufficiently many truth-value assignments, defined by 2-valued homomorphisms, to generate the quantum statistics for the propositions in the extended structure as measures over these different truth-value assignments.

If we want to prove a strong 'no go' theorem that excludes the possibility of a non-orthodox assignment of determinate values to observables (perhaps via underlying 'hidden variables'), then the constraints on such a value assignment should be as weak as possible. The problem considered here, however, involves characterizing the maximal sublattices of $\mathcal{L}$ that *allow* an interpretation of the associated observables as determinate for a given quantum state. The constraints imposed should reflect, roughly, the strongest 'classicality' conditions one can get away with, consistent with such an interpretation. From this standpoint, it seems appropriate to require that $\mathscr{D}(e,R)$ is a sublattice of $\mathscr{L}$ (rather than a partial Boolean subalgebra, in which operations corresponding to the conjunction and disjunction of propositions are defined only for compatible propositions represented by commuting projection operators), and that the possible truth-value assignments are defined by 2-valued lattice homomorphisms on $\mathscr{D}(e,R)$ (rather than Boolean homomorphisms on $\mathscr{D}(e,R)$, or equivalently, 2-valued homomorphisms on $\mathscr{D}(e,R)$ considered as a partial Boolean algebra: maps onto $\{0,1\}$ that reduce to 2-valued lattice homomorphisms only on each Boolean sublattice of $\mathscr{D}(e,R)$).

In the following two sections, I show that the determinate sublattices $\mathscr{D}(e,R)$ are uniquely characterized as follows: In an n-dimensional Hilbert space $\mathscr{H}_n$, suppose R has $m \leq n$ distinct eigenspaces r_i and the rays $e_{r_i} = (e \vee r_i^{\perp}) \wedge r_i$, $i = 1, \ldots, k \leq m$, are the nonzero projections of the state e onto these eigenspaces. The determinate sublattice $\mathscr{D}(e,R)$ of $\mathscr{L}$ is then the sublattice $\mathscr{L}_{e_{r_1}e_{r_2}\ldots e_{r_k}}$ generated by the k orthogonal rays e_{r_i} and all the rays in the subspace $(e_{r_1} \vee e_{r_2} \vee \ldots \vee e_{r_k})^{\perp}$ orthogonal to the k-dimensional subspace spanned by the e_{r_i} (shown in figure 4.1).

The sublattice $\mathscr{L}_{e_{r_1}e_{r_2}\ldots e_{r_k}}$ can also be characterized as $\{e_{r_i}, i = 1, \ldots, k\}'$, the commutant[90] in $\mathscr{L}$ of $\{e_{r_i}, i = 1, \ldots, k\}$, or as $\{p: e_{r_i} \leq p \text{ or } e_{r_i} \leq p^{\perp}, i = 1, \ldots, k\}$. If R has a continuous spectrum,[91] a rigorous treatment would be conceptually equivalent to partitioning the (infinite-dimensional) Hilbert space into orthogonal subspaces corresponding to a partition of the spectrum of R into small non-overlapping intervals, and taking the limit of the determinate sublattices obtained by successively finer partitions.

Physically, $\mathscr{L}_{e_{r_1}e_{r_2}\ldots e_{r_k}}$ contains the projections with values strictly correlated to the values of R when the system is in the state e. The full set of (not necessarily idempotent) observables associated with $\mathscr{L}_{e_{r_1}e_{r_2}\ldots e_{r_k}}$ includes any observable whose eigenspaces are spanned by rays in $\mathscr{L}_{e_{r_1}e_{r_2}\ldots e_{r_k}}$. The set of maximal observables includes any maximal observable with k eigenvectors in the directions e_{r_i}, $i = 1, \ldots, k$.

As an illustration, consider the case where the preferred determinate observable R is the unit observable I. Since the only eigenspace of I is the whole Hilbert space $\mathscr{H}$, and the projection of the state e onto $\mathscr{H}$ is just e, it follows that $\mathscr{D}(e,I) = \mathscr{L}_e = \{p: e \leq p \text{ or } e \leq p^{\perp}\}$. The propositions that belong to $\mathscr{L}_e$ are represented by all the subspaces that either include the ray e or are orthogonal to e. These are just the propositions assigned probability 1 or 0 by e: the propositions that are determinate according to the orthodox Dirac–von Neumann interpretation when the system is in the state e. Without the projection postulate, the orthodox interpretation is a 'no collapse' interpretation in the above sense, with $R = I$.

[90] An operator belongs to the commutant { }′ of { } if and only if it commutes with every operator in { }.
[91] See the appendix, section A.1.

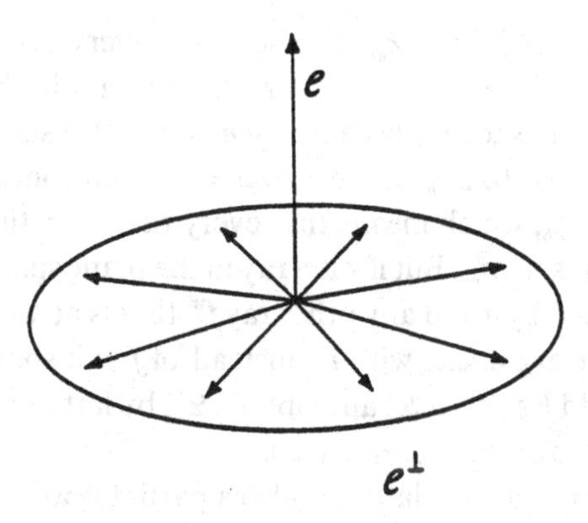

Figure 4.2. Generators of the orthodox lattice $\mathscr{L}_e$: the state e and all the rays in the subspace $e^\perp$.

One can picture the determinate sublattice $\mathscr{L}_e$ as an 'umbrella,'[92] generated by the state e as the handle, and all the rays in the subspace $e^\perp$ as the spines as shown in figure 4.2. (The ray e and the rays orthogonal to e are the atoms of $\mathscr{L}_e$.) The determinate sublattice $\mathscr{D}(e,R)$, for $R \neq I$, are 'generalized umbrellas' $\mathscr{L}_{e_{r_1}e_{r_2}\ldots e_{r_k}}$, with a set of handles $\{e_{r_i}, i = 1,\ldots,k\}$ given by the nonzero projections of e onto the eigenspaces of R, and the spines represented by all the rays in the orthogonal complement of the subspace spanned by the handles. (The ray e and the rays in $(e_{r_1} \vee {}_{r_2} \ldots \vee e_{r_k})^\perp$ are the atoms of $\mathscr{L}_{e_{r_1}e_{r_2}\ldots e_{r_k}}$.) Note that if e is an eigenstate of R, the determinate sublattice $\mathscr{D}(e,R)$ reduces to the orthodox determinate sublattice $\mathscr{L}_e$.

If the Hilbert space $\mathscr{H}$ is more than 2-dimensional,[93] there is one and only one 2-valued homomorphism on $\mathscr{L}_e$: the homomorphism generated by mapping the handle e onto 1 and all the other atoms in $\mathscr{L}_e$ onto 0. (In the case of a 'generalized umbrella' with k handles, there are exactly k 2-valued homomorphisms on $\mathscr{L}_{e_{r_1}e_{r_2}\ldots e_{r_k}}$: the homomorphisms generated by mapping each of the handles e_{r_i}, $i = 1,\ldots,k$, onto 1 and the remaining atoms in $\mathscr{L}_{e_{r_1}e_{r_2}\ldots e_{r_k}}$ onto 0.) It is easy to show that $\mathscr{L}_e$ is maximal as a determinate sublattice (if $\mathscr{H}$ is more than 2-dimensional). If we add anything to $\mathscr{L}_e$, lattice closure generates the lattice $\mathscr{L}(\mathscr{H})$ of all subspaces of $\mathscr{H}$, and there are no 2-valued homomorphisms on $\mathscr{L}(\mathscr{H})$.

To see this, first note that if $\mathscr{H}$ is more than 2-dimensional and we add a multi-dimensional subspace $b \notin \mathscr{L}_e$ to $\mathscr{L}_e$ and generate a lattice $\mathscr{L}_{eb}$ by closure, then every ray in b belongs to $\mathscr{L}_{eb}$. For any ray $f \leq b$ that is not orthogonal to e, $e \vee f \in \mathscr{L}_{eb}$ (because $e \vee f = e \vee g$, for some ray g orthogonal to e, and $e \vee g \in \mathscr{L}_{eb}$ because $e \in \mathscr{L}_e$ and $g \in \mathscr{L}_e$). So $(e \vee f) \wedge b = f \in \mathscr{L}_{eb}$.

Suppose, then, we add a subspace $b \notin \mathscr{L}_e$ to $\mathscr{L}_e$ and generate a lattice $\mathscr{L}_{eb}$ by closure. If b is a ray, then $b^\perp \in \mathscr{L}_{eb}$ is multi-dimensional. So it suffices to consider adding a multi-dimensional subspace b. It follows that every ray in b belongs to $\mathscr{L}_{eb}$. Consider a

[92] In previous publications (1992, b, 1993a), I characterized the determinate sublattices as 'fans.' The more visually suggestive term 'umbrella' was suggested by Richard Jeffrey.

[93] Problems with 2-dimensionality are addressed in the following section.

specific ray $f \leq b$. Since $f \in \mathscr{L}_{eb}$, $f^{\perp} \in \mathscr{L}_{eb}$, and so again every ray in $f^{\perp}$ belongs to $\mathscr{L}_{eb}$.

Now consider any ray h in $\mathscr{H}$ not already in $\mathscr{L}_{eb}$, and not in the plane spanned by f and e. The plane $e \vee h$ belongs to $\mathscr{L}_{eb}$, because $e \vee h = e \vee r$, for some ray r orthogonal to e. Also the plane $f \vee h$ belongs to $\mathscr{L}_{eb}$, because $f \vee h = f \vee s$, for some ray s orthogonal to f. So $(e \vee h) \wedge (f \vee h) = h \in \mathscr{L}_{eb}$, which means that every ray in $\mathscr{H}$ that is not in the plane spanned by f and e belongs to $\mathscr{L}_{eb}$. But if k is a ray in the plane spanned by f and e, then k is not in the plane spanned by e and any other ray $f^{\#}$ that is at an acute angle to f in the subspace b. So a similar argument with $f^{\#}$ instead of f will show that $k \in \mathscr{L}_{eb}$.

It follows that if we add $b \notin \mathscr{L}_e$ to $\mathscr{L}_e$ and obtain $\mathscr{L}_{eb}$ by lattice closure, then $\mathscr{L}_{eb}$ must contain every ray in $\mathscr{H}$, and so $\mathscr{L}_{eb} = \mathscr{L}(\mathscr{H})$.

As it turns out, a determinate sublattice is also a partial Boolean algebra. Recall that a partial Boolean algebra is the union of a family of Boolean algebras 'pasted together' in a certain way, such that the maximum and minimum elements of all the Boolean algebras are identified (and in general certain other elements as well), and such that, for every n-tuple of pairwise compatible elements (that is, pairs of elements belonging to the same Boolean algebra), there exists a Boolean algebra in the family containing the n elements. In the case of $\mathscr{L}_e$ this is obvious: In $\mathscr{H}_n$, the handle e and each subset of $n-1$ mutually orthogonal spines in $e^{\perp}$ generates a Boolean algebra, and $\mathscr{L}_e$ can be characterized as the union of elements in this family of Boolean algebras. So the Boolean algebras in $\mathscr{L}_e$ all have the handle e in common. In $\mathscr{L}_{e_{r_1}e_{r_2}\ldots e_{r_k}}$, the Boolean algebras all have the handles e_{r_i}, $i = 1, \ldots, k$ in common. Note that the 2-valued lattice homomorphisms on $\mathscr{L}_e$ and on $\mathscr{L}_{e_{r_1}e_{r_2}\ldots e_{r_k}}$ are also Boolean homomorphisms, or 2-valued homomorphisms on the corresponding partial Boolean algebras.

Bell (1966) and also Bohm and Bub (1966b) objected to a 'no go' theorem of Jauch and Piron (1963), which required that a truth-value assignment h to the lattice of quantum propositions (equivalently, the probabilities assigned by dispersion-free states) should satisfy the constraint (a consequence of axiom 4° in their numbering) that

$$h(p \wedge q) = 1 \text{ if } h(p) = h(q) = 1$$

for any propositions p, q (even incompatible propositions represented by noncommuting projection operators). The objection was that, to reproduce the quantum statistics, the constraint should be required to hold only for expectation values generated by distributions over the hidden variables corresponding to quantum states, but not necessarily for arbitrary hidden variable distributions (in particular, not for the truth-value assignments themselves).

Now, there exist 2-valued lattice homomorphisms on the determinate sublattices $\mathscr{L}_{e_{r_1}e_{r_2}\ldots e_{r_k}}$, so axiom 4° holds for these sublattices. The assumption that fails is Jauch and Piron's axiom 5, which requires that any non-null proposition is assigned the value 1 (that is, 'true') by some dispersion-free state. This is certainly a reasonable requirement on the full lattice $\mathscr{L}$: if a proposition is assigned the value 0 by every dispersion-free state, it will have zero probability for every quantum state (represented as a measure over dispersion-free states), hence will be orthogonal to every quantum state, and so can only be the null proposition. However, a similar argument does not apply to 'no collapse' interpretations that select proper sublattices of $\mathcal{L}$. In fact, there are non-null propositions in $\mathcal{L}_{e_{r_1}e_{r_2}\ldots e_{r_k}}$ (propositions in $(e_{r_1} \vee e_{r_2} \vee \ . \ . \ . \ \vee e_{r_k})^{\perp}$) that have zero probability in the state e and are assigned the value 0 by all 2-valued homomorphisms on $\mathcal{L}_{e_{r_1}e_{r_2}\ldots e_{r_k}}$. This is possible because $\mathcal{L}_{e_{r_1}e_{r_2}\ldots e_{r_k}}$ is always a *proper* sublattice of $\mathcal{L}$, constructed on the basis of the quantum state e. So the Jauch and Piron argument for axiom 5 fails for the 'no collapse' interpretations considered here.

Similarly, the Kochen and Specker argument, that the existence of hidden variables requires a Boolean embedding of the full partial Boolean algebra of idempotent observables of a quantum mechanical system, is called into question. If the lattice operations in $\mathcal{L}_{e_{r_1}e_{r_2}\ldots e_{r_k}}$ are confined to compatible elements (corresponding to commuting projection operators), then $\mathcal{L}_{e_{r_1}e_{r_2}\ldots e_{r_k}}$ can be regarded as a partial Boolean algebra. There exist 2-valued partial Boolean homomorphisms on $\mathcal{L}_{e_{r_1}e_{r_2}\ldots e_{r_k}}$ (equivalent to what I called 'Boolean homomorphisms' on the lattice in section 1.3), in fact sufficiently many to generate the probabilities defined by the quantum state for the propositions in $\mathcal{L}_{e_{r_1}e_{r_2}\ldots e_{r_k}}$, but insufficiently many to provide an embedding of $\mathcal{L}_{e_{r_1}e_{r_2}\ldots e_{r_k}}$ into a Boolean algebra. The proposal here is that the determinate sublattices $\mathcal{L}_{e_{r_1}e_{r_2}\ldots e_{r_k}}$, for suitable observables R, provide a class of perfectly viable 'no collapse' interpretations of quantum mechanics, in which the functional relationship constraint is satisfied without a Boolean embedding.

In chapter 5, I show that the determinate sublattice of a composite system $S + M$ (representing a system and a measuring instrument), for the 'entangled' state of $S + M$ arising dynamically from a unitary evolution representing an ideal quantum mechanical measurement interaction, contains the propositions corresponding to the S-observable correlated with the pointer observable of M, if the pointer observable (or some observable correlated with the pointer observable) is taken as the preferred determinate observable R. Note that this determinate sublattice of $S + M$ is derived *without any reference to measurement as an unanalyzed operation*; that is, this natural description of the measurement process falls out as just a special instance of the linear Schrödinger quantum dynamics, without requiring the projection or 'collapse' of the quantum state to validate the determinateness of measured values, as in the orthodox interpretation. (This is perhaps evident from the statement of the uniqueness theorem: $\mathcal{L}_{e_{r_1}e_{r_2}\ldots e_{r_k}}$ is generated from the nonzero projections of the quantum state e onto the eigenspaces of R.)

Measuring instruments are open systems, in interaction with their environments, which are typically systems with very many degrees of freedom. So we need to take account of the fact that even if the quantum state of $S + M$, immediately after an ideal measurement interaction, is a pure state that takes the form of a linear superposition of eigenstates of the measured observable correlated with eigenstates of the pointer observable of M, the interaction with the environment E will virtually instantaneously transform the state of the composite system $S + M + E$ to a linear superposition that entangles states of $S + M$ with states of E. I show, also in chapter 5, that this entanglement does not affect the function of R as a measurement pointer, if the instrument–environment interaction Hamiltonian commutes with R.

The actual value of the preferred determinate observable R must evolve in time in such a way as to conform to the dynamical evolution of possibilities and probabilities tracked by the quantum state as a dynamical state. I exploit a proposal by Vink (1993), following Bell (1987, pp. 176–7), to formulate a specific stochastic equation of motion for the evolution of determinate R values, for a fixed R, and I show that the effect of typical instrument–environment interactions in our world is to rapidly reduce the probability of an anomalous jump between different R-values after a measurement interaction, if R is the pointer observable. So environmental 'monitoring' to use Zurek's term (1993b, p. 290) guarantees that if the pointer observable is designated as a preferred always-determinate observable, it will exhibit the kind of stability we require of a pointer in a measurement interaction: if r_i is the determinate value of the pointer after a measurement, the probability of a spontaneous jump to r_j, where $j \neq i$, will almost instantaneously approach zero. Moreover, this feature does not apply to any other observable R' that does not commute with R.

Different choices for the preferred determinate observable R correspond to different 'no collapse' interpretations of quantum mechanics. I illustrate this in chapters 6 and 7 with Bohm's causal interpretation, a generalized version of the modal interpretation derived from earlier formulations by Kochen and by Dieks, and Bohr's complementarity interpretation. By contrast with complementarity, Bohm's interpretation and the modal interpretation are observer-free 'no collapse' interpretations. Bohm adopts position in configuration space as a fixed preferred determinate observable, while the preferred determinate observable of the modal interpretation is not fixed but varies with the quantum state.

4.2 A uniqueness theorem for 'no collapse' interpretations

We want to characterize the maximal sublattices, $\mathscr{D}(e,R)$, of the lattice of subspaces, $\mathscr{L}$, of an n-dimensional Hilbert space, $\mathscr{H}$, on which there exist sufficiently many property states, specified by 2-valued homomorphisms, to construct a probability space, with respect to which the Born probabilities generated by the quantum state e can be recovered as measures over subsets of property states. Since we suppose that $\mathscr{D}(e,R)$ is fixed by the state e, the preferred observable R, and the lattice structure of $\mathscr{H}$, we require also that $\mathscr{D}(e,R)$ is invariant under lattice automorphisms that preserve e and R.

This motivates the following statement of the theorem:

Consider a quantum system S in a (pure) quantum state represented by a ray e in an n-dimensional Hilbert space $\mathscr{H}$, and an observable R with $m \leq n$ distinct eigenspaces r_i.[94] Let $e_{r_i} = (e \vee r_i^{\perp}) \wedge r_i$, $i = 1, \ldots, k \leq m$, denote the nonzero projections of the ray e

[94] For convenience, I use the symbol r_i to denote the ith eigenspace of R here, rather than the corresponding eigenvalue.

onto the eigenspaces r_i. Then $\mathscr{D}(e,R) = \mathscr{L}_{e_{r_1}e_{r_2}\ldots e_{r_k}} = \{p: e_{r_i} \leq p \text{ or } e_{r_i} \leq p^{\perp}, i = 1, \ldots, k\}$ is the unique maximal sublattice of $\mathscr{L}(\mathscr{H})$ satisfying the following three conditions:

(1) *Truth and probability* (*TP*): $\mathscr{D}(e,R)$ is an ortholattice admitting sufficiently many 2-valued homomorphisms, h: $\mathscr{D}(e,R) \rightarrow \{0,1\}$, to recover all the (single and joint) probabilities assigned by the state e to mutually compatible sets of elements $\{p_i\}_{i \in I}$, $p_i \in \mathscr{D}(e,R)$, as measures on a Kolmogorov probability space $(X, \mathscr{F}, \mu)$, where X is the set of 2-valued homomorphisms on $\mathscr{D}(e,R)$, $\mathscr{F}$ is a field of subsets of X, and

$$\mu(\{h: h(p_i) = 1, \text{ for all } i \in I\}) = \text{tr}(e\prod_{i \in I} p_i).$$

(2) *R-preferred* (*R-PREF*): the eigenspaces r_i of R belong to $\mathscr{D}(e,R)$.

(3) *e,R-definability* (*DEF*): for any $e \in \mathscr{H}$ and observable R of S defined on $\mathscr{H}$, $\mathscr{D}(e,R)$ is invariant under lattice automorphisms that preserve e and R.

Proof: The strategy of the proof is to show, firstly, that if $p \in \mathscr{D}(e,R)$, then for any e_{r_i}, $i = 1, \ldots, k$, either $e_{r_i} \leq p$ or $e_{r_i} \leq p^{\perp}$, and secondly, that the sublattice $\mathscr{L}_{e_{r_1}e_{r_2}\ldots e_{r_k}} = \{p: e_{r_i} \leq p \text{ or } e_{r_i} \leq p^{\perp}, i = 1, \ldots, k\}$ satisfies the conditions of the theorem. Maximality then requires that $\mathscr{D}(e,R) = \mathscr{L}_{e_{r_1}e_{r_2}\ldots e_{r_k}}$.

Consider an arbitrary subspace p of $\mathscr{H}$, and suppose $p \in \mathscr{D}(e,R)$. Clearly, if r_i is included in p or in $p^{\perp}$, for any i, then $e_{r_i} \leq p$ or $e_{r_i} \leq p^{\perp}$.

Suppose this is not the case. That is, suppose some r_i is neither included in p nor in $p^{\perp}$. Then either the intersection of r_i with p or the intersection of r_i with $p^{\perp}$ is nonzero. (For, by the condition TP, there must be a 2-valued homomorphism mapping r_i onto 1, and this homomorphism would have to map both p and $p^{\perp}$ onto 0 if $r_i \wedge p = r_i \wedge p^{\perp} = 0$, which contradicts TP.) Assume that the intersection of r_i with p is nonzero (a similar argument applies if we assume that the intersection of r_i with $p^{\perp}$ is nonzero). Call this intersection b.

If p belongs to $\mathscr{D}(e,R)$, then $b = r_i \wedge p$ also belongs to $\mathscr{D}(e,R)$ (by the lattice closure assumption of TP, since r_i belongs to $\mathscr{D}(e,R)$). Now, either b is orthogonal to e_{r_i} in r_i, or b is skew to e_{r_i} in r_i. I shall show that in either case e_{r_i} must belong to $\mathscr{D}(e,R)$. It follows that $e_{r_i} \leq p$ or $e_{r_i} \leq p^{\perp}$, because if $e_{r_i} \nleq p$ and $e_{r_i} \nleq p^{\perp}$, then – since e_{r_i} is a ray – e_{r_i} must be skew to both p and $p^{\perp}$; that is, $e_{r_i} \wedge p = 0$ and $e_{r_i} \wedge p^{\perp} = 0$. But this contradicts TP, because there must be a 2-valued homomorphism mapping e_{r_i} onto 1, and this homomorphism would have to map both p and $p^{\perp}$ onto 0.

So, if r_i is included in p or in $p^{\perp}$ then $e_{r_i} \leq p$ or $e_{r_i} \leq p^{\perp}$, and if r_i is not included in p or in $p^{\perp}$, then e_{r_i} must belong to $\mathscr{D}(e,R)$, from which it follows that $e_{r_i} \leq p$ or $e_{r_i} \leq p^{\perp}$.

To see that e_{r_i} must belong to $\mathscr{D}(e,R)$ if (i) $b = r_i \wedge p$ is orthogonal to e_{r_i} in r_i, or (ii) b is skew to e_{r_i} in r_i, consider each of these cases in turn.

(i) Suppose b is orthogonal to e_{r_i} in r_i; that is, $b \leq e_{r_i}'$ (where the $'$ here denotes the orthogonal complement in r_i). If r_i is 1-dimensional, $e_{r_i} = r_i$ and it follows immediately from R-PREF that $e_{r_i} \in \mathscr{D}(e,R)$. If r_i is 2-dimensional, $b = e_{r_i}'$ and $e_{r_i} = r_i \wedge b^{\perp}$, so $e_{r_i} \in \mathscr{D}(e,R)$ by lattice closure (that is, by TP), because both r_i and $b^{\perp}$ belong to $\mathscr{D}(e,R)$. If r_i is more than 2-dimensional, consider all lattice automorphisms U that are rotations about e_{r_i} in r_i in r_i and the identity in $r_i^{\perp}$. Such rotations preserve e and R, because they preserve the eigenspaces of R and the projections of e onto the eigenspaces of R. It follows that $U(b) \in \mathscr{D}(e,R)$, by DEF. There are clearly sufficiently many rotations to generate a set of elements $U(b)$ whose span is e_{r_i}'. So $e_{r_i}' \in \mathscr{D}(e,R)$, and hence $e_{r_i} \in \mathscr{D}(e,R)$ by lattice closure.

(ii) Suppose b is skew to e_{r_i} in r_i. Consider first the case that the eigenspace r_i is more than 2-dimensional. In that case, we may suppose that b is not a ray (or else we apply the following argument to b'). The subspace b can therefore be represented as the span of a subspace c orthogonal to e_{r_i} in r_i, and the ray d, the projection of the ray e_{r_i} onto b. Consider a lattice automorphism U that is a reflection through the hyperplane $e_{r_i} \vee c$ in r_i and the identity in $r_i^{\perp}$. (So U preserves e_{r_i} and c, but not d.) As in (i) above, $U(b) \in \mathscr{D}(e,R)$ because U preserves e and R, and $b \wedge U(b) = c$, so $c \in \mathscr{D}(e,R)$. We can now consider rotations of c about e_{r_i} in r_i as in (i) to show that $e_{r_i}' \in \mathscr{D}(e,R)$, and hence that $e_{r_i} \in \mathscr{D}(e,R)$.

The argument (ii) fails if r_i is 2-dimensional. If b is skew to e_{r_i} in r_i and r_i is 2-dimensional, then b must be a ray in r_i; that is, c is the null subspace and $d = b$. The automorphism U reduces to a reflection through e_{r_i} in r_i. Suppose b is not at a 45° angle to e_{r_i}. The four rays, b, b', $U(b)$, $U(b)'$ would all have to belong to $\mathscr{D}(e,R)$ if b belongs to $\mathscr{D}(e,R)$, but this contradicts TP. A 2-valued homomorphism would have to map one of the rays b or b' onto 1, and one of the rays $U(b)$ or $U(b)'$ onto 1, and hence the intersection of these two rays – the null subspace – would also have to be mapped onto 1. So b cannot be skew to e_{r_i} in r_i, unless b is at a 45° angle to e_{r_i}. In this case, reflecting b through e_{r_i} yields b', and there is no contradiction with TP.

So, if r_i is 2-dimensional (and r_i is not included in p or $p^{\perp}$), we could conclude from this argument only that $e_{r_i} \leq p$ or $e_{r_i} \leq p^{\perp}$, or that $b \leq p$ or $b \leq p^{\perp}$, where b is a ray at a 45° angle to e_{r_i} in r_i. In the original Bub-Clifton (1966) formulation of the proof, a further 'weak separability' condition was introduced to exclude this one anomalous possibility, that the determinate sublattice $\mathscr{D}(e,R)$ might contain propositions p such that $b \leq p$ or $b \leq p^{\perp}$, where b is at a 45° angle to e_{r_i} in r_i

That such a condition is not required was pointed out by Sheldon Goldstein (private communication). Suppose r_i is 2-dimensional. Consider a lattice automorphism U that is the identity in $r_i^{\perp}$, preserves e_{r_i} in r_i, and maps b onto $U(b) \neq b$. For example, suppose we represent e_{r_i} by the vector $\begin{pmatrix}1\\0\end{pmatrix}$ in r_i and take the action of U on r_i as represented by the matrix $\begin{pmatrix}i & 0\\0 & 1\end{pmatrix}$. If b is represented by a vector $\begin{pmatrix}b_1\\b_2\end{pmatrix}$ in r_i, then:

$$\begin{pmatrix}i & 0\\0 & 1\end{pmatrix}\begin{pmatrix}1\\0\end{pmatrix} = \begin{pmatrix}i\\0\end{pmatrix} = i\begin{pmatrix}1\\0\end{pmatrix}$$

$$\begin{pmatrix}i & 0\\0 & 1\end{pmatrix}\begin{pmatrix}b_1\\b_2\end{pmatrix} = \begin{pmatrix}ib_1\\b_2\end{pmatrix} \neq k\begin{pmatrix}b_1\\b_2\end{pmatrix}$$

for any constant k. So U maps e_{r_i} onto the same ray but maps b onto a different ray, whether or not b is at a 45° angle to e_{r_i}. The four rays b, b', $U(b)$, $U(b)'$ would all have to belong to $\mathscr{D}(e,R)$ if b belongs to $\mathscr{D}(e,R)$, which contradicts TP, by the argument above.

We have now established that if $p \in \mathscr{D}(e,R)$ then either $e_{r_i} \leq p$ or $e_{r_i} \leq p^{\perp}$, for all $i = 1, \ldots, k$. The final stage of the proof involves showing that the set of *all* such elements $\{p : e_{r_i} \leq p \text{ or } e_{r_i} \leq p^{\perp}, i = 1, \ldots, k\} = \mathscr{L}_{e_{r_1}e_{r_2}\ldots e_{r_k}}$ satisfies the three conditions TP, R-PREF, and DEF. Maximality then requires that $\mathscr{D}(e,R) = \mathscr{L}_{e_{r_1}e_{r_2}\ldots e_{r_k}}$.

To see that TP is satisfied, we need to show that $\mathscr{L}_{e_{r_1}e_{r_2}\ldots e_{r_k}}$ is closed under the ortholattice operations, that there exist 2-valued homomorphisms on $\mathscr{L}_{e_{r_1}e_{r_2}\ldots e_{r_k}}$, and that the single and joint probabilities assigned by e to mutually compatible sets of elements $\{p_i\}_{i \in I}$ in $\mathscr{L}_{e_{r_1}e_{r_2}\ldots e_{r_k}}$ can be recovered as measures on a Kolmogorov probability space $(X, \mathscr{F}, \mu)$, where X is the set of 2-valued homomorphisms on $\mathscr{L}_{e_{r_1}e_{r_2}\ldots e_{r_k}}$.

Consider closure first. If $p \in \mathscr{L}_{e_{r_1}e_{r_2}\ldots e_{r_k}}$, then either $e_{r_i} \leq p$ or $e_{r_i} \leq p^{\perp}$, for all $i = 1, \ldots, k$. So, $p^{\perp} \in \mathscr{L}_{e_{r_1}e_{r_2}\ldots e_{r_k}}$ because either $e_{r_i} \leq p^{\perp}$ or $e_{r_i} \leq (p^{\perp})^{\perp} = p$. To show that $p \wedge q \in \mathscr{L}_{e_{r_1}e_{r_2}\ldots e_{r_k}}$ and $p \vee q \in \mathscr{L}_{e_{r_1}e_{r_2}\ldots e_{r_k}}$ if $p \in \mathscr{L}_{e_{r_1}e_{r_2}\ldots e_{r_k}}$ and $q \in \mathscr{L}_{e_{r_1}e_{r_2}\ldots e_{r_k}}$, we need to show that $e_{r_i} \leq p \wedge q$ or $e_{r_i} \leq (p \wedge q)^{\perp}$, and $e_{r_i} \leq p \vee q$ or $e_{r_i} \leq (p \vee q)^{\perp}$, for all $i = 1, \ldots, k$. If $p \in \mathscr{L}_{e_{r_1}e_{r_2}\ldots e_{r_k}}$ and $q \in \mathscr{L}_{e_{r_1}e_{r_2}\ldots e_{r_k}}$ then, for each $i = 1, \ldots, k$, $e_{r_i} \leq p$ or $e_{r_i} \leq p^{\perp}$, and $e_{r_i} \leq q$ or $e_{r_i} \leq q^{\perp}$. So either (i) $e_{r_i} \leq p$ and $e_{r_i} \leq q$, in which case $e_{r_i} \leq p \wedge q$ and $e_{r_i} \leq p \leq p \vee q$, or (ii) $e_{r_i} \leq p$ and $e_{r_i} \leq q^{\perp}$, in which case $e_{r_i} \leq p \leq p \vee q$ and $e_{r_i} \leq q^{\perp} \leq p^{\perp} \vee q^{\perp} = (p \wedge q)^{\perp}$, or (iii) $e_{r_i} \leq p^{\perp}$ and $e_{r_i} \leq q$, in which case $e_{r_i} \leq q \leq p \vee q$ and $e_{r_i} \leq p^{\perp} \leq p^{\perp} \vee q^{\perp} = (p \wedge q)^{\perp}$, or (iv) $e_{r_i} \leq p^{\perp}$ and $e_{r_i} \leq q^{\perp}$, in which case $e_{r_i} \leq p^{\perp} \wedge q^{\perp} = (p \vee q)^{\perp}$ and $e_{r_i} \leq p^{\perp} \leq p^{\perp} \vee q^{\perp} = (p \wedge q)^{\perp}$.

To show the existence of 2-valued homomorphisms on $\mathscr{L}_{e_{r_1}e_{r_2}\ldots e_{r_k}}$, consider the k maps h_i: $\mathscr{L}_{e_{r_1}e_{r_2}\ldots e_{r_k}} \to \{0,1\}$ defined as follows: For any element $p \in \mathscr{L}_{e_{r_1}e_{r_2}\ldots e_{r_k}}$, $h_i(p) = 1$ if $e_{r_i} \leq p$ and $h_i(p) = 0$ if $e_{r_i} \leq p^{\perp}$. Each h_i is a 2-valued homomorphism on $\mathscr{L}_{e_{r_1}e_{r_2}\ldots e_{r_k}}$. Clearly, $h_i(0) = 0$ and $h_i(I) = 1$, because $e_{r_i} \leq I = 0^{\perp}$. If $e_{r_i} \leq p$, then $h_i(p) = 1$ and $h_i(p^{\perp}) = 0$, because $p = (p^{\perp})^{\perp}$. Similarly, if $e_{r_i} \leq p^{\perp}$, then $h_i(p^{\perp}) = 1$ and $h_i(p) = 0$. So $h_i(p) = 1 - h_i(p^{\perp})$. If $e_{r_i} \leq p$ and $e_{r_i} \leq q$, so that $h_i(p) = 1$ and $h_i(q) = 1$, then $e_{r_i} \leq p \wedge q$, so $h_i(p \wedge q) = 1$. If $e_{r_i} \leq p$ and $e_{r_i} \leq q^{\perp}$, so that $h_i(p) = 1$ and $h_i(q) = 0$, or $e_{r_i} \leq p^{\perp}$ and $e_{r_i} \leq q$, so that $h_i(p) = 0$ and $h_i(q) = 1$, then $e_{r_i} \leq p^{\perp} \vee q^{\perp} = (p \wedge q)^{\perp}$, so that $h_i(p \wedge q) = 0$. If $e_{r_i} \leq p^{\perp}$ and $e_{r_i} \leq q^{\perp}$, so that $h_i(p) = h_i(q) = 0$, then $e_{r_i} \leq p^{\perp} \vee q^{\perp} = (p \wedge q)^{\perp}$, so that $h_i(p \wedge q) = 0$. So $h_i(p \wedge q) = h_i(p)h_i(q)$. Since $p \vee q = (p^{\perp} \wedge q^{\perp})^{\perp}$, it follows that $h_i(p \vee q) = h_i(p) \vee h_i(q) = h_i(p) + h_i(q) - h_i(p)h_i(q)$. So each h_i is a homomorphism.

Note that the k 2-valued homomorphisms h_i on $\mathscr{L}_{e_{r_1}e_{r_2}\ldots e_{r_k}}$ are in general the *only* 2-valued homomorphisms on $\mathscr{L}_{e_{r_1}e_{r_2}\ldots e_{r_k}}$. Since any 2-valued homomorphism that maps any ray in $(e_{r_1} \vee e_{r_2} \vee \ldots \vee e_{r_k})^{\perp}$ onto 1 must map each of the rays e_{r_i}, $i = 1, \ldots, k$, onto 0, it follows that any such homomorphism must map one of each orthogonal $(n-k)$-tuple of rays in $(e_{r_1} \vee e_{r_2} \vee \ldots \vee e_{r_k})^{\perp}$ onto 1 and the remaining members of the $(n-k)$-tuple onto 0. The infimum of any two rays that are mapped onto 1 must also be mapped onto 1, but this is the zero element. So no 2-valued homomorphism can map any of the rays in $(e_{r_1} \vee e_{r_2} \vee \ldots \vee e_{r_k})^{\perp}$ onto 1 – except, of course, when $\dim((e_{r_1} \vee e_{r_2} \vee \ldots \vee e_{r_k})^{\perp}) = 1$, that is, when $n = k + 1$. In all other cases $(n > k + 1)$, it follows that the only 2-valued homomorphisms on $\mathscr{L}_{e_{r_1}e_{r_2}\ldots e_{r_k}}$ are the homomorphisms h_i, $i = 1, \ldots, k$, where each h_i maps the ray e_{r_i} onto 1 and every other ray in $\mathscr{L}_{e_{r_1}e_{r_2}\ldots e_{r_k}}$ onto 0 (since all the other rays in $\mathscr{L}_{e_{r_1}e_{r_2}\ldots e_{r_k}}$ are orthogonal to e_{r_i}).

To generate the probabilities assigned by e to mutually compatible sets of elements $\{p_i\}_{i\in I}$, $p_i \in \mathscr{L}_{e_{r_1}e_{r_2}\ldots e_{r_k}}$, on the Kolmogorov probability space, the 2-valued homomorphism that maps e_{r_i} onto 1 (more precisely, the corresponding singleton subset) is assigned measure $\mathrm{tr}(ee_{r_i})$, for $i = 1, \ldots, k$. (The homomorphism that maps $(e_{r_1} \vee e_{r_2} \vee \ldots \vee e_{r_k})^\perp$ onto 1, assuming $\dim((e_{r_1} \vee e_{r_2} \vee \ldots \vee e_{r_k})^\perp) = 1$, is assigned measure 0.) To see that $\mu(\{h: h(p_1) = h(p_2) = \ldots = 1\}) = \mathrm{tr}(ep_1p_2\ldots)$, first note that $p_1p_2\ldots = p_1 \wedge p_2 \wedge \ldots$ for compatible projections p_i, and so the product $p_ip_2\ldots$ defines an element, call it p, that also belongs to $\mathscr{L}_{e_{r_1}e_{r_2}\ldots e_{r_k}}$, since $\mathscr{L}_{e_{r_1}e_{r_2}\ldots e_{r_k}}$ is closed under the ortholattice operations. Furthermore, $h(p_1) = h(p_2) = \ldots = 1$ if and only if $h(p_1 \wedge p_2 \wedge \ldots) = h(p) = 1$, since h is a lattice homomorphism. So it suffices to show that, for a general element $p \in \mathscr{L}_{e_{r_1}e_{r_2}\ldots e_{r_k}}$, $\mu(h: h(p) = 1) = \mathrm{tr}(ep)$.

Now any $p \in \mathscr{L}_{e_{r_1}e_{r_2}\ldots e_{r_k}}$ can be expressed as $p = e_{r_i} \vee e_{r_j} \vee \ldots \vee q$, for some $i, j, \ldots, q$, where q, is the projection operator onto a subspace (possibly the zero subspace) orthogonal to *all* the e_{r_i} (that is, $q \in (e_{r_i} \vee e_{r_2} \vee \ldots \vee e_{r_k})^\perp$). If $h(p) = 1$, $h(e_{r_i} \vee e_{r_j} \vee \ldots \vee q) = h(e_{r_i}) \vee h(e_{r_j}) \vee \ldots \vee h(q) = 1$. So:

$$\begin{aligned}\mu(\{h: h(p) = 1\}) &= \mu(\{h: h(e_{r_i}) \vee h(e_{r_j}) \vee \ldots \vee h(q) = 1\}) \\ &= \mu(\{h: h(e_{r_i}) = 1\}) + \mu(\{h: h(e_{r_j}) = 1\}) + \ldots + \mu(\{h{:}h(q) = 1\})\end{aligned}$$

because $\mu(\{h: h(q) = 1\}) = 0$, and $h(e_{r_i}) \neq h(e_{r_j})$ if $i \neq j$. It follows that:

$$\begin{aligned}\mu(\{h: h(p) = 1\}) &= \mathrm{tr}(ee_{r_i}) + \mathrm{tr}(ee_{r_j}) + \ldots + \mathrm{tr}(eq) \\ &= \mathrm{tr}(e(e_{r_i} + e_{r_j} + \ldots + q))\end{aligned}$$

because $\mu(\{h: h(e_{r_i}) = 1\}) = \mathrm{tr}(ee_{r_i})$ and $\mathrm{tr}(eq) = 0$, and hence:

$$\begin{aligned}\mu(\{h: h(p) = 1\}) &= \mathrm{tr}(e(e_{r_i} \vee e_{r_j} \vee \ldots \vee q)) \\ &= \mathrm{tr}(ep)\end{aligned}$$

since e_{r_i}, e_{r_j}, $\ldots$, and q are mutually orthogonal.

R-PREF is satisfied by construction. Since $e_{r_i} \in \mathscr{L}_{e_{r_1}e_{r_2}\ldots e_{r_k}}$, $i = 1, \ldots, k$, and every ray in $(e_{r_1} \vee e_{r_2} \vee \ldots \vee e_{r_k})^\perp$ belongs to $\mathscr{L}_{e_{r_1}e_{r_2}\ldots e_{r_k}}$, it follows that every ray in e_{r_i}' (the orthocomplement of e_{r_i} in r_i), $i = 1, \ldots, k$, belongs to $\mathscr{L}_{e_{r_1}e_{r_2}\ldots e_{r_k}}$. So $r_i \in \mathscr{L}_{e_{r_1}e_{r_2}\ldots e_{r_k}}$, $i = 1, \ldots, k$, by lattice closure (established above). The remaining $m - k$ elements r_i also belong to $\mathscr{L}_{e_{r_1}e_{r_2}\ldots e_{r_k}}$, because each of these is represented by a subspace orthogonal to e, and hence by a subspace in $(e_{r_1} \vee e_{r_2} \vee \ldots \vee e_{r_k})^\perp$. But since every ray in $(e_{r_1} \vee e_{r_2} \vee \ldots \vee e_{r_k})^\perp$ belongs to $\mathscr{L}_{e_{r_1}e_{r_2}\ldots e_{r_k}}$, every subspace of $(e_{r_1} \vee e_{r_2} \vee \ldots \vee e_{r_k})^\perp$ belongs to $\mathscr{L}_{e_{r_1}e_{r_2}\ldots e_{r_k}}$.

DEF is satisfied because lattice automorphisms that preserve e and R automatically preserve the nonzero projections of e onto the eigenspaces r_i of R, the rays e_{r_i}, $i = 1, \ldots, k$, that generate $\mathscr{L}_{e_{r_1}e_{r_2}\ldots e_{r_k}}$.

5

Quantum mechanics without observers: I

Many people must have thought along the following lines. Could one not just promote *some* of the 'observables' of the present quantum theory to the status of beables? The beables would then be represented by linear operators in the state space. The values which they are allowed to *be* would be the eigenvalues of those operators. For the general state the probability of a beable *being* a particular value would be calculated just as was formerly calculated the probability of *observing* that value.

Bell (1987, p. 41)

5.1 Avoiding the measurement problem

The analysis in chapter 4 shows that the determinate sublattices of the lattice of projection operators or subspaces, $\mathscr{L}(\mathscr{H})$, of a Hilbert space $\mathscr{H}$, representing the propositions ('yes–no' experiments) of a quantum mechanical system are just the sublattices $\mathscr{D}(e,R) = \mathscr{L}_{e_{r_1}e_{r_2}\ldots e_{r_k}} = \{p\text{: } e_{r_i} \leq p \text{ or } e_{r_i} \leq p^{\perp},\ i = 1, \ldots, k\}$. These are the maximal sub-collections of quantum propositions that can be true or false, given the quantum state e and a preferred observable R, subject to certain constraints that essentially require these sub-collections to be lattices determined by e and R on which sufficiently many truth valuations exist to recover the usual quantum statistics. The set of maximal observables associated with $\mathscr{L}_{e_{r_1}e_{r_2}\ldots e_{r_k}}$ includes any maximal observable with k eigenvectors in the directions e_{r_i}, $i = 1, \ldots, k$. The set of nonmaximal observables includes any nonmaximal observable that is a function of one of these maximal observables. So all the observables whose eigenspaces are spanned by rays in $\mathscr{L}_{e_{r_1}e_{r_2}\ldots e_{r_k}}$ are determinate, given e and R.

The uniqueness theorem characterizes a class of 'no collapse' interpretations of quantum mechanics, where each interpretation involves the selection of a particular preferred determinate observable, and hence the selection, via the quantum state at a particular time, of a particular determinate sublattice with respect to which the probabilities defined by the quantum state have the usual epistemic significance. That is, the quantum probabilities defined on the determinate sublattice can be understood as measures of ignorance about the actual values of observables associated with propositions in the determinate sublattice – the actual values of the preferred determinate observable and other observables that become determinate, given the quantum state and the preferred observable.

As I showed in section 4.1, the orthodox interpretation without the projection postulate is a 'no collapse' interpretation in the sense of the theorem. On the orthodox interpretation, an observable has a determinate value if and only if the state of the system is an eigenstate of the observable. Equivalently, the propositions that are determinately true or false of a system are the propositions represented by subspaces that either include the ray representing the state of the system, or are orthogonal to the state – the propositions assigned probability 1 or 0 by the state. The orthodox interpretation can therefore be formulated as the proposal that the preferred determinate observable is the unit observable I, and that $\mathscr{D}(e,I) = \mathscr{L}_e$ is the determinate sublattice of a system in the state e, where $\mathscr{L}_e = \{p\colon e \leq p \text{ or } e \leq p^{\perp}\}$.

From the standpoint of the uniqueness theorem, it is the choice of the preferred determinate observable as the unit observable that leads to the measurement problem. For a composite system $S + M$ in an entangled state of the form $|\psi\rangle = \Sigma_i c_i |a_i\rangle|r_i\rangle$ (suppressing the degeneracy index of the eigenstates of the pointer observable R), neither R-propositions nor A-propositions belong to the sublattice $\mathscr{L}_e$, where $e = e(|\psi\rangle)$ is the ray spanned by the vector $|\psi\rangle$. In order to avoid the problem, Dirac and von Neumann assume the projection postulate, that unitary evolution is suspended in the case of a measurement interaction, and that the state of $S + M$ is projected onto the ray spanned by one of the vectors $|a_i\rangle|r_i\rangle$, for some i, with probability $|c_i|^2$. For on the orthodox interpretation, it is only in such a state that the observables A and R have determinate values.

There is nothing in the mathematical structure of quantum mechanics that forces the choice of the preferred determinate observable R as the unit observable I, and indeed there is every reason to avoid this choice because it leads to the measurement problem. 'No collapse' interpretations that seek to solve this problem represent alternative proposals for choosing R. For such interpretations, there is no measurement problem if R plays the rôle of a pointer observable in all measurement interactions, or an observable with respect to which the pointer observable is defined as a function of R by an appropriate 'coarse-graining.'

To see this, consider a model quantum mechanical universe consisting of two systems, S and M, associated with a Hilbert space $\mathscr{H}_S \otimes \mathscr{H}_M$. An ideal ('first kind') measurement interaction between S and M, say a dynamical evolution of the quantum state of the composite system $S + M$ described by a unitary transformation that correlates eigenstates $|a_i\rangle$ of an S-observable A with eigenstates $|r_i\rangle$ of a (nonmaximal) M-observable R, will result (after a certain time) in a state represented by a unit vector of the form $|\psi\rangle = \Sigma_i c_i |a_i\rangle|r_i\rangle$ (assuming initial pure states for S and M). If we take R, or more precisely $I_S \otimes R$, as the preferred determinate observable, the projections of the ray $e(|\psi\rangle)$ onto the eigenspaces of $I_S \otimes R$ are the rays e_{r_i} spanned by the unit vectors

$|a_i\rangle|r_i\rangle$. So for this state, the determinate sublattice contains propositions represented by the projection operators $a_i \otimes I_M$, where a_i here represents the projection operator onto the 1-dimensional subspace in $\mathscr{H}_S$ spanned by the unit vector $|a_i\rangle$. That is, the determinate sublattice contains propositions corresponding to the eigenvalues of A and also, by stipulation, propositions corresponding to pointer positions, represented by the projection operators $I_S \otimes r_i$.

It follows that the propositions corresponding to the observable correlated with an appropriate pointer observable, in the entangled state arising from a unitary transformation representing an ideal quantum mechanical measurement interaction, turn out to be determinately true or false. And this is *derived* from structural features of the Hilbert space representation of the states and observables of a quantum mechanical system, without appealing to a special 'collapse' or projection process, or invoking any privileged status for measurement that transcends these structural features. So the uniqueness theorem justifies the interpretation of the probabilities defined by a quantum state for the eigenvalues of an observable A as 'the probabilities of finding the different possible eigenvalues of A in a measurement of A,' where a measurement is represented as a dynamical process that yields determinate values for A.

A question that now arises concerns the evolution of property states. Since we want measurements to be reliable, we have to consider how this evolution affects the pointer observable after a measurement interaction. It would not do, for example, if the pointer reading jumped anomalously between different values immediately after becoming correlated with values of the measured observables. Further questions concern the analysis of non-ideal measurements, and what happens to the pointer if the measuring instrument is an open system interacting with its environment, which is bound to be the case for any system we are actually able to use as a measuring instrument in our quantum universe. I take up these questions in the following sections of this chapter. It turns out that typical instrument–environment interactions that occur in our universe ensure the stability of pointers after ideal or non-ideal measurements, for a suitable choice of the preferred determinate observable. So the possibility of reliable measuring instruments depends on a contingent feature of our quantum universe.

5.2 The evolution of property states

The property states defined by 2-valued homomorphisms (with nonzero measure) on the determinate sublattice $\mathscr{L}_{e_{r_1}e_{r_2}\ldots e_{r_k}}$ of a quantum universe will evolve in time, just as the property states defined by 2-valued homomorphisms on the Boolean lattice of a classical universe evolve in time. The quantum property states are in 1–1 correspondence with the rays $e_{r_i} \in \mathscr{L}_{e_{r_1}e_{r_2}\ldots e_{r_k}}$, hence with the eigenspaces r_i of R, and assign the same value to e_{r_i} and r_i, for all i. Since the evolution of these states is completely determined by the evolution of e and of R, we want an equation of motion for the actual values of R that will preserve the distribution of R-values specified by the quantum state

e, as e evolves unitarily in time according to Schrödinger's time-dependent equation of motion.

The solution to this problem is implicit in a proposal by Bell (1987, pp. 176–7) for constructing stochastic trajectories for fermion number density, regarded as an always-determinate discrete observable or 'beable' for quantum field theory. Bell's equation of motion generalizes the deterministic evolution of position in configuration space (a continuous observable) in Bohm's 'causal' interpretation of quantum mechanics to the stochastic evolution of a discrete observable. I shall follow a formulation by Vink (1993).

For a single particle of mass M, with no internal degrees of freedom (where configuration space is just the 3-dimensional coordinate space of the particle), Bohm extracts two real equations from Schrödinger's time-dependent complex equation of motion for the wave function in configuration space, taking $H = -(\hbar^2/2M)\nabla^2 + V$:

$$i\hbar\frac{\partial\psi}{\partial t} = -\frac{\hbar^2}{2M}\nabla^2\psi + V\psi,$$

by substituting $\psi = R\exp(iS/\hbar)$:[95]

$$\frac{\partial S}{\partial t} + \frac{(\nabla S)^2}{2M} + V - \frac{\hbar^2}{2M}\frac{\nabla^2 R}{R} = 0 \tag{5.1}$$

$$\frac{\partial R^2}{\partial t} + \nabla\cdot\left(\frac{R^2\nabla S}{M}\right) = 0 \tag{5.2}$$

Equation (5.2), derived from the imaginary part of the Schrödinger equation, can be expressed as a continuity equation for an ensemble of positions in configuration space, with a probability density $\rho = R^2 = |\psi|^2$ and a probability current vector $\boldsymbol{J} = \rho\nabla S/M = (\hbar/M)\mathrm{Im}(\psi^*\nabla\psi)$:

$$\frac{\partial\rho}{\partial t} + \nabla\cdot\boldsymbol{J} = 0 \tag{5.3a}$$

The continuity equation (5.3a) guarantees that if $\rho = |\psi|^2$ initially, ρ will remain equal to $|\psi|^2$ at all times.

Note that each element of the ensemble represents a possible position of the particle in the one-particle case, and a possible configuration of the entire n-particle system in the n-particle case. In the $3n$-dimensional configuration space of n particles ($k = 1, \ldots, n$):

$$\boldsymbol{J}_k = \frac{\rho\nabla_k S}{M_k} = \frac{\hbar}{M_k}\mathrm{Im}(\psi^*\nabla_k\psi)$$

$$\frac{\partial\rho}{\partial t} + \sum_{k=1}^{n}\nabla_k\cdot\boldsymbol{J}_k = 0 \tag{5.3b}$$

[95] The symbol 'R' here for the real part of the wave function ψ should not, of course, be confused with the symbol 'R' for the preferred determinate observable.

Vink considers an arbitrary complete set of commuting observables R^i ($i = 1, 2, \ldots, I$), with simultaneous eigenvectors $|r^1_{n^1}, r^2_{n^2}, \ldots, r^I_{n^I}\rangle$, where the $n^i = 1, 2, \ldots, N^i$ label the discrete eigenvalues of R^i. Suppressing the index i for notational convenience, these eigenstates are written as $|r_n\rangle$. We can regard the $|r_n\rangle$ as the different eigenvectors of a maximal observable R. Then each R^i is a function of R.

The time evolution of the state vector is given by the Schrödinger equation of motion:

$$i\hbar\frac{\mathrm{d}|\psi(t)\rangle}{\mathrm{d}t} = H|\psi(t)\rangle$$

or

$$i\hbar\frac{\mathrm{d}\langle r_n|\psi\rangle}{\mathrm{d}t} = \langle r_n|H\psi\rangle = \sum_m \langle r_n|H|r_m\rangle\langle r_m|\psi\rangle$$

in the R-representation.

The Schrödinger equation yields a continuity equation for the probability density $P_n(t) = |\langle r_n|\psi(t)\rangle|^2$ in the R-representation (analogous to (5.3a)):

$$\frac{\mathrm{d}P_n}{\mathrm{d}t} - \frac{1}{\hbar}\sum_m J_{nm} = 0 \tag{5.4}$$

where the components of the probability current matrix $J_{nm}(t)$ – proportional to the net flow of probability from r_m to r_n in time $\mathrm{d}t$ – are defined as:

$$J_{nm}(t) = 2\mathrm{Im}(\langle\psi(t)|r_n\rangle\langle r_n|H|r_m\rangle\langle r_m|\psi(t)\rangle) \tag{5.5}$$

For the nonmaximal (degenerate) observables R^i, the probability densities P^i_n and the current matrices J^i_{nm} are defined by summing over the remaining indices. For example,

$$P^i_n = \sum_u |\langle r^i_{n^i}, u|\psi\rangle|^2,$$

where $u = r^{i'}_{mi'}$ for $i' \neq i$, and similarly for the current matrices J^i_{nm}.

We want an equation of motion for the evolution of the actual values of the discrete observable R (and of the nonmaximal R^i) consistent with the continuity equation. The evolution will have to be stochastic to preserve the distribution of discrete R-values specified by the quantum state e, as e evolves in time. For suppose that at some time t_1 the quantum state $e(t_1)$ is an eigenstate of R with the eigenvalue r_i, for some i, so that the determinate sublattice is $\mathscr{L}_e(t_1) = \mathscr{L}_{e_{r_i}}(t_1)$. At t_1, the state $e(t_1)$ assigns unit probability to the 2-valued homomorphism that maps the proposition represented by the ray $e_{r_i}(t_1)$ onto 1 (or equivalently, $e(t_1)$ assigns unit probability to the value r_i of R). At time $t_2 > t_1$, if $e(t_2)$ is no longer an eigenstate of R, the determinate sublattice will take the form $\mathscr{L}_{e_{r_1}e_{r_2}\ldots e_{r_k}}(t_2)$, and each of the k rays $e_{r_i}(t_2)$ (or equivalently, each of the k corresponding determinate values of R) will be assigned nonzero probability by $e(t_2)$. A deterministic evolution would map the ray $e_{r_i}(t_1)$ onto some particular ray $e_{r_j}(t_2)$, or equivalently, the equivalently, the value r_i would be mapped onto some particular value r_j. So there would be determinate R-values assigned nonzero probability by the quantum state e at time t_2 that could not have become determinate through the temporal evolution of the determinate R-value at time t_1.

The argument is, in effect, that the support of the quantum state $e(t_1)$ in R-space – the determinate values of R assigned nonzero probability by $e(t_1)$ – can be a finite set with smaller cardinality than the support of $e(t_2)$ in R-space. So if the determinate values of R at t_1 (one of which is the actual value at t_1) evolve deterministically, there would have to be determinate values of R at t_2, with nonzero probability specified by the quantum state $e(t_2)$, that could not have evolved deterministically from any of the determinate R-values at t_1. For continuous R, this argument fails, because the support of e in R-space is always a continuum of values, for any quantum state e: there are, strictly, no eigenstates of R in the continuous case.[96]

Suppose the discontinuous jumps in the actual value of R are governed by transition probabilities $T_{mn}\mathrm{d}t$, where $T_{mn}\mathrm{d}t$ denotes the probability of a jump from value r_n to value r_m in the time interval dt. The transition matrix gives rise to time-dependent probability distributions of R-values. It follows that the rate of change of the probability density $P_n(t)$ for r_n must satisfy a master equation:

$$\frac{\mathrm{d}P_n(t)}{\mathrm{d}t} = \sum_m (T_{nm}P_m - T_{mn}P_n) \tag{5.6}$$

From the Schrödinger equation, $\dfrac{\mathrm{d}P_n(t)}{\mathrm{d}t}$ also satisfies the continuity equation (5.4).

To reconcile (5.6) with (5.4), it suffices to find a transition matrix T, with $T_{mn} \geq 0$, that satisfies the equation:

$$\frac{J_{nm}}{\hbar} = T_{nm}P_m - T_{mn}P_n \tag{5.7}$$

for a given P and J. Since $J_{mn} = -J_{nm}$ (hence $J_{nn} = 0$), this equation yields $n(n-1)/2$ equations for the n^2 elements of T. So there are many solutions. Bell's choice (1987, p. 176) for $n \neq m$ was:

$$\begin{aligned} T_{nm} &= \frac{J_{nm}}{\hbar P_m}, \text{ if } J_{nm} > 0 \\ T_{nm} &= 0, \text{ if } J_{nm} \leq 0 \end{aligned} \tag{5.8}$$

For $n = m$, T_{nn} is fixed by the normalization condition $\Sigma_m T_{mn}\mathrm{d}t = 1$.

Bohm interpreted equation (5.1), derived from the real part of the Schrödinger equation, as a Hamilton–Jacobi equation for the motion of a particle under the influence of a potential function V and an additional 'quantum potential' $(\hbar^2/2m)(\nabla^2 R/R)$.

[96] I owe this argument to Rob Clifton.

The possible trajectories of the particle are given by the solutions to the 'guiding equation':

$$\frac{\mathrm{d}\boldsymbol{x}}{\mathrm{d}t} = \frac{\boldsymbol{J}}{\rho} = \frac{\nabla S}{M} = \frac{\hbar}{M}\,\mathrm{Im}\left(\frac{\nabla\psi}{\psi}\right) = \frac{\hbar}{2\mathrm{i}M}\,\frac{\psi^*\nabla\psi - \psi\nabla\psi^*}{|\psi|^2} \tag{5.9a}$$

That is, the particle trajectories are given by the integral curves of a velocity field defined by the gradient of the phase S. So the trajectories $\boldsymbol{x}(t)$ depend on the wave function ψ as a function of the position coordinates of the particle.

For a composite system consisting of n particles, the system is represented by a point in the $3n$-dimensional configuration space of the n particles, and the velocity of this point $(\boldsymbol{x}_1, \ldots, \boldsymbol{x}_n)$ depends on the wave function as a function on configuration space:

$$\frac{\mathrm{d}\boldsymbol{x}_k}{\mathrm{d}t} = \frac{\boldsymbol{J}_k}{\rho} = \frac{\nabla_k S}{M_k} = \frac{\hbar}{M_k}\,\mathrm{Im}\left(\frac{\nabla_k\psi}{\psi}\right) = \frac{\hbar}{2\mathrm{i}M_k}\,\frac{\psi^*\nabla_k\psi - \psi\nabla_k\psi^*}{|\psi|^2} \tag{5.9b}$$

Vink shows that Bell's solution for the transition matrix leads to Bohm's theory in the continuum limit, when R is position in configuration space. For example, consider a single particle of mass M on a 1-dimensional lattice. Let $R = x = n\varepsilon$, with $n = 1, 2, \ldots, N$, where ε is the distance between neighbouring lattice sites. Schrödinger's equation for the wave function $\psi(n\varepsilon,t) = \langle x_n|\psi(t)\rangle = \langle n\varepsilon|\psi(t)\rangle$ on this discrete x-space (with an appropriately discretized second derivative)[97] becomes:

$$\begin{aligned} i\hbar\frac{\partial}{\partial t}\psi(n\varepsilon,t) &= -\frac{\hbar^2}{2M\varepsilon^2}\left[\psi(n\varepsilon+\varepsilon,t) + \psi(n\varepsilon-\varepsilon,t) - 2\psi(n\varepsilon,t)\right] + V(n\varepsilon)\psi(n\varepsilon,t) \\ &= \sum_m H_{nm}\psi(m\varepsilon,t) \\ &= \sum_m\left[-\frac{\hbar^2}{2M\varepsilon^2}(\delta_{n+1,m} + \delta_{n-1,m} - 2\delta_{nm}) + V(n\varepsilon)\delta_{nm}\right]\psi(m\varepsilon,t) \end{aligned} \tag{5.10}$$

Suppressing the time coordinate, the probability density P_n and the current matrix J_{nm} can be expressed as:

$$\begin{aligned} P_n &= |\psi(n\varepsilon)|^2 \\ J_{nm} &= 2\mathrm{Im}[\psi^*(n\varepsilon)H_{nm}\psi(m\varepsilon)] \\ &= -\frac{\hbar^2}{M\varepsilon^2}\mathrm{Im}[\psi^*(n\varepsilon)\psi(m\varepsilon)\delta_{n+1,m} + \psi^*(n\varepsilon)\psi(m\varepsilon)\delta_{n-1,m}] \end{aligned} \tag{5.11a}$$

The potential function $V(x)$ drops out of the expression for J_{nm} because it contributes a term $\psi^*(n\varepsilon)V(n\varepsilon)\delta_{nm}\psi(m\varepsilon)$, which is zero if $m \neq n$ (obviously, since the matrix of the potential energy is diagonal in the position representation), and real if $m = n$. Since $\psi^*(n\varepsilon)\psi(m\varepsilon)\delta_{n+1,m} = \psi^*(m\varepsilon - \varepsilon)\psi(m\varepsilon)\delta_{m,n+1}$ (because these expressions are only non-

[97] The first partial derivative, $\partial/\partial x$, of a function f on the discrete lattice can be computed as $[f(n\varepsilon + \mathrm{e}) - f(n\varepsilon)]/\varepsilon = \Sigma_m[(\delta_{n+1,m} - \delta_{nm})f(m\varepsilon)]/\varepsilon$, or as $[f(n\varepsilon) - f(n\varepsilon - \varepsilon)]/\varepsilon = \Sigma_m[(\delta_{nm} - \delta_{n-1,m})f(m\varepsilon)]/\varepsilon$. To compute the second partial derivative, differentiation is applied symmetrically.

zero if $m = n + 1$, in which case they are equal) and $\psi^*(n\varepsilon)\psi(m\varepsilon)\delta_{n-1,m} = \psi^*(m\varepsilon + \varepsilon)\psi(m\varepsilon)\delta_{m,n-1}$ (because these expressions are only nonzero if $m = n - 1$, in which case they are equal), J_{nm} can be expressed as:

$$J_{nm} = -\frac{\hbar^2}{M\varepsilon^2}\mathrm{Im}[\psi^*(m\varepsilon + \varepsilon)\psi(m\varepsilon)\delta_{m,n-1} + \psi^*(m\varepsilon - \varepsilon)\psi(m\varepsilon)\delta_{m,n+1}] \quad (5.11\text{b})$$

So $J_{nm} = 0$ unless $m = n - 1$ or $m = n + 1$; that is, the Hamiltonian induces nearest-neighbour transitions only:

$$J_{21} = -\frac{\hbar^2}{M\varepsilon^2}\mathrm{Im}[\psi^*(2\varepsilon)\psi(\varepsilon)],\ J_{12} = -\frac{\hbar^2}{M\varepsilon^2}\mathrm{Im}[\psi^*(\varepsilon)\psi(2\varepsilon)]$$

$$J_{32} = -\frac{\hbar^2}{M\varepsilon^2}\mathrm{Im}[\psi^*(3\varepsilon)\psi(2\varepsilon)],\ J_{23} = -\frac{\hbar^2}{M\varepsilon^2}\mathrm{Im}[\psi^*(2\varepsilon)\psi(3\varepsilon)], \text{ etc.} \quad (5.12)$$

Note that this is a feature of discretized position, for potentials $V(x)$ that are functions of position. If we take the preferred always-determinate observable R as discretized momentum, for example, J_{nm} (and hence T_{nm}) can be nonzero between distant lattice sites, and will depend on the specific form of the potential function $V(x)$. Since x becomes $i\hbar\frac{\partial}{\partial p}$ in the momentum representation, if $V(x) = e^x$, say, the potential in the momentum representation will take the form

$$1 + \sum_j \frac{(i\hbar)^j}{j!}\frac{\partial^j}{\partial p^j}$$

(expanding e^x as $1 + \Sigma_j x^j/j!$) and can induce transitions between arbitrarily distant lattice sites through nonzero terms in the transition matrix T_{nm} for arbitrary m, n. From (5.10), we see that the matrix of the discretized Laplacian operator, the second partial derivative $\partial^2/\partial x^2$ in the discrete x representation, takes the form $(\delta_{n+1,m} + \delta_{n-1,m} - 2\delta_{nm})/\varepsilon^2$. The matrix of the fourth partial derivative $\partial^4/\partial x^4$, for example, takes the form $(\delta_{n,m+2} + \delta_{n,m-2} - 4\delta_{n,m+1} - 4\delta_{n,m-1} + 6\delta_{nm})/\varepsilon^4$. Evidently, this is also the form of the matrix of the fourth partial derivative of momentum in the momentum representation. The term $[(i\hbar)^j/j!]\partial^4/\partial p^4$ in the expansion of the potential $V(x) = e^x$ in the momentum representation can therefore introduce nonzero terms in the matrix J_{nm}, and hence in the transition matrix T_{nm}, for $m = n - 1, n + 1$, as well as for $m = n - 2, n + 2$. Higher order derivatives will introduce nonzero off-diagonal terms in T_{nm} for arbitrary m, n. The same thing can, of course, occur for position as the preferred observable, if the potential is velocity dependent.

Bohm points out in his reply (1953a) to Epstein (1952) that for a system like the hydrogen atom, where $V(x)$ is proportional to $1/x$, the potential in the momentum representation cannot even be expressed as a convergent series of operators $i\hbar\ \partial/\partial p$. Taking the preferred always-determinate observable R as momentum is not at all the same thing as taking R as position, and then transforming the continuity equation (5.4), the master equation (5.6), and the solution (5.8) for the transition matrix T_{nm} via a canonical transformation from a position to a momentum representation. The latter procedure simply yields a more complicated redescription of a 'no collapse' interpretation with position as the preferred observable.

Assuming that ε is small and the wave function is sufficiently smooth on this lattice scale, J_{nm} in (5.11b) reduces, to first order in ε, to:

$$J_{nm} = \frac{\hbar}{M\varepsilon}[S'(m\varepsilon)P_m\delta_{m,n-1} - S'(m\varepsilon)P_m\delta_{m,n+1}], \tag{5.13}$$

where S is the phase. Bell's solution (5.8) yields:

$$T_{nm} = \frac{S'(m\varepsilon)}{M\varepsilon}\delta_{m,n-1}, \; S'(m\varepsilon) \geq 0$$

$$T_{nm} = -\frac{S'(m\varepsilon)}{M\varepsilon}\delta_{m,n+1}, \; S'(m\varepsilon) \leq 0 \tag{5.14}$$

So as the distance between lattice sites tends to zero, the particle either remains at a given site or jumps to the next site in a particular direction. For positive $S'(m\varepsilon)$ the particle tends to jump from site m to site $m+1$ with probability $[|S'(m\varepsilon)|dt]/M\varepsilon$, and for negative $S'(m\varepsilon)$ the particle tends to jump from site m to $m-1$ with the same probability (the probability of a jump in the opposite direction tending to zero as $\varepsilon \to 0$). The probability of a jump depends on the absolute value of the gradient of the phase of the wave function at the lattice position, and the direction in which the particle jumps depends on whether the gradient of the phase is positive or negative at that lattice position. In other words, whether the particle jumps to the next lattice site, and which direction it jumps, depends stochastically only on where the particle is on the lattice, through a feature of the wave function at the lattice position of the particle.

As the distance between lattice sites tends to zero, the stochastic trajectories become smoother and smoother, because the stochastic transitions are between nearest-neighbours in the lattice only. Since each jump is over a distance ε, the average displacement in a time interval dt is:

$$dx = \frac{S'(x)dt}{M}$$

and hence the velocity of the particle is:

$$\frac{dx}{dt} = \frac{S'(x)}{M}$$

As $\varepsilon \to 0$, $S' \to \partial_x S$, and so in the continuum limit:

$$\frac{dx}{dt} = \frac{\partial_x S}{M}$$

(where the partial derivative here explicitly reflects the time dependence of the phase, $S(x,t)$). Vink shows that the dispersion in the average displacement vanishes in the limit as $\varepsilon \to 0$, and so the discontinuous stochastic trajectories approach the continuous deterministic Bohmian trajectories in the continuum limit.

Other solutions for the transition matrix induce jumps between distant sites, so continuous deterministic trajectories are not recovered, even in the continuum limit. Nelson's stochastic dynamics (1966) is characterized by one such solution. In a 'no collapse' interpretation of quantum mechanics based on position in configuration space as a preferred, always-determinate observable R, the Bohmian trajectories appear to be unique, as the deterministic continuum limit of a stochastic evolution for a discretized R.[98]

The discrete version of Bohmian mechanics provides an explanation for the origin of the Born probabilities that is not available in the continuous version. Suppose the initial quantum state $|\psi(0)\rangle$ is an eigenstate of the discrete preferred observable R with eigenvalue r_0:

$$|\psi(0)\rangle = |r_0\rangle$$

The determinate sublattice defined by the ray $e(|\psi(0)\rangle)$ and R is the sublattice $\mathscr{L}_{e_{r_0}}$ (that is, in this case, the determinate sublattice coincides with the determinate sublattice of the orthodox interpretation). There is only one 2-valued homomorphism on $\mathscr{L}_{e_{r_0}}$ (if $\mathscr{H}$ is more than 2-dimensional): the 2-valued homomorphism that assigns 1 to the atomic proposition represented by the ray e_{r_0} and 0 to every other atom in $\mathscr{L}_{e_{r_0}}$. So the initial probability distribution defined by the state $e(|\psi(0)\rangle)$, as an ignorance measure on the determinate sublattice, is just the distribution that assigns unit probability to this 2-valued homomorphism (or, more precisely, to the subset of all 2-valued homomorphisms on $\mathscr{L}_{e_{r_0}}$ containing this single 2-valued homomorphism). Continuity now guarantees that the stochastic transitions governed by the Bell–Vink equations of motion will produce a distribution of values for R at time t that is just the probability distribution defined by $|\psi(t)\rangle$.

If R is continuous, then the corresponding choice for the initial state would be a state that is sharply peaked about r_0 (again, because there are, strictly speaking, no R-eigenstates for a continuous observable R). But now, unless we make the additional statistical assumption that the values of R are initially distributed in accordance with the probability distribution of R-values defined by the Born rule for this initial state, continuity alone will not guarantee that the values of R at later times t will be distributed as defined by the Born rule for $|\psi(t)\rangle$. The nature and status of the statistical assumptions required to yield the Born probabilities in the continuous version of Bohmian mechanics is the subject of an ongoing debate. See Bohm (1953c), Bohm and Hiley (1993, chapter 9), Holland (1993b, chapter 3), Dürr, Goldstein, and Zanghì (1992a, b, c, 1993), and Valentini (1991a, b, 1996).

[98] For a discussion of the existence and uniqueness of Bohmian trajectories, see Berndl *et al.* (1993).

Taking the initial quantum state $|\psi(0)\rangle$ as an eigenstate of R is appropriate if we consider $|\psi(0)\rangle$ not as the initial state of the universe, but rather as the initial state of the instrument obtained after a (possibly non-ideal) measurement, when the state of the measured system + measuring instrument + environment has the form $|\Psi'(t)\rangle = \Sigma_i c_i'(t)|a_i'\rangle|r_i\rangle|\varepsilon_i(t)\rangle$ (where $\{|a_i'\rangle\}$ represent the states of the measured system and $\{|\varepsilon_i(t)\rangle\}$ represent environmental states – see sections 5.3 and 5.4). The pointer value r_0 (here the value at time t, when we re-set t to zero) is effectively associated with just one of the components $|a_i'\rangle|r_i\rangle|\varepsilon_i(t)\rangle$. This is because different environmental states approach orthogonality virtually instantaneously after a measurement, so the stochastic evolution of the pointer position depends effectively only on the component $|a_0'\rangle|r_0\rangle|\varepsilon_0(0)\rangle$ corresponding to the value r_0 of R at a time $t = 0$ very shortly after a measurement, which means that the instrument state is effectively $|r_0\rangle$. See the discussion of measurement in Bohmian mechanics in section 6.1.

From the perspective of the uniqueness theorem of chapter 4, the origin of the Bornprobabilities for the distribution of values of a discrete preferred observable R lies in the stochastic evolution of the actual property state or 2-valued homomorphism associated with the actual value of R at time t. The property state is required to evolve stochastically in such a way as to conform to the dynamically evolving possibility structure defined by the determinate sublattice, which is selected by the quantum state and R as a sublattice of the non-Boolean lattice $\mathscr{L}(\mathscr{H})$. The distribution problem, which arises for Bohm's choice of R as position in configuration space, appears to be an artefact of choosing a continuous observable as the preferred observable in a 'no collapse' interpretation. We can think of position in configuration space as the continuous limit of a sequence of discrete coarse-grained position observables, and for each of these discrete position observables there is no distribution problem.

5.3 Non-ideal measurements

In section 5.1, I characterized an ideal ('first kind') measurement in quantum mechanics as an interaction between two quantum systems, S (the system measured) and M (the measuring instrument), that induces a correlation between eigenstates $|a_i\rangle$ of an observable A of S (the observable measured) and eigenstates $|r_i\rangle$ of an observable R of M (the 'pointer' observable) in the state of the composite system $S + M$ after the interaction (hence a correlation between properties of S and pointer properties of M).

By a non-ideal measurement, I mean an interaction between S and M that is 'close to' an ideal measurement, in the sense that eigenstates of an S-observable A and eigenstates of the pointer observable R of M are 'almost correlated' in the state of the composite system $S + M$ after the interaction.

For example, suppose the Hilbert spaces of S and M are both 2-dimensional, and the initial state of S is a linear superposition of A-eigenstates, $|\alpha\rangle = c_1|a_1\rangle + c_1|a_2\rangle$. Suppose the initial state of M is $|r_1\rangle$. Then an ideal measurement of A is represented by the transition:

$$|\alpha\rangle|r_1\rangle = (c_1|a_1\rangle + c_2|a_2\rangle)|r_1\rangle \rightarrow |\psi\rangle = c_1|a_1\rangle|r_1\rangle + c_2|a_2\rangle|r_2\rangle \quad (5.15)$$

Of course, in any case of interest M will be a macroscopic system, or at any rate, an open system in interaction with its environment E. The above representation of a measurement is meant to be schematic. My purpose in this section is to bring out certain aspects of non-ideal measurements. I consider the rôle of the environment in the following section.

A non-ideal measurement involves a transition of the following sort:

$$\begin{aligned} |\alpha\rangle|r_1\rangle &= (c_1|a_1\rangle + c_2|a_2\rangle)|r_1\rangle \\ \rightarrow |\psi'\rangle &= c_{11}|a_1\rangle|r_1\rangle + c_{12}|a_1\rangle|r_2\rangle + c_{21}|a_2\rangle|r_1\rangle + c_{22}|a_2\rangle|r_2\rangle \end{aligned} \quad (5.16)$$

where $|c_{12}|^2$ and $|c_{21}|^2$ are small compared to $|c_{11}|^2$ and $|c_{22}|^2$, and $|c_{11}|^2 \approx |c_1|^2$ and $|c_{22}|^2 \approx |c_2|^2$.

The state $|\psi'\rangle$ can be expressed as:

$$\begin{aligned} |\psi'\rangle &= (c_{11}|a_1\rangle + c_{21}|a_2\rangle)|r_1\rangle + (c_{12}|a_1| + c_{22}|a_2\rangle)|r_2\rangle \\ &= c_1'|a_1'\rangle|r_1\rangle + c_2'|a_2'\rangle|r_2\rangle, \end{aligned} \quad (5.17)$$

where the states $|a_1'\rangle = c_{11}/c_1'|a_1\rangle + c_{21}/c_1'|a_2\rangle$ and $|a_2'\rangle = c_{12}/c_2'|a_1\rangle + c_{22}/c_2'|a_2\rangle$, with $|c_1'|^2 = |c_{11}|^2 + |c_{21}|^2$ and $|c_2'|^2 = |c_{12}|^2 + |c_{22}|^2$, are non-orthogonal *relative states* of S (relative to the eigenstates of the pointer observable R of M), 'close to' the eigenstates of the measured observable A. So, a non-ideal measurement can be regarded as correlating eigenstates of R with relative states of S that are non-orthogonal linear superpositions of A-eigenstates, rather than A-eigenstates. Since these relative states of S are non-orthogonal, they do not correspond to the eigenstates of any S-observable.

Alternatively:

$$\begin{aligned} |\psi'\rangle &= |\alpha_1\rangle(c_{11}|r_1\rangle + c_{12}|r_2\rangle) + |a_2\rangle(c_{21}|r_1\rangle + c_{22}|r_2\rangle) \\ &= c_1''|a_1\rangle|r_1'\rangle + c_2''|a_2\rangle|r_2'\rangle, \end{aligned} \quad (5.18)$$

where the states $|r_1'\rangle = c_{11}/c_1''|r_1\rangle + c_{12}/c_1''|r_2\rangle$ and $|r_2'\rangle = c_{21}/c_2''|r_1\rangle + c_{22}/c_2''|r_2\rangle$, with $|c_1''|^2 = |c_{11}|^2 + |c_{12}|^2$ and $|c_2''|^2 = |c_{21}|^2 + |c_{22}|^2$, are non-orthogonal relative states of M (relative to the eigenstates of A), 'close to' the eigenstates of the pointer observable R. So, a non-ideal measurement can also be regarded as correlating eigenstates of the measured observable A of S with relative states of M that are non-orthogonal linear superpositions of pointer eigenstates. Again, since these relative states of M are non-orthogonal, they do not correspond to the eigenstates of any M-observable.

In the general case, a non-ideal measurement is characterized by the transition:

$$\begin{aligned} |\alpha\rangle|r_0\rangle = (\sum_i c_i|a_1\rangle)|r_o\rangle \rightarrow |\psi'\rangle &= \sum_{kj} c_{kj}|a_k\rangle|r_j\rangle \\ &= \sum_j c_j'|a_j'\rangle|r_j\rangle \end{aligned} \quad (5.19)$$

$$= \sum_k c_k''|a_k\rangle|r_k'\rangle \tag{5.20}$$

where $|r_0\rangle$ is the initial 'ready state' of M (suppressing the degeneracy index of the pointer eigenstates). The relative states $|a_j'\rangle$ of S are:

$$|a_j'\rangle = \sum_k \frac{c_{kj}}{c_j'}|a_k\rangle \tag{5.21}$$

with $|c_{j'}|^2 = \Sigma_k|c_{kj}|^2$, and the relative states $|r_k'\rangle$ of M are:

$$|r_k'\rangle = \sum_j \frac{c_{kj}}{c_k''}|r_j\rangle \tag{5.22}$$

with $|c_k''|^2 = \Sigma_j|c_{kj}|^2$. Note that we could have $c_j' = c_j$ (so $|c_j|^2 = \Sigma_k|c_{kj}|^2$), for all j, as a special case of (5.19), or $c_k'' = c_k$ (so $|c_k|^2 = \Sigma_j|c_{kj}|^2$), for all k, as a special case of (5.20).

A non-ideal measurement interaction might, for example, be induced by an interaction Hamiltonian of the form:

$$H_{\text{int}} = g(t)\, A \otimes V,$$

where V is an M-observable conjugate to R (that is, $RV - VR = iI$, taking $\hbar = 1$), and $g(t)$ is a coupling constant that is nonzero only during the time interval from $t = 0$ to $t = T$, when the interaction is 'switched on.' If $g(t)$ is sufficiently large so that the total Hamiltonian can be approximated by H_{int} during this time interval, the measurement is said to be 'impulsive.' For an impulsive measurement, Schrödinger's time-dependent equation reduces to:

$$\mathrm{i}\frac{\mathrm{d}|\psi\rangle}{\mathrm{d}t} = H_{\text{int}}|\psi\rangle = g(t)\, A \otimes V\, |\psi\rangle$$

during the interaction, and has the solution:

$$|\psi(t)\rangle = \exp\left(\int -ig(t)A \otimes V\mathrm{d}t\right) |\psi(0)\rangle$$

If $g(t) = g$ during the interaction, then

$$|\psi(t)\rangle = \exp(-igtA \otimes V)\, |\psi(0)\rangle$$

for $0 \leq t \leq T$. If $|\psi(0)\rangle = \Sigma_k c_k|a_k\rangle|r_0\rangle$, where $|r_0\rangle$ is the initial 'ready state' of M, then

$$\exp(-igtA \otimes V)\, |\psi(0)\rangle = \sum_k c_k|a_k\rangle \exp(-igta_k V)|r_0\rangle$$

So, after the measurement interaction at time T:

$$|\psi(T)\rangle = \sum_k c_k|a_k\rangle|r_k'\rangle,$$

where $|r_k'\rangle = \exp(-igTa_k V)|r_0\rangle$.

The shifted states $|r_k'\rangle$ will generally not be mutually orthogonal, and hence not eigenstates of R. For example, suppose R represents position, Q, and V momentum, P. In the position representation:

$$\langle q|\psi(T)\rangle = \sum_k c_k|a_k\rangle \exp(-gTa_k\frac{\partial}{\partial q})\phi(q)$$

$$= \sum_k c_k|a_k\rangle\phi(q - gTa_k)$$

If $\phi_0 = \phi(q)$ is a narrow Gaussian wave packet symmetric about 0, the wave functions $\phi(q - gTa_k)$ are narrow Gaussians, symmetric about gTa_k, $k = 1, 2, \ldots$. They are relatively localized if gT is sufficiently large, so that the separation between peaks is very much greater than the width of the Gaussians. But their tails overlap, and so long as the overlap between the tails is nonzero, the wave functions are not orthogonal.

Quantum mechanical measurements are typically non-ideal. For example, a Stern–Gerlach measurement of the spin of a spin-$\frac{1}{2}$ particle, in which the position of the particle itself (or a coarse-grained position) acts as the pointer R, is non-ideal in this sense. In a measurement of spin in the z-direction, the particle, represented initially by a relatively localized wave packet moving in the x-direction, enters a region between two magnets and is subjected to the influence of an inhomogeneous magnetic field that gives the particle a momentum in the positive or negative z-direction, depending on whether the spin of the particle in the z-direction, σ_z, is 'up' (+) or 'down' (−). So the position of the particle in the z-direction, after the particle has left the magnetic field, indicates the value of the spin component:

$$(c_+|z+\rangle + c_-|z-\rangle)\phi(q) \rightarrow c_+|z+\rangle\phi(q - gt) + c_-|z-\rangle\phi(q + gt), \quad (5.23)$$

where $|z+\rangle$ and $|z-\rangle$ represent the 'up' and 'down' spin states of the particle, respectively.

Here $gt = \mu t(\partial H_z/\partial z)_{z=0}$ to a first order approximation, where $\mu = -e/2mc$ is the magnetic moment of the particle and H_z is the component of the magnetic field in the z-direction. The two wave packets $\phi_+ = \phi(q - gt)$ and $\phi_- = \phi(q + gt)$, centered at gT and $-gT$, respectively, at time T, begin to spread out after time T but continue to move in the positive and negative z-directions. If the separation between the peaks is much greater than the width of the packets (and this depends essentially on the time t and the gradient of the magnetic field in the z-direction), one would like to say that this analysis shows that a relatively rough observation of the position of the particle in the z-direction reveals the value of the spin component of the particle in the z-direction. But strictly speaking, these packets will always have infinitely long tails, so they never *completely* separate in the configuration space of the particle. They are not orthogonal wave functions, and hence not eigenfunctions of R. While there is a correlation between the packet moving up and $|z+\rangle$ and the packet moving down and $|z-\rangle$, there is no strict correlation between z-position and the eigenvalue, ± 1, of σ_z.

In an observer-free 'no collapse' interpretation with z-position as the preferred determinate observable R, the trajectory of the particle in R-space (that is, the position of the particle in the z-direction) follows one particular packet, with negligible probability (in any time period during which the packets maintain their macroscopic separation) of an anomalous transition to an R-value associated with a different packet. This is because (i) the stochastic equation of motion for the evolution of the actual value of R induces transitions between neighbouring R-values only, so (ii) in order for the actual value of R to jump between values of R associated with macroscopically separated packets in R-space, the stochastic trajectories would have to go through many nearest-neighbour transitions, and (iii) the wave function in R-space is close to zero between the two packets, so the trajectory of R would have to move through a region in which the transition matrix is effectively zero. Which particular packet is relevant – which trajectory the particle follows in R-space – depends entirely on the initial R-value of the particle when it enters the magnetic field. In the following section, I show that the probability of a jump between different R-values associated with the two packets is reduced still further almost instantaneously after the particle interacts with its environment.

A Stern–Gerlach measurement is (essentially) non-ideal because it is the *position* of the particle (more precisely, a coarse-grained position that distinguishes between an 'up' region and a 'down' region), that functions as the pointer observable, but the interaction results in a correlation between spin eigenstates and wave functions – non-orthogonal relative states of M expressible as linear superpositions (integrals) over 'position eigenstates' (δ-functions of position),[99] rather than 'position eigenstates.'

One might think that a non-ideal measurement can be understood as 'close to' an ideal measurement in the following sense: In any real measurement, there is always some probability of the measuring instrument malfunctioning and producing a faulty reading. A measurement interaction between the system and a 'good' instrument might yield an almost perfect correlation between A-eigenvalues and R-eigenvalues, but any real instrument will induce an interaction with S that results in a final state of the form $|\psi'\rangle$ with significant error terms like $c_{12}|a_1\rangle|r_2\rangle$ and $c_{21}|a_2\rangle|r_1\rangle$. The closer the probabilities $|c_{12}|^2$ and $|c_{21}|^2$ are to zero, the closer the non-ideal measurement approaches an error-free measurement.

Now, presumably a measuring instrument produces a faulty reading if the instrument registers r_1 when the value of A is a_2, or r_2 when the value of A is a_1. But on the orthodox interpretation, the probabilities $|c_{12}|^2$ and $|c_{21}|^2$ refer to *the outcomes of measurements on the composite system $S + M$ by a second measuring instrument.* Just as we cannot maintain, for a system S in the state $|\alpha\rangle = c_1|a_1\rangle + c_2|a_2\rangle$, that A has some determinate but unknown value, and that $|c_i|^2$ represents the probability that A has

[99] See the appendix, section A.3.

the value a_i, so we cannot maintain, on the orthodox interpretation, that A and R have determinate but unknown values in the state $|\psi'\rangle = c_{11}|a_1\rangle|r_1\rangle + c_{12}|a_1\rangle|r_2\rangle + c_{21}|a_2\rangle|r_1\rangle + c_{22}|a_2\rangle|r_2\rangle$, and that $|c_{ij}|^2$ represents the probability that A has the value a_i and R has the value r_j. The orthodox interpretation does not license the interpretation of $|c_{12}|^2$ and $|c_{21}|^2$ as error probabilities, in the sense that $|c_{12}|^2$ and $|c_{21}|^2$ represent, respectively, the probability that A has the value a_1 when the pointer R reads r_2, and the probability that A has the value a_2 when the pointer R reads r_1.

But quite apart from the orthodox interpretation, we cannot interpret a non-ideal measurement as an 'error-prone' ideal measurement in this sense, because such an interpretation would in general be contradictory.

To see this, consider the case where the Hilbert spaces of S and M are 3-dimensional. A non-ideal measurement can be regarded as correlating non-orthogonal linear superpositons of eigenstates of A with pointer eigenstates:

$$\begin{aligned} |\psi'\rangle &= \sum_{i,j=1}^{3} c_{ij}|a_i\rangle|r_j\rangle \\ &= \sum_{k=1}^{3} c_k'|a_k'\rangle|r_k\rangle \end{aligned} \tag{5.24a}$$

where

$$c_k'|a_k'\rangle = \sum_{i=1}^{3} c_{ik}|a_i\rangle \tag{5.25a}$$

So three non-orthogonal states are correlated with eigenstates of R. These three non-orthogonal states are 'almost orthogonal' if the 'error terms' are small. But these non-orthogonal states are not unique. We could equally well express $|\psi'\rangle$ as:

$$\begin{aligned} |\psi'\rangle &= \sum_{i,j=1}^{3} d_{ij}|b_i\rangle|r_j\rangle \\ &= \sum_{k=1}^{3} d_k'|b_k'\rangle|r_k\rangle \end{aligned} \tag{5.24b}$$

where

$$d_k'|b_k'\rangle = \sum_{i=1}^{3} d_{ik}|b_i\rangle \tag{5.24b}$$

and the states $|b_i\rangle$ are eigenstates of some obsevable B 'close to' A in the Hilbert space. I shall show that there exist families of observables 'close to' A, such that the set of maps selecting one and only one eigenstate in each orthogonal triple of eigenstates (corresponding to the attribution of a particular determinate value to the observable) cannot be assigned weights consistent with the interpretation of $|c_{ij}|^2, |d_{ij}|^2, \ldots, i \neq j$, as error probabilities. So in a non-ideal measurement in which the final state of the system plus measuring instrument is $|\psi'\rangle$ and the pointer reading is r_k, we canot maintain that A actually has the value a_k with high probability, and the weights $|c_{kj}|^2$, for $j \neq k$, represent error probabilities that A actually has some other value a_j, and similarly for other observables $B, C, \ldots$ 'close to' A.

Consider the following three orthogonal triples (not necessarily normalized), defined relative to the orthogonal triple $\{|a_1\rangle, |a_2\rangle, |a_3\rangle\}$:[100]

$$\begin{aligned} |b_1\rangle &= |a_1\rangle + p|a_3\rangle \\ |b_2\rangle &= |a_2\rangle \\ |b_3\rangle &\perp |b_1\rangle, |b_2\rangle \end{aligned}$$

$$\begin{aligned} |c_1\rangle &= |b_1\rangle + \varepsilon^{-1}p|a_2\rangle = |a_1\rangle + \varepsilon^{-1}p|a_2\rangle + p|a_3\rangle \\ |c_2\rangle &= \varepsilon p|a_2\rangle - p|a_3\rangle \\ |c_3\rangle &\perp |c_1\rangle, |c_2\rangle \end{aligned}$$

$$\begin{aligned} |d_1\rangle &= |b_1\rangle + \eta^{-1}p|a_2\rangle = |a_1\rangle + \eta^{-1}p|a_2\rangle + p|a_3\rangle \\ |d_2\rangle &= \eta p|a_2\rangle - p|a_3\rangle \\ |d_3\rangle &\perp |d_1\rangle, |d_2\rangle \end{aligned}$$

Now:

$$\begin{aligned} |w_1\rangle &= |c_1\rangle + |c_2\rangle = |a_1\rangle + (\varepsilon + \varepsilon^{-1})p|a_2\rangle \\ |w_2\rangle &= |d_1\rangle + |d_2\rangle = |a_1\rangle + (\eta + \eta^{-1})p|a_2\rangle \end{aligned}$$

So $|w_1\rangle$ is orthogonal to $|w_2\rangle$ if

$$(\varepsilon + \varepsilon^{-1})(\eta + \eta^{-1})p^2 = -1$$

or

$$(\eta + \eta^{-1})p = \frac{-1}{(\varepsilon + \varepsilon^{-1})p}$$

Let $k = 1/(\varepsilon + \varepsilon^{-1})p$. Then the condition for orthogonality is:

$$\eta^2 p + k\eta + p = 0$$

This equation has two real solutions, $\eta_+, \eta_- = (\eta_+)^{-1}$, for each value of $k \geq 2p$, or equivalently, for values of ε satisfying the inequality

$$\varepsilon^2 - \frac{\varepsilon}{2p^2} + 1 \leq 0$$

The function $\varepsilon^2 - (\varepsilon/2p^2) + 1$ will have negative values for values of ε between the two real solutions of the quadratic equation $\varepsilon^2 - (\varepsilon/2p^2) + 1 = 0$. This equation has two real solutions $\varepsilon_+, \varepsilon_- = (\varepsilon_+)^{-1}$, for values of $p \leq \frac{1}{2}$. So, if $p \leq \frac{1}{2}$, there are values of ε

[100] The following construction is inspired by Bell's version of the Kochen and Specker 'no go' theorem, presented as a corollary to Gleason's theorem. See Bell (1966).

between ε_+ and ε_- for which there exist values of η (real solutions to the quadratic equation for η), such that $|w_1\rangle = |c_1\rangle + |c_2\rangle$ and $|w_2\rangle = |d_1\rangle + |d_2\rangle$ are orthogonal vectors, both lying in the $|a_1\rangle$, $|a_2\rangle$ plane. Hence $|w_1\rangle$, $|w_2\rangle$, $|w_3\rangle = |a_3\rangle$ form a fifth orthogonal triple, in addition to the $|a\rangle$-triples, $|b\rangle$-triples, $|c\rangle$-triples, and $|d\rangle$-triples.

For examples, suppose $p = 0.01$. Then $\varepsilon_+ \approx 5000$ and $\varepsilon_- \approx 0.0002$. Take $\varepsilon = 2500$. The two values of η are $\eta_+ \approx 0.267\,942$ and $\eta_- \approx 3.732\,05$. So $(\varepsilon + \varepsilon^{-1})p = 25.000\,004$ and $(\eta + \eta^{-1})p \approx 0.039\,9999$, and we have:

$$|b_1\rangle = |a_1\rangle + 0.01|a_3\rangle$$
$$|b_2\rangle = |a_2\rangle$$
$$|b_3\rangle \perp |b_1\rangle, |b_2\rangle$$

$$|c_1\rangle = |a_1\rangle + 0.000\,004|a_2\rangle + 0.01|a_3\rangle$$
$$|c_2\rangle = 25|a_2\rangle - 0.01|a_3\rangle$$
$$|c_3\rangle \perp |c_1\rangle, |c_2\rangle$$

$$|d_1\rangle \approx |a_1\rangle + 0.037\,3205|a_2\rangle + 0.01|a_3\rangle$$
$$|d_2\rangle \approx 0.002\,679\,42|a_2\rangle - 0.01|a_3\rangle$$
$$|d_3\rangle \perp |d_1\rangle, |d_2\rangle$$

$$|w_1\rangle = |c_1\rangle + |c_2\rangle = |a_1\rangle + 25.000\,004|a_2\rangle$$
$$|w_2\rangle \approx |d_1\rangle + |d_2\rangle = |a_1\rangle - 0.039\,999|a_2\rangle$$
$$|w_3\rangle = |a_3\rangle$$

These four orthogonal triples of vectors are all 'close to' the $|a\rangle$-triple, in the sense that they are all related by small rotations relative to the $|a\rangle$-triple. (The largest angles are between the $|d\rangle$-triple and the $|a_2\rangle$-triple: the angle between $|d_1\rangle$ and $|a_1\rangle$ is about 3°, and the angle between $|d_2\rangle$ and $-|a_3\rangle$ is about 15°. The angles between $|b\rangle$ and $|a_1\rangle$ and between $|c_1\rangle$ and $|a_1\rangle$, are less than 1°.)

Suppose a non-ideal measurement correlates an 'almost orthogonal' triple of vectors $\{|a_1'\rangle,|a_2'\rangle,|a_3'\rangle\}$ with indicator observable readings, and the $|a'\rangle$-triple is 'close to' the orthogonal $|a\rangle$-triple (for example, suppose $|a_1'\rangle$ bisects the angle between $|a_1\rangle$ and $|d_1\rangle$). The $|a'\rangle$-triple will then also be 'close to' the four orthogonal triples defined by small rotations relative to the $|a\rangle$-triple. Now, the non-ideal measurement cannot be interpreted as yielding simultaneous 'error-prone' measurements of the observables associated with all five (normalized) orthogonal triples as eigenvectors. For the 'error probabilities' would then have to be represented as measures over maps that select one and only one vector in each orthogonal triple (corresponding to the actual value of the associated observable). But any such map that selects $|a_k\rangle$ will also have to select $|b_k\rangle$, for $k = 1,2,3$. For suppose the selection is defined by a map, v, that assigns a 1 or a 0 to the vectors. If two vectors in a plane are assigned a 0, then any vector in the plane is assigned a 0 (because the orthogonal vector would have to be assigned a 1). So if $v(|a_1\rangle) = 1$ and $v(|b_1\rangle) = 0$, then $v(|a_2\rangle) = v(|a_3\rangle) = 0$, and hence

$v(|c_1\rangle) = v(|b_1\rangle + \varepsilon^{-1}p|a_2\rangle) = 0$ and $v(|c_2\rangle) = v(\varepsilon p|a_2\rangle - p|a_3\rangle) = 0$. It follows that $v(|w_1\rangle) = v(|c_1\rangle + |c_2\rangle) = 0$. Similarly, $v(|w_2\rangle) = v(|d_1\rangle + |d_2\rangle) = 0$. But $v(|w_3\rangle) = v(|a_3\rangle) = 0$. So the map v violates the requirement that $v(|w_i\rangle) = 1$ for some (particular) i, if $v(|a_1\rangle) = 1$ and $v(|b_1\rangle) = 0$. But if $v(|a_1\rangle) = 0$ and $v(|b_1\rangle) = 1$, then $v(|b_2\rangle) = v(|b_3\rangle) = 0$, and since $v(|a_2\rangle) = v(|b_2\rangle) = 0$, $v(|a_3\rangle) = 1$. An analogous argument shows that this is impossible. That is, if $v(|a_3\rangle) = 1$, then we must have $v(|b_3\rangle) = 1$. So $v(|a_1\rangle) = 1$ if and only if $v(|b_1\rangle) = 1$, and hence $v(|a_3\rangle) = 1$ if and only if $v(|b_3\rangle) = 1$ (because $|a_2\rangle = |b_2\rangle$).

Since the maps represent all the different outcomes of the non-ideal measurement, all the different ways in which indicator values r_1, r_2, r_3 are coupled with (presumed definite) values for A, B, C, D, and W, it follows that the values a_i for A and b_i for B, for $i = 1, 2, 3$, would always have to co-occur together. But this is inconsistent with the fact that the 'error probabilities' for A and B have different values. Obviously, a similar construction is possible for higher dimensional Hilbert spaces, so the artificiality of the schematic representation of a non-ideal measurement here does not compromise the generality of the argument.[101]

If any one of the observables A, B, C, D, W is taken as having a determinate value in virtue of the non-ideal measurement, then all these observables must have determinate values: there is no non-arbitrary way of distinguishing one of these observables as the observable that is 'really' measured in the interaction. But since all these observables cannot have determinate values simultaneously, by the above argument, none of the observables – indeterminate before the measurement, on the orthodox interpretation – can come to have a determinate value as a result of the non-ideal measurement.

Since all real measurements in quantum mechanics are essentially non-ideal, an observable that is indeterminate in a certain quantum state cannot become determinate as the result of a non-ideal measurement. What a quantum mechanical measurement does is to reproduce, more or less precisely, the probabilities associated with an observable of the measured system in the probabilities of the pointer readings of the measuring instrument. An interpretation of quantum mechanics that goes beyond the orthodox interpretation and 'solves the measurement problem' must, at a minimum, guarantee that the pointer readings are determinate in any measurement process – that the probabilities of observables of the measured system are reproduced in the distribution of determinate pointer readings. The more accurately the distribution of pointer readings reproduces the probabilities of measured system observables, the closer the measurement is to an ideal measurement. In the limiting case where the distribution of pointer readings exactly matches the probabilities of a system observable, we can regard the observable as having a value (because the set of such observables will all be representable as functions of a single observable). So we can

[101] Note that Bell's objection (1966) to his own version of the Kochen and Specker argument does not apply to the argument here, which involves a single simultaneous measurement of several (in principle, infinitely many) 'unsharp' observables, and the possibility of a certain interpretation of this measurement.

regard the limiting case of an ideal measurement as a dynamical process in virtue of which a measured observable comes to have a determinate value.

The methodological significance of an ideal measurement in quantum mechanics is similar to that of other ideal elements introduced in physics; for example, an ideal gas, or the straight line motion of a body not under the action of any forces in a Newtonian universe. It is only relative to the theoretical possibility of an ideal measurement that we can understand a non-ideal measurement as a *measurement* – as reproducing, more or less precisely, the probabilities of an observable of the measured system, where a precise reproduction of the probabilities in an ideal measurement would yield a value of the measured observable correlated with the value of the pointer reading.

So what we require for an observer-free 'no collapse' interpretation of quantum mechanics, is the determinateness, for every quantum state, of an appropriate observable or set of observables, subject to the constraints of the 'no go' theorems, such that every measurement process can be represented *dynamically* in the theory (with the standard linear 'no collapse' dynamics) as resulting in a distribution of values for a determinate pointer observable of the instrument that reproduces, more or less precisely, the probabilities defined by the quantum state for the measured observable. In the following section, I consider the interaction between a measuring instrument M and its environment E and show how typical environmental interactions that occur in our universe allow the possibility of determinate observables that can function as suitable pointers in measurement interactions.

5.4 Environmental 'monitoring'

An ideal measurement interaction between a system S, initially in a state $|\alpha\rangle = \Sigma_i c_i |a_i\rangle$, and a measuring instrument M, initially in an eigenstate $|r_0\rangle$ of a pointer observable R (suppressing the degeneracy index), results in the transition, after a certain time, of the form:

$$|\alpha\rangle|r_0\rangle \rightarrow |\psi\rangle = \sum_i c_i |a_i\rangle|r_i\rangle \tag{5.26}$$

A non-ideal measurement results in a transition of the form:

$$|\alpha\rangle|r_0\rangle \rightarrow |\psi'\rangle = \sum_{kj} c_{kj} |a_k\rangle|r_j\rangle = \sum_j c_j' |a_j'\rangle|r_j\rangle, \tag{5.27}$$

where the relative states of S, $|a_j'\rangle = \Sigma_k (c_{kj}/c_j')|a_k\rangle$ are 'close to' eigenstates $|a_j\rangle$ of A and $|c_j'|^2 = \Sigma_k |c_{kj}|^2 \approx |c_j|^2$. (See (5.19) and (5.21).) Equivalently:

$$|\alpha\rangle|r_0\rangle \rightarrow |\psi'\rangle = \sum_{kj} c_{kj} |a_k\rangle|r_j\rangle = \sum_k c_k'' |a_k\rangle|r_k'\rangle, \tag{5.28}$$

where the relative states of M, $|r_k'\rangle = \Sigma_j(c_{kj}/c_k'')|r_j\rangle$ are 'close to' eigenstates $|r_k\rangle$ of R and $|c_k''|^2 = \Sigma_j|c_{kj}|^2 \approx |c_k|^2$. (See (5.20) and (5.22).)

Consider first, for simplicity, an ideal measurement of a spin-$\frac{1}{2}$ system initially in the state:

$$|\alpha\rangle = \tfrac{1}{\sqrt{2}}|z+\rangle + \tfrac{1}{\sqrt{2}}|z-\rangle, \tag{5.29}$$

where $|z+\rangle$ and $|z-\rangle$ are the two eigenstates of σ_z (that is, S is initially in the state $|\alpha\rangle = |x+\rangle$, the 'up' eigenstate of σ_x). Suppose S interacts with a second system M, whose states are also represented on a 2-dimensional Hilbert space, so that $S+M$ ends up, after a certain time, in an entangled state of the form:

$$|\psi\rangle = \tfrac{1}{\sqrt{2}}|z+\rangle|r_{z+}\rangle + \tfrac{1}{\sqrt{2}}|z-\rangle|r_{z-}\rangle, \tag{5.30}$$

We can think of M as a two-state quantum measuring instrument, with $|r_{z+}\rangle$ and $|r_{z-}\rangle$ representing the two eigenstates of a pointer observable R_z of M.

The state $|\psi\rangle$ has the form of a biorthogonal decomposition, which is non-unique if and only if the squares of the absolute values of the coefficients are all equal.[102] Since the coefficients of $|\psi\rangle$ are both equal to $\frac{1}{\sqrt{2}}$, $|\psi\rangle$ can also be expressed, for example, as the biorthogonal decomposition:

$$|\psi\rangle = \tfrac{1}{\sqrt{2}}|x+\rangle|r_{x+}\rangle + \tfrac{1}{\sqrt{2}}|x-\rangle|r_{x-}\rangle, \tag{5.31}$$

where R_x has the eigenstates:

$$|r_{x+}\rangle = \tfrac{1}{\sqrt{2}}|r_{z+}\rangle + \tfrac{1}{\sqrt{2}}|r_{z-}\rangle$$
$$|r_{x-}\rangle = -\tfrac{1}{\sqrt{2}}|r_{z+}\rangle + \tfrac{1}{\sqrt{2}}|r_{z-}\rangle$$

So certain entangled states of $S+M$ can define a correlation between σ_z-values and the values of an observable R_z of M, and at the same time a similar correlation between σ_x-values and the values of an observable R_x of M, where R_x does not commute with R_z. If we want to regard $|\psi\rangle$ as representing a state of affairs where the 'instrument' M has measured the value of σ_z, then M must equally be regarded as measuring the value of σ_x.

Consider, now, an interaction between M and a third system E, whose states are also represented on a 2-dimensional Hilbert space, that results in the transition, after a certain time, to a pure state of the form:

$$|\Psi\rangle = \tfrac{1}{\sqrt{2}}|z+\rangle|r_{z+}\rangle|\varepsilon_{z+}\rangle + \tfrac{1}{\sqrt{2}}|z-\rangle|r_{z-}\rangle|\varepsilon_{z-}\rangle, \tag{5.32}$$

where $|\varepsilon_{z+}\rangle$ and $|\varepsilon_{z-}\rangle$ are two orthogonal states that span the Hilbert space of E. The form of the state $|\Psi\rangle$ as a triorthogonal decomposition is *unique*: if the interaction with the 'environment' system E preserves the correlation between σ_z-eigenstates and eigenstates of R_z expressed by the biorthogonal decomposition (5.30), then:

$$|\Psi\rangle \neq \tfrac{1}{\sqrt{2}}|x+\rangle|r_{x+}\rangle|\varepsilon_{x+}\rangle + \tfrac{1}{\sqrt{2}}|x-\rangle|r_{x-}\rangle|\varepsilon_{x-}\rangle \tag{5.33}$$

[102] This follows from the biorthogonal decomposition theorem. See appendix, section A.5.

for *any* two distinct states $|\varepsilon_{x+}\rangle \neq |\varepsilon_{z+}\rangle$ and $|\varepsilon_{x-}\rangle \neq |\varepsilon_{z-}\rangle$ of E (orthogonal or non-orthogonal). So the correlation between σ_x-eigenstates and eigenstates of R_x is lost after the interaction with the system E. This follows from the triorthogonal decomposition theorem, which guarantees the uniqueness of the form (5.32) if the sets of states $\{|z+\rangle, |z-\rangle\}$, $\{|r_{z+}\rangle, |r_{z-}\rangle\}$, and $\{|\varepsilon_{z+}\rangle, |\varepsilon_{z-}\rangle\}$ are mutually orthogonal in their respective Hilbert spaces. The triorthogonal decomposition theorem generalizes to a 'tridecompositional' theorem, which guarantees the uniqueness of the form (5.32) if the sets of states $\{|z+\rangle, |z-\rangle\}$ and $\{|r_{z+}\rangle, |r_{z-}\rangle\}$ are even linearly independent (and not necessarily orthogonal) in the Hilbert spaces of S and M, respectively, and $|\varepsilon_{z+}\rangle, |\varepsilon_{z-}\rangle$ are *any* states that do not span the same ray in the Hilbert space of E (that is, noncollinear states that do not differ by a phase difference only). Finally, the tridecompositional theorem generalizes to an n-decompositional theorem, so if E is a composite system consisting of n subsystems, each with its own 2-dimensional Hilbert space, and the interaction preserves the correlations between eigenstates of σ_z and R_z expressed by a biorthogonal decomposition of $|\psi\rangle$ with respect to states $|\varepsilon_{z+}\rangle, |\varepsilon_{z-}\rangle$ that are n-product states, it will not preserve any other correlations defined by other biorthogonal decompositions of $|\psi\rangle$. (For a proof of the theorem, see the following section.)

Apart from inessential phase differences in the pure states, this is essentially the example considered by Zurek (1981) to illustrate the origin of 'environment-induced decoherence.' Zurek considers a reversible Stern–Gerlach experiment, in which the Stern–Gerlach magnets are arranged to first split and then recombine a beam of spin-$\frac{1}{2}$ particles. A bistable atom that interacts in a suitable way with the spin functions as a quantum measuring instrument if it is inserted along one of the two possible trajectories in the apparatus. Two states, $|r_{z+}\rangle$ and $|r_{z-}\rangle$ of the atom M, become correlated with spin states $|z+\rangle$ and $|z-\rangle$ of a spin-$\frac{1}{2}$ particle S passing through the reversible Stern–Gerlach apparatus, so that when the particle leaves the apparatus, the particle + atom pair (the system $S + M$) is in the state:

$$|\psi\rangle = (-\tfrac{i}{\sqrt{2}}|z+\rangle|r_{z+}\rangle + \tfrac{1}{\sqrt{2}}|z-\rangle|r_{z-}\rangle)\phi(q,t),$$

where $\phi(q,t)$ is a narrow wave packet representing the position of the particle after it leaves the apparatus. Ignoring the spatio-temporal component of the total state, this is the same as the state $|\psi\rangle$ in equation (5.30) above, except for the phase relations between the component product vectors in the biorthogonal decomposition (the squares of the absolute values of the coefficients are the same: $\frac{1}{2}$).

Zurek notes that the atom M contains information about the spin of the particle S in different directions in the reversible Stern–Gerlach apparatus, because we can express $|\psi\rangle$ as a biorthogonal decomposition with respect to σ_y, or for that matter with respect to σ_x, for an appropriate M-observable R:

$$|\psi\rangle = -\tfrac{i}{\sqrt{2}}|x+\rangle|r+\rangle - \tfrac{i}{\sqrt{2}}|x-\rangle|r-\rangle,$$

where the eigenstates of R are:

$$|r+\rangle = \tfrac{1}{\sqrt{2}}|r_{z+}\rangle + \tfrac{1}{\sqrt{2}}|r_{z-}\rangle$$
$$|r-\rangle = \tfrac{1}{\sqrt{2}}|r_{z+}\rangle - \tfrac{1}{\sqrt{2}}|r_{z-}\rangle$$

After the spin-$\frac{1}{2}$ particle S has passed through the reversible Stern–Gerlach apparatus (when the upper and the lower beams have recombined), we can choose to measure either the observable R_z or the observable R on the atom M, and so obtain information about the spin of S in either the z-direction or the x-direction (even though the apparatus is set up to measure z-spin and not x-spin).

This is no longer possible if the atom M interacts with the environment E in a specific way. Zurek takes the environment of $S + M$ as another bistable atom. The interaction Hamiltonian between M and E is assumed to commute with the observable R_z, and not the observable R. After a certain time, the state of $S + M + E$ is represented by the triorthogonal decomposition:

$$|\Psi\rangle = -\tfrac{1}{\sqrt{2}}|z+\rangle|r_{z+}\rangle|\varepsilon_{z+}\rangle + \tfrac{1}{\sqrt{2}}|z-\rangle|r_{z-}\rangle|\varepsilon_{z-}\rangle$$

for orthogonal states $|\varepsilon_{z+}\rangle$ and $|\varepsilon_{z-}\rangle$, which is the same as the state $|\Psi\rangle$ in equation (5.32), except for the phase relations between the two components.

Zurek remarks (1981, p. 1522):

Because of the correlations with the environment, knowing the state of the apparatus [the systems M] in the [$|r+\rangle$, $|r-\rangle$] basis does not suffice any more to determine the state of the system. Part of the information about the state of the spin has been 'transferred' from the apparatus to the environment. And both the environment and the apparatus are correlated with [$|z+\rangle$] or [$|z-\rangle$] states of the spin.

We can therefore conclude that when the environment atom is present and interacts with the apparatus via [the appropriate interaction Hamiltonian] . . . , then the apparatus–spin system will retain perfect correlation in only one product basis [$|z+\rangle|r_{z+}\rangle$, $|z-\rangle|r_{z-}\rangle$] of the direct-product space. Hence, [$|r_{z+}\rangle$, $|r_{z-}\rangle$] is the pointer basis of the apparatus, which will eventually appear on the diagonal of the density matrix obtained by tracing out 'environmental degrees of freedom,' i.e., the state of the environment atom. Measurements made by the apparatus on a spin eigenfunction along any other direction are to some degree obliterated by the interaction with the environment. In particular, no information about the orientation of the spin in the direction of the [x]-axis can be derived from the state of the apparatus alone.

There are two distinct features of environmental interaction that are alluded to here. Firstly, Zurek points out that a *specific* interaction between a measuring instrument M and its environment E, an interaction characterized by a Hamiltonian that commutes with a certain M-observable designated as the pointer observable, will preserve the correlations between eigenstates of a particular S-observable and eigenstates of the pointer observable, established by an initial ideal measurement interaction between S and M, and will not preserve correlations between any other S-observable and M-observable (not commuting with the 'preferred' observables selected by the environment) that might arise through the measurement interaction. This feature of composite systems depends on the triorthogonal decomposition theorem, a special case of the tridecomposition theorem (or n-decomposition theorem), and generalizes to non-ideal measurements, because the tridecomposition theorem requires only that the states of S correlated with the pointer states of M should be linearly independent rather than strictly orthogonal (which will be the case for the non-orthogonal S-states 'close to' the eigenstates of the measured S-observable correlated with pointer states in a non-ideal measurement).

The second feature concerns the diagonalization of the reduced density matrix of the system $S + M$ in the $(S + M)$-basis selected (via the triorthogonal decomposition theorem) by the environmental interaction in the case of an ideal measurement, or the diagonalization of the density matrix by 'tracing out environmental degrees of freedom.' In the case of a non-ideal measurement, or when the environmental states are non-orthogonal, the diagonalization is approximate in the $(S + M)$-basis selected by the tridecompositional theorem. Zurek's claim is that the pointer observable is picked out as effectively determinate by the interaction with the environment, *because* this environmental 'monitoring' of the measuring instrument leads to almost instantaneous 'decoherence' or loss of interference between different pointer states: an effective 'collapse' of the state onto a tensor product of a pointer state and correlated measured observable eigenstate.

In chapter 8, I argue that environment-induced decoherence, in the sense of a loss of interference terms exhibited by the reduced density operator of $S + M$ in a certain representation, obtained by 'tracing out' the environment, does not resolve the measurement problem. But the fact that 'the environment acts, in effect, as an observer continuously monitoring certain preferred observables which are selected mainly by the system–environment interaction hamiltonians' (Zurek, 1993b, p. 290) *is* crucially important to an interpretation of quantum mechanics able to account for measurements in a satisfactory way. It is not decoherence that is relevant, but the tridecomposition or n-decomposition theorem, and the stochastic evolution induced by environmental monitoring for the actual values of an appropriately selected preferred determinate observable, in the sense of the uniqueness theorem of chapter 4. What we require for an interpretation of quantum mechanics is the stipulation of a *suitable* preferred determinate observable or set of observables that will yield an interpretation of the processes we regard as measurements in terms of the distributions of determinate pointer readings. The sort of environmental monitoring that happens to occur in our universe guarantees the existence of such preferred determinate observables, and also narrows the range of possible choices. So it is a contingent matter that measurements are possible in our quantum universe. But environmental monitoring cannot make an indeterminate observable determinate through the process of decoherence.

Consider again the non-ideal measurement interaction described by equation (5.28a):

$$|\alpha\rangle|r_0\rangle \to |\psi'(t)\rangle = \sum_{kj} c_{kj}(t)|a_k\rangle|r_j\rangle = \sum_j c_j'(t)|a_j'(t)\rangle|r_j\rangle,$$

where I have explicitly indicated the time dependence of the states $|a_j'(t)\rangle$ 'close to' the eigenstates of the measured observable A, the 'almost orthogonal' relative states of the pointer states $|r_j\rangle$. If the interaction between the measuring instrument and the environment E is induced by an interaction Hamiltonian that commutes with the pointer observable R, the state of the composite system $S + M + E$ will evolve to a tridecomposition:

$$|\alpha\rangle|r_0\rangle|\varepsilon_0\rangle \to |\psi'(t)\rangle|\varepsilon_0\rangle \to |\Psi'(t)\rangle = \sum_j c_j'(t)|a_j'(t)\rangle|r_j\rangle|\varepsilon_j(t)\rangle \tag{5.34}$$

Sophisticated models of the instrument–environment interaction treat the environment as a collection of harmonic oscillators or as a quantum field. Here the state $|\varepsilon_0\rangle$ is some initial state of the environment E, and the states $|\varepsilon_j(t)\rangle$ are non-orthogonal states that almost instantaneously approach orthogonality, typically on a time scale that is less than 10^{-23} seconds for systems that are well isolated from their environments, and orders of magnitude less than that for the typical macrosystems we use as measuring instruments (see section 8.1, and Zurek, 1991, p. 41).

This environmental monitoring has two important consequences for the possibility of measurements. Firstly, the correlation between pointer states and relative states of the measured system S, brought about by a non-ideal measurement, is preserved. By the tridecomposition theorem, since the states $|a_j'(t)\rangle$ are linearly independent even if non-orthogonal, and the pointer states are orthogonal, this preservation of measurement correlations occurs only for a pointer observable R that commutes with the instrument–environment interaction Hamiltonian. Of course, the same thing will occur for an ideal measurement, where the states $|a_j\rangle$ of S are orthogonal. It follows that if we select R as the preferred determinate observable, or an observable in terms of which R is defined by coarse-graining, then the distribution of pointer readings in an ideal or non-ideal measurement interaction will continue to reproduce, more or less precisely, the distribution of values of the measured observable specified by the original quantum state of the system, even after the instrument interacts with the environment. With any other choice for the preferred determinate observable, the instrument–environment interaction will almost instantaneously destroy correlations between the values of a system observable and values of the preferred observable induced by an initial measurement interaction.

Secondly, environmental monitoring will have the effect of rapidly reducing the probability of an anomalous jump between different pointer readings after a non-ideal measurement interaction, resulting from the stochastic evolution of the actual value of R, if we select R as the preferred determinate observable, or an observable in terms of which R is defined by coarse-graining.

To see this, recall the stochastic equation of motion for the preferred determinate observable R given in section 5.2. The jumps in R-values are governed by transition probabilities $T_{mn}dt$, where $T_{mn}dt$ denotes the probability of a jump from value r_n to value r_m in time dt. From equation (5.8):

$$T_{nm} = \frac{J_{nm}}{\hbar P_m}, \text{ if } J_{nm} > 0, \text{ and } T_{nm} = 0, \text{ if } J_{nm} \leq 0$$

For a composite system $S + M + E$, the pointer R is nonmaximal in the Hilbert space $\mathcal{H}_S \otimes \mathcal{H}_M \otimes \mathcal{H}_E$ of $S + M + E$. After a non-ideal measurement between S and M, $J_{nm}(t)$ takes the form (from (5.5), for a nonmaximal R):

$$J_{nm}(t) = 2\mathrm{Im} \sum_{uu'vv'} (\langle\Psi'(t)|\gamma_{unv}\rangle\langle\gamma_{unv}|H|\gamma_{u'mv'}\rangle\langle\gamma_{u'mv'}|\Psi'(t)\rangle), \tag{5.35}$$

where $|\gamma_{unv}\rangle = |a_u\rangle|r_n\rangle|\eta_v\rangle$ and $\{|a_u\rangle\}$, $\{|r_n\rangle\}$, $\{|\eta_v\rangle\}$ are orthonormal basis sets in the Hilbert spaces $\mathcal{H}_S$, $\mathcal{H}_M$, and $\mathcal{H}_E$, respectively (suppressing the degeneracy index of R-eigenstates in the R-eigenspaces of $\mathcal{H}_M$), and the summation is over degeneracy indices in the Hilbert spaces $\mathcal{H}_S$ and $\mathcal{H}_E$. Here $|\Psi'(t)\rangle = \Sigma_j c_j'(t)|a_j'(t)\rangle|r_j\rangle|\varepsilon_j(t)\rangle$ and H is the total Hamiltonian after the measurement interaction (with the interaction between S and M 'switched off'):

$$\begin{aligned} H &= H_{S+M} \otimes I_E + H_{int} \\ &= H_S \otimes I_M \otimes I_E + I_S \otimes H_M \otimes I_E + H_{int}, \end{aligned}$$

where H_{int} is the Hamiltonian of the interaction between M and the environment E.

Now, H_{int} commutes with R, and so is diagonal in the R-indices m, n. It follows that H_{int} does not induce transitions between different R-values, because $\langle\gamma_{unv}|H|\gamma_{u'mv'}\rangle = 0$ for $n \neq m$. Similarly, the Hamiltonian $H_S \otimes I_M \otimes I_E$ does not contribute to the probability of a transition between different R-values. The matrix element

$$\langle\Psi'(t)|\gamma_{unv}\rangle\langle\gamma_{unv}|I_S \otimes H_M \otimes I_E|\gamma_{u'mv'}\rangle\langle\gamma_{u'mv'}|\Psi'(t)\rangle$$

becomes

$$\begin{aligned} &\delta_{uu'}\delta_{vv'}\langle\Psi'(t)|\gamma_{unv}\rangle\langle r_n|H_M|r_m\rangle\langle\gamma_{u'mv'}|\Psi'(t)\rangle \\ &= \delta_{uu'}\delta_{vv'}c_n'^* c_m'\langle a_n'(t)|a_u\rangle\langle a_{u'}|a_m'(t)\rangle\langle\varepsilon_n(t)|\eta_v\rangle\langle\eta_{v'}|\varepsilon_m(t)\rangle\langle r_n|H_M|r_m\rangle \end{aligned}$$

Summing over the indices u, u', v, v' yields:

$$J_{nm} = 2\mathrm{Im}(c_n'^* c_m'\langle a_n'(t)|a_m'(t)\rangle\langle\varepsilon_n(t)|\varepsilon_m(t)\rangle\langle r_n|H|r_m\rangle) \tag{5.36a}$$

or (explicitly indicating the degeneracy of R-eigenstates in the Hilbert space $\mathcal{H}_M$):

$$J_{nm} = 2\mathrm{Im}[\sum_{pq} c_n'^* c_m'\langle a_n'(t)|a_m'(t)\rangle\langle\varepsilon_n(t)|\varepsilon_m(t)\rangle\langle r_n, p|H|r_m, q\rangle] \tag{5.36b}$$

Since $\langle\varepsilon_n(t)|\varepsilon_m(t)\rangle$ decays to zero almost instantaneously if $m \neq n$, the transition probability of a jump between different pointer values becomes negligible almost instantaneously, even though $\langle a_n'(t)|a_m'(t)\rangle \neq 0$. That is, the transition probability, which is proportional to $\langle a_n'(t)|a_m'(t)\rangle$ after a non-ideal measurement interaction between S and M, and is 'close to' zero if the relative states $|a_j'\rangle$ are 'almost orthogonal,' is rapidly reduced by the term $\langle \varepsilon_n(t)|\varepsilon_m(t)\rangle$ after the instrument–environment interaction. (Evidently, the probability of a jump between different pointer readings is zero after an ideal measurement, where $\langle a_n(t)|a_m(t)\rangle = 0$.)

Suppose we take R as (discretized) position in configuration space, and the pointer observable as a nonmaximal position observable in $\mathscr{H}_M$ related to R by an appropriate coarse-graining. A non-ideal measurement between S and M can be represented (via (5.28)) as correlating eigenstates of a measured observable A of S with 'almost-orthogonal' states of M (as in the representation of the Stern–Gerlach measurement in the previous section, where the two sharply peaked wave functions $\phi_+ = \phi(q - gt)$ and $\phi_- = \phi(q + gt)$ in (5.23) can be regarded as linear superpositions of position eigenstates). As we saw in section 5.2, for potentials that are functions of position (and not nth order derivatives of position), the Hamiltonian of $S + M$ induces only nearest-neighbour transitions in R-space, so the probability of a jump between R-values associated with different pointer readings (which correspond to macroscopically distinct R-values) is negligible. Position enjoys a privileged status as a preferred determinate observable, because a position pointer is stable, in the sense that anomalous jumps in pointer readings after a measurement are highly improbable, even without environmental monitoring. This feature of position is a consequence of a contingent fact about our universe: the ubiquity of potential functions of a certain form for the systems we use as measuring instruments. If the interaction with the environment is induced by an interaction Hamiltonian that commutes with position, the very small probability of an anomalous pointer-reading jump is reduced virtually instantaneously by many orders of magnitude to negligible levels.

Schematically, in terms of a discretized position observable R, the state of $S + M + E$ at time $t = 0$ immediately after a non-ideal measurement can be expressed as:

$$|\Psi'(0)\rangle = \sum_k c_k''|a_k\rangle|r_k'(t)\rangle|\varepsilon(0)\rangle,$$

where the 'almost-orthogonal' states $|r_k'(t)\rangle$ (linear superpositions of R-eigenstates, $|r_i\rangle$) correspond to the macroscopically separated Gaussians in equation (5.23), and $|\varepsilon(0)\rangle$ is the state of the environment at $t = 0$. After the interaction with the environment, if the interaction Hamiltonian commutes with R, the state of $S + M + E$ takes the form:

$$|\Psi(t)\rangle = \sum_k c_k''|a_k\rangle|\chi_k(t)\rangle, \tag{5.37}$$

where the states $|\chi_k(t)\rangle$ are entangled states of $M + E$ that rapidly approach orthogonality.

As before, the only component of the total Hamiltonian that induces transitions between different R-values after the measurement is the component $I_S \otimes H_M \otimes I_E$. The matrix element

$$\langle \Psi'(t)|\gamma_{unv}\rangle\langle\gamma_{unv}|I_S \otimes H_M \otimes I_E|\gamma_{u'mv'}\rangle\langle\gamma_{u'mv'}|\Psi'(t)\rangle \tag{5.38}$$

becomes

$$\begin{aligned}&\delta_{uu'}\delta_{vv'}\langle\Psi'(t)|\gamma_{unv}\rangle\langle r_n|H_M|r_m\rangle\langle\gamma_{u'mv'}|\Psi'(t)\rangle\\ &= \delta_{uu'}\delta_{vv'}c_u''^{*}c_{u'}''\langle\chi_u(t)|r_n\rangle|\eta_v\rangle\langle\eta_{v'}|\langle r_m|\chi_{u'}(t)\rangle\langle r_n|H_M|r_m\rangle\end{aligned}$$

Summing over the degeneracy indices u, u', v, v' in the Hilbert spaces $\mathcal{H}_S$ and $\mathcal{H}_E$ yields:

$$J_{nm} = 2\mathrm{Im}\left[\sum_u |c_u''|^2\left(\sum_v \langle\chi_u(t)|r_n\rangle|\eta_v\rangle\langle\eta_v|\langle r_m|\chi_u(t)\rangle\right)\langle r_n|H_M|r_m\rangle\right] \tag{5.39}$$

Since the entangled $(M + E)$-states take the form $|\chi_u(t)\rangle = \Sigma_j d_{u_j}|r_j\rangle|\varepsilon_{u_j}(t)\rangle$ after environmental monitoring, where the $|\varepsilon_{u_j}(t)\rangle$ are environmental states that rapidly approach orthogonality, J_{nm} can be expressed as:

$$\begin{aligned}J_{nm} &= 2\mathrm{Im}\left[\sum_u |c_u''|^2\left(\sum_v d_{u_n}{}^{*}d_{u_m}\langle\varepsilon_{u_n}(t)|\eta_v\rangle\langle\eta_v|\varepsilon_{u_m}(t)\rangle\right)\langle r_n|H_M|r_m\rangle\right]\\ &= 2\mathrm{Im}\left(\sum_u |c_u''|^2 d_{u_n}{}^{*}d_{u_m}\langle\varepsilon_{u_n}(t)|\varepsilon_{u_m}(t)\rangle\langle r_n|H_M|r_m\rangle\right)\end{aligned} \tag{5.40}$$

Here the indices m and n refer to eigenstates of the discretized position observable R, not to different pointer values. The term $\langle r_n|H_M|r_m\rangle$ induces nearest-neighbour transitions in the value of R only, and we see that the probability of these transitions is reduced by the factor $\langle\varepsilon_{u_n}(t)|\varepsilon_{u_m}(t)\rangle$, which decays to zero almost instantaneously if $m \neq n$.

The choice of a preferred determinate observable R for an observer-free 'no collapse' interpretation of quantum mechanics that can account for measurements is constrained by the nature of the interaction between open systems and their environments in our universe. In chapter 6, I examine two interpretations that adopt very different proposals for the preferred determinate observable: Bohm's causal interpretation, which designates position in configuration space as a preferred always-determinate observable, and the modal interpretation (in a version generalized from earlier formulations by Kochen, 1985, and by Dieks, 1988, 1989a), in which the preferred determinate observable is time dependent and derived from the quantum state. From the above analysis, Bohm's choice would appear to be a viable option, because open systems typically interact with their environments through interaction potentials that depend on distance. Such interactions will preserve the correlations generated by ideal or non-ideal measurements for suitably coarse-grained position pointers, and also ensure the virtual impossibility of anomalous jumps in pointer readings after a measurement. But a more coarse-grained choice for the preferred determinate observable such as fermion number density might also be an option. As Bell (1987, p. 175) pointed out: 'The distribution of fermion number in the world certainly includes the positions of instruments, instrument pointers, ink on paper, . . . and much much more.'

5.5 Proof of the tridecompositional theorem

By the biorthogonal decomposition theorem, the representation of a state $|\psi\rangle \in \mathcal{H} \otimes \mathcal{H}'$ in the form:

$$|\psi\rangle = \sum_i c_i |u_i\rangle |v_i\rangle$$

always exists for some basis sets $\{|u_i\rangle\} \in \mathcal{H}$ and $\{|v_i\rangle\} \in \mathcal{H}'$, and is unique if and only if the $|c_i|^2$ are all distinct. The representation of a state $|\psi\rangle \in \mathcal{H} \otimes \mathcal{H}' \otimes \mathcal{H}''$ as a triorthogonal decomposition:

$$|\psi\rangle = \sum_i c_i |u_i\rangle |v_i\rangle |w_i\rangle$$

does not always exist, but if such a decomposition does exist, it is unique without any condition on the coefficients. The proof of uniqueness for triorthogonal decompositions appears in Elby and Bub (1994), and follows from the uniqueness of the more general 'tridecompositions,' also proved there. See Clifton (1995a) for an elegant reformulation and strengthening of the result. The following proof incorporates some of the features of both proofs.

Lemma 1: If $\{|u_i\rangle\}$ and $\{|u_j'\rangle\}$ are two linearly independent sets of unit vectors (not necessarily orthogonal) in a Hilbert space $\mathcal{H}_1$, and $|\psi\rangle \in \mathcal{H}_1 \otimes \mathcal{H}_2$ takes the form:

$$|\psi\rangle = \sum_i c_i |u_i\rangle |v_i\rangle = \sum_j d_j |u_j'\rangle |v_j'\rangle$$

for any sets of unit vectors $\{|v_i\rangle\}$ and $\{|v_j'\rangle\}$ in $\mathcal{H}_2$, then the sets $\{|v_i\rangle\}$ and $\{|v_j'\rangle\}$ span the same subspace in $\mathcal{H}_2$.

Proof: Denote the subspace of $\mathcal{H}_2$ spanned by the set of vectors $\{|v_i\rangle\}$ by $\mathcal{V}$. For any vector $v^\perp$ orthogonal to $\mathcal{V}$, we have:

$$\begin{aligned} \langle v^\perp|\psi\rangle &= \sum_i c_i \langle v^\perp|v_i\rangle |u_i\rangle \\ &= 0 \\ &= \sum_j d_j \langle v^\perp|v_j'\rangle |u_j'\rangle \end{aligned}$$

Since the $|u_i'\rangle$ are linearly independent and the d_j are nonzero, it follows that:

$$\langle v^\perp|v_j'\rangle = 0, \text{ for all } j,$$

which means that each vector $|v_j'\rangle$ is orthogonal to every vector $v^\perp$ in $\mathcal{V}$, and so the vectors $\{|v_j'\rangle\}$ span the subspace $\mathcal{V}$.

Lemma 2: If $\{|u_i\rangle\}$ is a linearly independent set of unit vectors in a Hilbert space $\mathscr{H}_1$, and $|\psi\rangle \in \mathscr{H}_1 \otimes \mathscr{H}_2$ can be represented as:

$$|\psi\rangle = \sum_i c_i|u_i\rangle|v_i\rangle = \sum_j d_j|u_j\rangle|v_j'\rangle,$$

for any sets of unit vectors $\{|v_i\rangle\}$ and $\{|v_i'\rangle\}$ in $\mathscr{H}_2$, then $c_i|v_i\rangle = d_i|v_i'\rangle$, for all i (that is, the 'relative states' $\{c_i|v_i\rangle\}$ in $\mathscr{H}_2$, with respect to a given set of linearly independent unit vectors $\{|u_i\rangle\}$ in $\mathscr{H}_1$, are unique).

Proof: Supposc:

$$|\psi\rangle = \sum_i c_i|u_i\rangle|v_i\rangle = \sum_j d_j|u_j\rangle|v_j'\rangle,$$

For any vector $|v\rangle \in \mathscr{H}_2$, we have:

$$\langle v|\psi\rangle = \sum_i c_i\langle v|v_i\rangle|u_i\rangle = \sum_j d_j\langle v|v_j'\rangle|u_j\rangle$$

and hence:

$$\sum_i (c_i\langle v|v_i\rangle - d_i\langle v|v_i'\rangle)|u_i\rangle = 0$$

By linear independence of the $\{|u_i\rangle\}$:

$$c_i\langle v|v_i\rangle = d_i\langle v|v_i'\rangle, \text{ for all } i,$$

and and since this equality holds for any vectors $|v\rangle \in \mathscr{H}_2$, it follows that $c_i|v_i\rangle = d_i|v_i'\rangle$, for all i.

Lemma 3: If $\{|u_i\rangle\}$ is a linearly independent set of vectors in $\mathscr{H}_1$ and $\{|v_i\rangle\}$ is any set of mutually noncollinear unit vectors in $\mathscr{H}_2$ (any set of unit vectors such that no two vectors span the same ray in $\mathscr{H}_2$), then no product vector $|\psi\rangle|\phi\rangle \in \mathscr{H}_1 \otimes \mathscr{H}_2$ can be represented nontrivially in the form:

$$|\psi\rangle|\phi\rangle = \sum_i c_i|u_i\rangle|v_i\rangle$$

Proof: For any vector $|\phi^\perp\rangle \in \mathscr{H}_2$ orthogonal to $|\phi\rangle$, taking the partial scalar product with $|\psi\rangle|\phi\rangle$ yields:

$$\langle\phi^\perp|\psi\rangle|\phi\rangle = 0 = \sum_i c_i\langle\phi^\perp|v_i\rangle|u_i\rangle$$

By linear independence of the $\{|u_i\rangle\}$:

$$\langle\phi^{\perp}|v_i\rangle = 0, \text{ for all } i,$$

and since $|\phi^{\perp}\rangle$ is any vector orthogonal to $|\phi\rangle$, this means that each vector $|v_i\rangle$ is orthogonal to every vector that is orthogonal to $|\phi\rangle$, which is only possible if $|v_i\rangle = k_i|\phi\rangle$, for all i, contrary to the assumption that the product $|\psi\rangle|\phi\rangle$ can be represented as a nontrivial sum.

Corollary: If $\{|u_i\rangle\}$ and $\{|v_i\rangle\}$ are linearly independent sets of unit vectors in $\mathscr{H}_1$ and $\mathscr{H}_2$, respectively, and $\{|w_i\rangle\}$ is any set of mutually noncollinear unit vectors in $\mathscr{H}_3$, then no product vector $|\psi\rangle|\phi\rangle|\xi\rangle \in \mathscr{H}_1 \otimes \mathscr{H}_2 \otimes \mathscr{H}_3$ can be represented nontrivially in the form:

$$|\psi\rangle|\phi\rangle|\xi\rangle = \sum_i c_i|u_i\rangle|v_i\rangle|w_i\rangle$$

Evidently, the result extends to tensor products of n vectors, for any n.

Tridecompositional uniqueness theorem: If $\{|u_i\rangle\} \in \mathscr{H}_1$ and $\{|v_i\rangle\} \in \mathscr{H}_2$ are linearly independent sets of unit vectors in their respective Hilbert spaces, and $\{|w_i\rangle\} \in \mathscr{H}_3$ is any set of mutually noncollinear unit vectors, then the representation of $|\psi\rangle \in \mathscr{H}_1 \otimes \mathscr{H}_2 \otimes \mathscr{H}_3$ in the form of a tridecomposition:

$$|\psi\rangle = \sum_i c_i|u_i\rangle|v_i\rangle|w_i\rangle$$

is unique if it exists.

Proof: Suppose:

$$|\psi\rangle = \sum_i c_i|u_i\rangle|v_i\rangle|w_i\rangle$$

$$= \sum_j d_j|u_j'\rangle|v_j'\rangle|w_j'\rangle$$

where $\{|u_j'\rangle\} \in \mathscr{H}_1$ and $\{|v_j'\rangle\} \in \mathscr{H}_2$ are also linearly independent sets of unit vectors, and $\{|w_j'\rangle\} \in \mathscr{H}_3$ is any set of mutually noncollinear unit vectors, and the two representations differ nontrivially.

The sets $\{|u_i\rangle|w_i\rangle\}$ and $\{|u_i'\rangle|w_i'\rangle\}$ are linearly independent if $\{|u_i\rangle\}$ and $\{|u_i'\rangle\}$ are linearly independent. By Lemma 1 applied to $|\psi\rangle$ as a linear superposition of tensor biproducts with respect to the subspaces $\mathscr{H}_2$ and $\mathscr{H}_1 \otimes \mathscr{H}_3$, the linearly independent sets $\{|v_i\rangle\}$ and $\{|v_i'\rangle\}$ span the same subspace. It follows that we can represent $|v_j'\rangle$ as:

$$|v_j'\rangle = \sum_k e_{jk}|v_k\rangle$$

So $|\psi\rangle$ can be represented as:

$$|\psi\rangle = \sum_k \sum_j d_j e_{jk} |u_j'\rangle |v_k\rangle |w_j'\rangle$$

By Lemma 2 (again applied to the representation of $|\psi\rangle$ as a sum of tensor biproducts of vectors from the Hilbert spaces $\mathscr{H}_2$ and $\mathscr{H}_1 \otimes \mathscr{H}_3$), since:

$$|\psi\rangle = \sum_i c_i |u_i\rangle |v_i\rangle |w_i\rangle$$

$$= \sum_k \sum_j d_j e_{jk} |u_j'\rangle |v_k\rangle |w_j'\rangle$$

it follows that:

$$c_i |u_i\rangle |w_i\rangle = \sum_j d_j e_{ji} |u_j'\rangle |w_j'\rangle,$$

which is impossible by Lemma 3, since the product vector $|u_i\rangle|w_i\rangle$ would then be representable nontrivially in the form:

$$|u_i\rangle |w_i\rangle = \sum_j \frac{d_j e_{ji}}{c_j} |u_j'\rangle |w_j'\rangle$$

with respect to the linearly independent set $\{|u_j'\rangle\}$ in $\mathscr{H}_2$.

Corollary: Evidently, a similar result holds for the uniqueness of n-decompositions, if they exist.

Corollary: The uniqueness of triorthogonal decompositions follows for the special case where the linearly independent sets $\{|u_i\rangle\}$ and $\{|v_i\rangle\}$ are orthogonal.

QED

6

Quantum mechanics without observers: II

> Not all 'observables' can be given beable status, for they do not all have simultaneous eigenvalues, i.e. do not all commute. It is important to realize therefore that most of these 'observables' are entirely redundant. What is essential is to be able to define the positions of things, including the positions of instrument pointers or (the modern equivalent) of ink on computer output.
>
> *Bell (1987, p. 175)*

6.1 Bohmian mechanics

On Bohm's 1952 hidden variable theory or 'causal' interpretation of quantum mechanics – now often referred to as 'Bohmian mechanics' – a quantum world consists of particles that always have determinate positions, and a wave function ψ interpreted as a 'guiding field' that propagates in the configuration space of the particles according to Schrödinger's equation of motion. The particles change their positions over time according to the deterministic equation of motion (5.9b), which reflects the evolution of ψ in configuration space.

As Bell puts it (1987, p. 128; italics in the original):

No one can understand this theory until he is willing to think of ψ as a real objective field rather than just a 'probability amplitude'. Even though it propagates not in 3-space but in 3N-space.

This field is (p. 128) 'just as "real" and "objective" as say the fields of classical Maxwell theory – although its action on the particles [via equation (5.9b)] is rather original.' So the only real change in a Bohmian quantum world is the change in ψ as a field in configuration space, and the change in the positions of the particles. The only real properties are the position properties of particles, and these positions change as the particles are pushed and pulled in various ways by the evolving ψ-field. Other properties, such as spin properties, play no rôle in the dynamical evolution of the trajectory in configuration space, which depends entirely on the initial configuration space position and the ψ-field. On this view, what we call a 'measurement' of a spin component of an electron in a Stern–Gerlach measurement, say, simply catalogues a certain characteristic movement of a particle in the presence of an inhomogeneous magnetic field, which reflects a particular way in which a certain sort of ψ-field can evolve in configuration space and guide the motion of particles. This suffices to account for all quantum phenomena, insofar as these phenomena can be characterized by changes in the positions of particles (or changes in macroscopic position variables defined as coarse-grained functions over fine-grained positions in configuration space: instrument readings, image densities on photographic plates, and so forth).

This is the formulation of Bohmian mechanics favoured by authors such as Bell (1987, pp. 117–38, 159–68), Dürr, Goldstein, and Zanghì (1990, 1992a,b,c, 1993), Albert (1992), Cushing (1994), and Maudlin (1995a). Other authors, such as Bohm, Schiller, and Tiomno (1955), Dewdney (1992), Holland and Vigier (1988), and Bohm and Hiley (1993) treat spin as well as position in configuration space as observables that always have determinate values. A spin-component observable can take determinate values that are not eigenvalues of spin in states that are not spin eigenstates, but the account of spin measurements shows that the measured value of spin is always an eigenvalue of spin. This treatment of spin could, in principle, be extended to other observables.

There is an alternative way of interpreting Bohmian mechanics: as an observer-free 'no collapse' interpretation in the sense of chapter 4, in which the preferred determinate observable R is fixed once and for all as position in configuration space. In section 5.2 I showed that the deterministic equation of motion (5.9a,b) for position in configuration space in Bohmian mechanics arises as the limit of a stochastic evolution of property states defined by 2-valued homomorphisms (with nonzero measure) on the determinate sublattice $\mathscr{L}_{e_{r_1}e_{r_2}\ldots e_{r_k}}$, generated by a discretized configuration observable R, as the quantum state e evolves dynamically in time. On this view, as e evolves dynamically in time, it determines, at each instant, a Bohmian determinate sublattice, defined by position in configuration space as the only preferred, always-determinate observable (apart from functions of position). For convenience, we can think of this observable as the continuum limit of a discretized position observable. While position in configuration space is always determinate, other observables are sometimes determinate and sometimes indeterminate, depending on the quantum state (understood as a dynamical state). The determinate observables at a particular time t are associated with the *possible* properties of the system at time t – the collection of properties that define the space of alternatives at t, some subcollection of which is actual at t. The dynamical evolution of the quantum state tracks the evolution of what is possible – not what is actual – as given by the properties in the dynamically evolving Bohmian determinate sublattice, and the evolution of the probabilities defined by the quantum state over these properties. The actual property state at time t, defined by a particular 2-valued homomorphism on the determinate sublattice at time t, evolves stochastically over time (or deterministically, in the continuum limit) so as to preserve the dynamical evolution of probabilities.

There is a difference between the ψ-field of quantum mechanics and the electromagnetic field of Maxwell's theory: the action of the ψ-field on the particles is indeed 'rather original.' It is not simply that the ψ-field propagates in $3N$-dimensional configuration space (for an N-particle system), rather than in ordinary 3-space. The more fundamental difference is that the change in the quantum state $|\psi\rangle$ manifests itself directly at a *modal* level – the level of possibility rather than actuality – through the determinate sublattice defined by $|\psi\rangle$ and position in configuration space as the preferred determinate observable. What changes when the electromagnetic field changes are, roughly, the actual pushes and pulls on the particles in 3-space. What changes when $|\psi\rangle$ changes is a sublattice of alternative possibilities, with probabilities defined over these possibilities. These possibilities and probabilities in turn constrain what actually occurs in a quantum world.

If we take the preferred observable R as continuous position in configuration space then, as we saw in section 5.3, all measurements with R as pointer are essentially non-ideal (because such measurement interactions correlate eigenstates of a measured observable with wave functions that are not literally 'position eigenstates,' but are only relatively localized in configuration space). But a Bohmian particle always has a determinate position in configuration space, and insofar as the particle position functions as a pointer in measurement interactions, we can take the distribution of particle positions in a spin component measurement, say, as reflecting, more or less precisely, the probabilities specified by the initial quantum state for the eigenvalues of the spin component. Moreover, we can understand a non-ideal measurement of spin as yielding an 'effective collapse' of the linear superposition of product states, representing the final state after the measurement interaction, to a product state of a relatively localized wave function in configuration space and a spin eigenstate.

To illustrate, consider the measurement of spin-related observables on Bohm's theory (following an analysis by Pagonis and Clifton, 1995).

Let $S_x^2, S_{x'}^2, S_y^2, S_{y'}^2, S_z^2$ represent the squared components of spin in the x, x', y, y', and z directions of a spin-1 particle, respectively, where x, y, z and x', y', z form two orthogonal triples of directions with the z-direction in common. Each of these observables has eigenvalues 1 and 0 (taking units in which $\hbar = 1$ and a spin component has eigenvalues -1, 0, and $+1$), corresponding respectively to a plane and a ray in the Hilbert space $\mathscr{H}_3$. The three 0-eigenrays of S_x^2, S_y^2, S_z^2 form an orthogonal triple in $\mathscr{H}_3$, and the three 0-eigenrays of $S_{x'}^2, S_{y'}^2, S_z^2$ form another orthogonal triple in $\mathscr{H}_3$, with the 0-eigenray of S_z^2 in common.[103]

Define the observables H and H' as:

$$\begin{aligned} H &= S_x^2 - S_y^2 \\ H' &= S_{x'}^2 - S_{y'}^2 \end{aligned} \tag{6.1}$$

The observables H and H' are maximal, with three eigenvalues, -1, 0, and $+1$, and incompatible (that is, the corresponding operators do not commute). The eigenvalues -1, 0, and $+1$ of H correspond to the orthogonal triple of eigenrays defined by the

[103] See the appendix, section A.4.

0-eigenrays of S_x^2, S_z^2, and S_y^2, respectively. The eigenvalues -1, 0, and $+1$ of H' correspond to the orthogonal triple of eigenrays defined by the 0-eigenrays of $S_{x'}^2$, S_z^2, and $S_{y'}^2$, respectively. The nonmaximal observable S_z^2 can be represented as a function of H and also as a function of H':

$$S_z^2 = H^2 = H'^2 \tag{6.2}$$

So S_z^2 can be measured via a measurement of H or of H'.

The observable H can be measured, in principle, by passing the particle through a suitable inhomogeneous electromagnetic field, which functions much like a Stern–Gerlach magnet for the measurement of spin (see Swift and Wright, 1980). The Hilbert space of the particle is a tensor product of an infinite-dimensional Hilbert space, $\mathscr{H}$, for the representation of spatial motion, and a 3-dimensional Hilbert space, $\mathscr{H}'$, for the representation of spin observables. In the position representation of $\mathscr{H}$, the interaction of the particle with the field will be governed by a Hamiltonian of the form:

$$H_{int} = -g(t)\mathrm{i}\frac{\partial}{\partial q}H$$

where $g(t)$ is a positive coupling constant that is nonzero only during the interaction, and q is a component of the particle's position. If we assume, as in the analysis of non-ideal measurements in section 5.3, that the measurement is impulsive and $g(t) = g$ during the interaction, the Schrödinger equation reduces to

$$\mathrm{i}\frac{\partial\psi}{\partial t} = H_{int}\psi$$

during the interaction, and has the solution:

$$\psi(q,t) = \exp(-g\frac{\partial}{\partial q}Ht)\psi(q,0),$$

where $\psi(q,t)$ represents the quantum state in the position representation in $\mathscr{H}$.[104] Taking the initial quantum state as

$$\psi(q,0) = \phi(q)\sum_n c_n|H = n\rangle,$$

where $\phi(q)$ is a narrow wave packet symmetric about $q = 0$, and $|H = n\rangle$ is an eigenstate of H (for $n = -1, 0, +1$), the state at any time t during the interaction is:

$$\psi(q,t) = \sum_n c_n\phi(q - gnt)|H = n\rangle \tag{6.3}$$

With a suitable choice for the coupling constant g, after a time $t \geq T$ (at the end of the

[104] So $\psi(q,t)$ represents the quantum state in a mixed notation that in general entangles wave functions in $\mathscr{H}$ with Hilbert space vectors in $\mathscr{H}'$.

interaction), gt will be significantly larger than the width of the packet $\phi(q)$, so that the overlap between adjacent wave packets $\phi(q - gnt)$ is negligible. As a result of the interaction, the particle's H-value will therefore effectively become correlated with the particle's q-position.

The Bohmian particle trajectories during the interaction are governed by the equation of motion for q:

$$\frac{\mathrm{d}q}{\mathrm{d}t} = \frac{J_q}{\rho},$$

where $\rho(q) = |\psi(q,t)|^2$ is the probability density, and $J_q = \psi^*(q,t)gH\psi(q,t)$ is the q-component of the probability current. So:

$$\frac{\mathrm{d}q}{\mathrm{d}t} = \frac{g\sum_n n|c_n|^2|\phi(q - gnt)|^2}{\sum_n |c_n|^2|\phi(q - gnt)|^2} \tag{6.4}$$

This equation can be solved to yield different trajectories for the different initial positions of the particle in the initial wave packet. By the continuity equation (5.3a), $\partial\rho/\partial t + \partial J_q/\partial q = 0$, the particle trajectories at time $t \geq T$ will be distributed over the positions $q = gnt$ with probabilities that are close to $|c_n|^2$, for $n = -1, 0, +1$ (to the extent that the overlap between adjacent wave packets is small). So q acts as a measurement pointer for H-values. A displacement of the particle from its initial position in the narrow packet $\phi(q)$ centered about $q = 0$ by an amount gnt as the particle leaves the field can be understood as a measurement of H with the outcome $H = n$.

In the discrete version of Bohm's theory, the state of the particle develops sharp peaks in R-space, with negligible overlap between peaks, where each peak is associated with the product of an H eigenstate and a function in R-space. The trajectory of the particle in R-space follows one particular peak, with negligible probability (in any time period during which the peaks remain sharp and maintain their macroscopic separation) of a transition to an R-value associated with a different peak. As in the discussion of the Stern–Gerlach measurement of spin as a non-ideal measurement in section 5.3 this is because the stochastic equation of motion for the actual value of R induces transitions between neighbouring R-values only, so a change of R-value associated with macroscopically separated sharp peaks would require many nearest-neighbour transitions through a region in which the transition matrix is effectively zero. The transition probabilities are almost instantaneously reduced still further by environmental monitoring, as shown in section 5.4. Since different H-eigenstates are correlated with values of R associated with different peaks, and all but one peak can be neglected for the evolution of R-values (the relevant peak depending on the initial R-value when the particle enters the field), the effect is as if the state collapses onto a tensor product of an eigenstate of H and a relatively localized wave function in configuration space.

Now, the same analysis for H' instead of H will yield an equation of motion for the particle trajectories in terms of a position coordinate q':

$$\frac{dq'}{dt} = \frac{J_{q'}}{\rho}$$

$$= \frac{g\sum_n n|c_n'|^2|\phi(q' - gnt)|^2}{\sum_n |c_n'|^2|\phi(q' - gnt)|^2}, \tag{6.5}$$

where the initial state of the particle is:

$$\psi(q',0) = \phi(q')\sum_n c_n'|H' = n\rangle$$

and q' is the position coordinate of the particle that becomes correlated with the value of H' during the interaction (in which the particle is passed through an electromagnetic field oriented in a direction suitable for a measurement of $H' = S_{x'}^2 - S_{y'}^2$ rather than $H = S_x^2 - S_y^2$). Assume, for simplicity, that the form of the initial position wave function is the same for q' as for q, and that the strength, g, and duration, T, of the interaction is the same for H' as for H.

Distinct possible particle trajectories cannot cross along the q-axis in an H-measurement, or the q'-axis in an H'-measurement, because the equation of motion is deterministic and velocity changes only as a function of position. This means that in an H-measurement, trajectories that end up, after time $t \geq T$, at one of the three possible final q-positions, $-gt$, 0, or gt, in that order, begin in one of three possible q-regions in the initial wave packet $\phi(q)$, ordered from negative to positive values of q. The relative sizes of the q-regions in an H-measurement will differ from the relative sizes of the q'-regions in an H'-measurement, because fractions $|c_{-1}|^2$, $|c_0|^2$, $|c_{+1}|^2$ of the initial q-positions (in an ensemble distributed according to $|\phi(q)|^2$) end up at the positions $q = -gt, q = 0, q = gt$, respectively (corresponding to the results $H = -1, 0, +1$ in an H-measurement), while *different* fractions $|c_{-1}'|^2$, $|c_0'|^2$, $|c_{+1}'|^2$ of the initial positions (in an ensemble distributed according to the same distribution function $|\phi(q')|^2$) end up at the positions $q' = -gt$, $q' = 0$, $q' = gt$, respectively (corresponding to the results $H' = -1, 0, +1$ in an H'-measurement). Note that $|c_n|^2 \neq |c_n'|^2$, for $n = -1, 0, +1$, unless $H = H'$. It follows that a given initial position of the particle will be affected differently by different interaction Hamiltonians, and so a measurement of S_z^2 via an H-measurement need not yield the same result as a measurement of S_z^2 via an H'-measurement, for the same initial position and quantum state of the particle.

In this sense, Bohm's theory violates the Kochen and Specker functional relationship constraint. If values are assigned to all observables of a spin-1 particle as the values that *would be* obtained on measurement (where a measurement of an observable is understood as an evolution of the quantum state of the particle to a form that correlates position values with values of the observable in the above sense) then, for some initial positions of the particle, it won't be the case that the value assigned to S_z^2 is equal to the square of the value assigned to H^2 and *also* equal to the square of the value assigned to H'^2. For example, suppose the initial position of the particle is such that a measurement of H would yield the value $+1$, while a measurement of H' would yield the value 0. If these values are assigned to H and H', then the functional relationship constraint requires that $S_z^2 = 1$ and also that $S_z^2 = 0$, since $S_z^2 = H^2 = H'^2$. Of course, this contradiction does not show any inconsistency in Bohmian mechanics. Rather, it shows that Bohmian mechanics is 'contextual,' in the sense that determinate values can be attributed to 'nonpreferred' observables like H, H', and S_z^2 only with respect to a specific determinate sublattice. Since the determinate sublattice depends on the quantum state, and the measurement of an observable involves the dynamical evolution of the quantum state to a certain form, the measured value of an observable like S_z^2 will depend on what other observables are measured together with S_z^2.

In a similar sense, Bohm's theory is nonlocal. Consider two spin-$\frac{1}{2}$ particles, S_1 and S_2, in the singlet spin state at time $t = 0$:

$$\psi(q_1,q_2,0) = \phi(q_1)\phi(q_2)(\tfrac{1}{\sqrt{2}}|+\rangle_1|-\rangle_2 - \tfrac{1}{\sqrt{2}}|-\rangle_1|+\rangle_2) \tag{6.6}$$

A measurement of spin in the z-direction on S_1 (via a Stern–Gerlach interaction at S_1 that correlates the position of S_1 with the z-spin of S_1) induces an evolution of the quantum state of the two-particle system to a form:

$$\psi(q_1,q_2,t) = [\tfrac{1}{\sqrt{2}}\phi(q_1 - gt)|+\rangle_1|-\rangle_2 - \tfrac{1}{\sqrt{2}}\phi(q_1 + gt)|-\rangle_1|+\rangle_2]\phi(q_2), \tag{6.7}$$

where the wave packets $\phi_+(q_1) = \phi(q_1 - gt)$ and $\phi_-(q_1) = \phi(q_1 + gt)$ for the relevant position coordinates of S_1 are separated with negligible overlap (cf. (5.23)). The q_1 position coordinate of the particle S_1 is effectively associated with just one of these wave packets. This means that the position of the two-particle system $S_1 + S_2$ in configuration space – the q_1, q_2 position (ignoring other position coordinates not affected by the Stern–Gerlach interaction at S_1) – can be associated with just one of the wave packets $\phi_+(q_1)\phi(q_2)$ or $\phi_-(q_1)\phi(q_2)$, assuming we can neglect the overlap. So in a subsequent measurement of the spin of S_2 via a Stern–Gerlach interaction at S_2, the evolution of q_2 will depend on the configuration space position of $S_1 + S_2$, which is effectively correlated either with the spin state $|+\rangle_1|-\rangle_2$ or with the spin state $|-\rangle_1|+\rangle_2$. It follows that $S_1 + S_2$ will behave in the S_2-interaction as if its quantum state has effectively collapsed to $\phi_+(q_1)\phi(q_2)|+\rangle_1|-\rangle_2$ or to $\phi_-(q_1)\phi(q_2)|-\rangle_1|+\rangle_2$, and so q_2 will evolve to a position associated with a z-spin value for S_2 that is oppositely correlated with the z-spin value associated with the final position of q_1. Evidently, then, the outcome of a spin measurement at S_2 will depend on the orientation of the Stern–Gerlach magnet that determines the type of spin measurement at S_1, because this will determine how the quantum state of $S_1 + S_2$ evolves into a linear superposition of two states with effectively disjoint supports in configuration space. The quantum state of $S_1 + S_2$ will evolve differently for a measurement of x-spin at S_1 rather than z-spin, say, and so will affect the evolution of q_2 differently in a subsequent spin measurement at S_2, as in the measurement of H and H' on the spin-1 system.[105]

There is another sense, though, in which both the functional relationship constraint and the locality condition are satisfied by Bohm's causal interpretation. Consider the spin-1 system again, and the quantum state:

$$\psi(q,t) = \sum_n c_n \phi(q - gnt)|H = n\rangle$$

at times $t \geq T$ after the H-measurement interaction. Both the observables H and S_z^2 belong to the determinate sublattice defined by the quantum state $\psi(q,t)$ for position as the preferred determinate observable, to the extent that the wave functions $\phi(q - gnt)$ can be regarded as orthogonal eigenfunctions. Of course, the wave function tails will always overlap to some small extent, so even if we partition the position observable Q to three disjoint ranges of values associated with the three peaks of the wave functions, the projections of the state $e(\psi(q,t))$ onto the three corresponding eigenspaces will not yield the rays e_{q_n} spanned by the vectors $\phi(q - gnt)\,|\,H = n\rangle$, for $n = -1, 0, +1$, but only rays e'_{q_n} that are 'close to' the rays e_{q_n} (for suitable values of t).

A more precise treatment would follow the analysis of non-ideal measurements in section 5.3. In their analysis of measurement, Bohm and Hiley (1984; 1993, pp. 97–133) simply treat wave functions like $\phi(q - gnt)$ as non-overlapping and effectively eigenfunctions of q, as in an ideal measurement with a discrete pointer Q. Then H-propositions and S_z^2-propositions belong to the lattice $\mathscr{L}_{e_{q_{-1}}e_{q_0}e_{q_{+1}}}$. By the uniqueness theorem, $\mathscr{L}_{e_{q_{-1}}e_{q_0}e_{q_{+1}}}$ is the maximal set of propositions that are determinately true or false for the state $\psi(q,t)$, taking Q as the preferred determinate observable, and contains propositions associated with the observables that are measured via the pointer position q. For the state $\psi(q,t)$, observables like H and S_z^2 associated with $\mathscr{L}_{e_{q_{-1}}e_{q_0}e_{q_{+1}}}$ satisfy the functional relationship constraint. The value assigned to S_z^2 is the square of the value assigned to H, for any q-value, and similarly for other observables associated with $\mathscr{L}_{e_{q_{-1}}e_{q_0}e_{q_{+1}}}$. ($H'$-propositions do not, of course, belong to $\mathscr{L}_{e_{q_{-1}}e_{q_0}e_{q_{+1}}}$.) So, if we consider the set of observables that are measured ideally by the evolution of the quantum state to a specific form in which distinct q-values become correlated with distinct values of these observables, for a discrete preferred determinate pointer observable Q, then the functional relationship constraint is satisfied for these observables. It follows that the locality condition is satisfied in *this* sense, for composite systems associated with tensor product Hilbert spaces, because locality is a special case of the functional relationship constraint.

Bohm and Hiley (1993, p. 41) characterize Bohmian mechanics as a reformulation of quantum mechanics in terms of 'beables':

> This theory is formulated basically in terms of what Bell has called 'beables' rather than of 'observables.' These beables are assumed to have a reality that is independent of being observed

[105] For an illuminating account of measurement in Bohm's theory, see Albert (1992) or Maudlin (1995a).

or known in any other way. The observables therefore do not have a fundamental significance in our theory but rather are treated as statistical functions of the beables that are involved in what is currently called a measurement.

For Bell (1987, p. 174), the 'beables' of a physical theory correspond to possible 'elements of reality,' in the terminology employed in the Einstein–Podolsky–Rosen argument.

In particular we will exclude the notion of 'observable' in favour of that of '*be*able.' The beables of the theory are those elements which might correspond to elements of reality, to things which exist. Their existence does not depend on 'observation.' Indeed observation and observers must be made out of beables.

Bohm and Hiley (1993, p. 120) take the beables of Bohmian mechanics as 'the overall wave function together with the coordinates of the particles,' and Bell (1987, p. 176) adopts a similar position in his extension of Bohmian mechanics to fermion fields: 'The lattice fermion numbers are the local beables of the theory, being associated with definite positions in space. The state vector $|t\rangle$ also we consider as a beable, although not a local one.' On the interpretation of Bohmian mechanics proposed here, though, the beables are the properties (or the corresponding idempotent observables) in the determinate sublattice defined at time t by the quantum state and position in configuration space as the preferred determinate observable. The difference between classical and quantum mechanics is that a classical world is characterized by a fixed Boolean lattice of beables for all time, while the non-Boolean lattice of beables for a quantum world evolves dynamically, so that different beables apply at different times, depending on the quantum state.

From the discussion in chapter 2, we know that a beable interpretation of quantum mechanics cannot satisfy all of Einstein's realist requirements, as encapsulated in his principles of separability and locality. Referring to the Einstein–Podolsky–Rosen argument, Einstein writes as follows (Schilpp, 1949, pp. 681–2):

Of the 'orthodox' quantum theoreticians whose position I know, Niels Bohr's seems to me to come nearest to doing justice to the problem. Translated into my own way of putting it, he argues as follows:

If the partial systems A and B form a total system which is described by its ψ-function $\psi/(AB)$, there is no reason why any mutually independent existence (state of reality) should be ascribed to the partial systems A and B viewed separately, *not even if the partial systems are spatially separated from each other at the particular time under consideration.* The assertion that, in this latter case, the real situation of B could not be (directly) influenced by any measurement taken on A is, therefore, within the framework of quantum theory, unfounded and (as the paradox shows) unacceptable.

By this way of looking at the matter it becomes evident that the paradox forces us to relinquish one of the following two assertions:

(1) the description by means of the ψ-function is *complete*

(2) the real states of spatially separated objects are independent of each other.

On the other hand, it is possible to adhere to (2), if one regards the ψ-function as the description of a (statistical) ensemble of systems (and therefore relinquishes (1)). However, this view blasts the framework of the 'orthodox quantum theory.'

The 'real state' or 'real situation' of a physical system here corresponds to Einstein's notion of the 'being-thus' of a system (Einstein, 1948, p. 320; 1971, p. 192). For a 'no collapse' interpretation in the sense of chapter 4, it corresponds to the properties selected by a particular 2-valued homomorphism on the determinate sublattice defined by the quantum state of the system and the preferred determinate observable. Each 2-valued homomorphism selects a particular value for the preferred observable, and also particular values for other determinate observables associated with the determinate sublattice. That is, each 2-valued homomorphism at time t selects a maximal set of determinate properties for the system, and some particular 2-valued homomorphism represents the 'being-thus' or actual properties of the system at time t.

The separability and locality principles can therefore be formulated as follows:

Separability: The actual properties ('real states') of spatially separated systems are independent of each other.

Locality: If two systems are spatially separated, then the actual properties ('real state') of one system cannot be directly influenced by any measurement on the other system.

As Einstein presents it, the issue of the completeness of quantum mechanics – the heart of the dispute between Einstein and Bohr – concerns the separability principle. The determinate sublattices preserve only a weak separability principle: the determinate properties (and hence the 'real states') of spatially separated systems are independent of each other (that is, each system is independently characterized by its own determinate sublattice), if and only if the quantum state of the composite system is *not* an entangled state (a linear superposition of product states) arising from past interaction between the systems. What the uniqueness theorem shows is that the possible 'completions' of quantum mechanics in Einstein's sense can be uniquely characterized and reduced to the choice of a fixed determinate observable, or a fixed beable. So, in fact, the option of preserving separability in the strong sense is excluded in a beable interpretation.

As we saw for Bohmian mechanics, the determinate sublattices of a 'no collapse' interpretation satisfy the Kochen and Specker functional relationship constraint in what might be termed an 'ontological' sense. That is, the values of the determinate observables associated with a determinate sublattice, as assigned by all 2-valued homomorphisms on the lattice, preserve the functional relationships satisfied by these observables, and hence preserve locality as a special case of the functional relation constraint. But in what might be termed a 'dynamical' sense, the 'no collapse' interpretations associated with determinate sublattices are nonlocal. If we understand an ideal measurement as an interaction that induces an evolution of the quantum state to a form that generates a determinate sublattice, in which certain observables become determinate, with determinate values correlated to the values of the preferred determinate observable functioning as measurement pointer, then an ideal measurement on a system S_1 can make determinate a property of a system S_2, spatially separated from S_1, that was not determinate before the measurement. In this sense, Einstein's locality principle is violated. The 'real states' of one system can be influenced by a measurement on another spatially separated system, through the alteration of what is possible – what can be – for the first system.

6.2 The modal interpretation

The idea behind a 'modal' interpretation of quantum mechanics is that quantum states, unlike classical states, constrain possibilities rather than actualities – which leaves open the question of whether one can introduce property states, or 'value states' in van Fraassen's terminology (1991, p. 275), that attribute values to (some) observables of the theory, or equivalently, truth values to the corresponding propositions. As van Fraassen puts it (1991, p. 279):

> In other words, the [quantum] state delimits what can and cannot occur, and how likely it is – it delimits possibility, impossibility, and probability of occurrence – but does not say what actually occurs.

Apart from van Fraassen's original version of the modal interpretation (1973, 1974, 1981, 1991), there are now a variety of other observer-free 'no collapse' interpretations of quantum mechanics that can be seen as modal in this sense; for example, the interpretations of Kochen (1985), Krips (1987), Healey (1989, 1995, 1996), Dieks (1988, 1989a, 1994a,b), Bub (1992a, 1994b), Clifton (1995b), Vermaas and Dieks (1995) and, of course, Bohmian mechanics. All these modal interpretations share with van Fraassen's interpretation the feature that an observable can have a determinate value even if the quantum state is not an eigenstate of the observable, so they preserve the linear, unitary dynamics for quantum states without requiring the projection postulate to validate the determinateness of pointer readings and measured observable values in quantum measurement processes. I have adopted this perspective in my analysis of determinate sublattices and the measurement problem in chapters 4 and 5.

The modal interpretations of Kochen, Dieks, and Healey exploit the biorthogonal (Schmidt) decomposition theorem to define value states. According to this theorem, any pure quantum state $|e\rangle$ of a system $S + S^*$ can be expressed in the form:

$$|e\rangle = \sum c_i |u_i\rangle |v_i\rangle \tag{6.8}$$

for some orthonormal set of vectors $\{|u_i\rangle\}$ in $\mathscr{H}(S)$ and some orthonormal set $\{|v_i\rangle\}$ in $\mathscr{H}(S^*)$. The decomposition is unique if and only if $|c_i|^2 \neq |c_j|^2$ for all $i \neq j$. In the non-degenerate case, the basic idea is to take the propositions that are determinately true or false for S in the quantum state $|e\rangle$ as the propositions represented by the Boolean algebra of projection operators generated by the set $\{P_{u_i}\}$. (Similar remarks apply to S^*, of course.) There are alternative proposals for the degenerate case, and for the ascription of properties to composite systems and their component subsystems. Healey, in particular, has a rather elaborate set of rules for property ascription to composite systems, involving biorthogonal decomposition as well as permissive and prohibitive composition rules.

Recent reformulations by Clifton (1995b) and Vermaas and Dieks (1995) exploit the uniqueness of the spectral resolution of the reduced state of a system S, obtained by partial tracing over the environment E of the system, to characterize the determinate properties of S at any time. In Clifton's formulation, if W is the reduced state of a system S (pure or mixed), the determinate properties of S are the properties represented by the projection operators in the set:

$$\mathscr{D}_{\text{modal}}(S) = \{P\colon PP_{w_i} = P_{w_i} \text{or } 0, \text{ for all } P_{w_i} \in SR(W)\}, \tag{6.9}$$

where $SR(W)$ is the set of projection operators defined by the spectral resolution of W (that is, the projection operators onto the eigenspaces corresponding to the nonzero eigenvalues of W). It is easy to see that $\mathscr{D}_{\text{modal}}(S)$ forms a sublattice of $\mathscr{L}(S)$. Note, in particular, that if W is a projection operator representing a pure state e, all the properties in the orthodox sublattice $\mathscr{L}_e$ belong to $\mathscr{D}_{\text{modal}}(S)$. The probability of a property P in $\mathscr{D}_{\text{modal}}(S)$ is defined as $\text{tr}(WP)$. If S is a composite system composed of subsystems, there are further rules for the joint probabilities of properties of different subsystems, where these probabilities are derived from the reduced states of the subsystems.

To see how the reformulation is related to the original formulation in terms of the biorthogonal decomposition theorem, consider a system composed of two subsystems in the pure state (6.8), where $|c_i|^2 \neq |c_j|^2$ if $i \neq j$. Taking the partial trace over the system S^* yields the density operator:

$$W = \sum_i |c_i|^2 P_{u_i}$$

representing the reduced state of S. Since the spectral resolution of W is unique, $SR(W) = \{P_{u_i}, \text{ for all } i\}$. So the properties of S selected via the biorthogonal decomposition theorem applied to the composite system $S + S^*$ are also selected by the new rule applied to the reduced state of S alone.

It is now straightforward to prove a 'recovery theorem' for this formulation of the modal interpretation, in terms of the framework for 'no collapse' interpretations provided by the uniqueness theorem of chapter 4 (see the following section). The determinate sublattices of S, as a subsystem of the system $S + E$, are the sublattices $\mathscr{D}(e, W \otimes I)|_S$, where '$|_S$' denotes the restriction of the sublattice to the Hilbert space $\mathscr{H}(S)$ of the subsystem S, e is the ray representing the quantum state of $S + E$, and W is the reduced state of e for the system S. Note that the preferred determinate observable $R = W \otimes I$ generates the determinate properties of S only. Different preferred determinate observables would be required to generate the determinate properties of subsystems (or supersystems) of S.

In this modal interpretation, the preferred determinate observable (for a given system) is not fixed as in Bohmian mechanics, but changes with time as the state e evolves. Bacciagaluppi (1996) and Bacciagaluppi and Dickson (1996) have proposed a stochastic evolution for property states in $\mathscr{D}_{\text{modal}}(S)$ analogous to the stochastic evolution in 'no collapse' interpretations with a fixed preferred determinate observable formulated in section 5.2. (See also Dieks, 1994a, b and Vermaas, 1996a for a different approach to the temporal evolution of property states in $\mathscr{D}_{\text{modal}}(S)$.) Bacciagaluppi and Hemmo (1996) and Bacciagaluppi, Elby, and Hemmo (1996) have shown how one can get a satisfactory analysis of quantum measurements in this modal framework, by exploiting the instrument–environment interaction to account for the determinateness of pointer readings and the determinateness of observers' beliefs in ideal or non-ideal measurements. I shall sketch the analysis by Bacciagaluppi and Hemmo of the rôle of environmental 'monitoring' in non-ideal measurements, and show how it relates to the similar discussion in section 5.4 for a 'no collapse' interpretation with a fixed preferred determinate observable.

After a non-ideal measurement between a system S, initially in a state $\Sigma_i c_i|a_i\rangle$, and a measuring instrument M, initially in a ready state $|r_0\rangle$, the state of $S + M$ takes the form (5.19):

$$|\psi'\rangle = \sum_j c_j'|a_j'\rangle|r_j\rangle,$$

where $|a_j'\rangle = \Sigma_k(c_{k_j}/c_j')|a_k\rangle$ are 'almost orthogonal' relative states of S, 'close to' the eigenstates $|a_j\rangle$ of an S-observable A, and $|c_j'|^2 = \Sigma_k|c_{k_j}|^2 \approx |c_j|^2$. Tracing over the system S yields the reduced density operator for M:

$$\begin{aligned} W &= \sum_k \langle a_k|\psi'\rangle\langle\psi'|a_k\rangle \\ &= \sum_k \langle a_k|\Big(\sum_{ij} c_i'c_j'^*|a_i'\rangle\langle a_j'| \otimes |r_i\rangle\langle r_j|\Big)|a_k\rangle \\ &= \sum_{ij} c_i'c_j'^*\Big(\sum_k \langle a_k|a_i'\rangle\langle a_j'|a_k\rangle\Big)|r_i\rangle\langle r_j| \\ &= \sum_{ij} c_i'c_j'^*\langle a_j'|a_i'\rangle|r_i\rangle\langle r_j| \end{aligned} \qquad (6.10)$$

(Note that $|a_i'\rangle = \Sigma_k\langle a_k|a_i'\rangle|a_k\rangle$, and so $\langle a_j'|a_i'\rangle = \langle a_j'|\Sigma_k\langle a_k|a_i'\rangle|a_k\rangle = \Sigma_k\langle a_k|a_i'\rangle\langle a_j'|a_k\rangle$.)

The diagonal terms in the matrix of W – the iith terms, for all i – are equal to $|c_i'|^2$ (since $\langle a_i'|a_i'\rangle = 1$). The off-diagonal terms – the ijth terms, for all $i \neq j$ – are equal to $c_i'c_j'^*\langle a_j'|a_i'\rangle$. Since the relative states $|a_i'\rangle$ are 'almost orthogonal,' the off-diagonal terms are close to zero. The question is: how close? If they are not close enough, the modal interpretation is subject to an objection raised by Albert and Loewer (1990, 1993). If the diagonal terms $|c_i'|^2$ are all almost equal and the off-diagonal terms are sufficiently small, W will be close to degeneracy. In such a case, the eigenvectors $|w_i\rangle$ of W can be very different from the eigenstates $|r_i\rangle$ of the pointer observable R. Since the determinate properties of M are derived from the projection operators P_{w_i} in the spectral resolution of W:

$$W = \sum_i w_i P_{w_i}$$

the pointer properties will not be determinate, nor will properties 'close to' the pointer properties be determinate.

Bacciagaluppi and Hemmo provide the following example to illustrate the point. The diagonal matrices:

$$W_1 = \begin{pmatrix} 0.99 & 0 \\ 0 & 0.01 \end{pmatrix}$$
$$W_2 = \begin{pmatrix} 0.501 & 0 \\ 0 & 0.499 \end{pmatrix}$$

have the same eigenvectors:

$$|+\rangle = \begin{pmatrix} 1 \\ 0 \end{pmatrix} \text{ and } |-\rangle = \begin{pmatrix} 0 \\ 1 \end{pmatrix}$$

but different eigenvalues. In terms of their spectral measures:

$$W_1 = 0.99|+\rangle\langle+| + 0.01|-\rangle\langle-|$$
$$W_2 = 0.501|+\rangle\langle+| + 0.499|-\rangle\langle-|$$

Although W_2 is close to degeneracy, with both eigenvalues close to 0.5, the eigenvalues of W_1 are very different. Now consider the matrices:

$$W_1' = \begin{pmatrix} 0.99 & 0.05 \\ 0.05 & 0.01 \end{pmatrix}$$
$$W_2' = \begin{pmatrix} 0.501 & 0.05 \\ 0.05 & 0.499 \end{pmatrix}$$

where W_1' and W_2' differ from W_1 and W_2, respectively, by the same small off-diagonal terms. The eigenvectors of W_1' are:

$$|+\rangle' = \begin{pmatrix} 0.9987 \\ 0.0508 \end{pmatrix} \text{ and } |-\rangle' = \begin{pmatrix} -0.0508 \\ 0.9987 \end{pmatrix}$$

and these are close to the eigenvectors $|+\rangle$ and $|-\rangle$ of W_1. But the eigenvectors of W_2' are:

$$|+\rangle'' = \begin{pmatrix} 0.7141 \\ 0.7000 \end{pmatrix} \text{ and } |-\rangle'' = \begin{pmatrix} -0.7141 \\ 0.7000 \end{pmatrix}$$

and these are clearly not at all close to the eigenvectors of W_2. In fact, $\langle +''|+\rangle = \langle +''|-\rangle = \langle -''|+\rangle = \langle -''|-\rangle \approx \frac{1}{\sqrt{2}}$, so the eigenvectors of W_2' are rotated by approximately 45° relative to the eigenvectors of W_2, even though the eigenvalues of W_2 and W_2' are close (both approximately 0.5).

After M interacts with its environment E, through an interaction Hamiltonian that commutes with the pointer observable R, the resulting state of $S + M + E$ takes the form:

$$|\Psi'(t)\rangle = \sum_i c_i'|\alpha_i\rangle|r_i\rangle|\varepsilon_i(t)\rangle$$

Tracing over $S + E$ yields the reduced density operator for M as:

$$W'(t) = \sum_{ij} c_i'c_j'^*\langle a_j'|a_i'\rangle\langle\varepsilon_j(t)|\varepsilon_i(t)\rangle|r_i\rangle\langle r_j| \tag{6.11}$$

The off-diagonal elements:

$$c_i'c_j'^*\langle a_j'|a_i'\rangle\langle\varepsilon_j(t)|\varepsilon_i(t)\rangle \tag{6.12}$$

are now reduced by the factor $\langle\varepsilon_j(t)|\varepsilon_i(t)\rangle$, which almost instantaneously decays to zero for $i \neq j$. A quantitative analysis (see Bacciagaluppi and Hemmo, 1996) shows that in typical cases this effect of environmental monitoring will very rapidly compensate for the effect of near degeneracy and ensure that the determinate properties of M are indistinguishable from the pointer properties, within the limits of vagueness associated with these properties.

The rôle of environmental monitoring in guaranteeing the determinateness of pointer properties in the modal interpretation depends on the very rapid decay of the off-diagonal elements of the reduced density matrix of M given by (6.12). For a 'no collapse' interpretation with a suitable fixed determinate observable (in virtue of which the pointer of M is determinate), environmental monitoring ensures the stability of pointer readings after a measurement: the actual value of the pointer observable, evolving stochastically, will not jump anomalously from one value to another.[106] This depends on the rapid decay of the off-diagonal elements J_{ji} in the current matrix for R given by equation (5.36a) (with an appropriate change in the indices):

$$2\mathrm{Im}(c_i'c_j'^*\langle a_j'|a_i'\rangle\langle\varepsilon_j(t)|\varepsilon_i(t)\rangle\langle r_j|H|r_i\rangle)$$

In both cases, the rapid decay of the environmental factor $\langle\varepsilon_j(t)|\varepsilon_i(t)\rangle$, for $i \neq j$, completely swamps any small effect of the non-orthogonality of the relative states $|a_i'\rangle$ that arise in non-ideal measurements.

Clifton (1996b) has pointed out that the modal interpretations of van Fraassen, Kochen, and Dieks, as well as the new formulation presented above, share an undesirable feature of 'perspectivalism.' If a system S is part of a composite system

[106] The question of stability in this sense is also relevant for the actual value of the pointer observable in the modal interpretation, which will evolve according to the stochastic modal dynamics.

$S + S^*$, then a property represented by a projection operator P on the Hilbert space $\mathscr{H}$ of S can be determinate, without the property represented by the projection operator $P \otimes I^*$ on the Hilbert space $\mathscr{H} \otimes \mathscr{H}^*$ of $S + S^*$ being determinate, and conversely. If we regard $P \otimes I^*$ as representing the same property, from the perspective of the composite system, that P represents from the perspective of the subsystem S, then whether or not this property obtains depends on whether we consider the question from the perspective of the whole composite system, or from the perspective of a subsystem.

As an example, suppose the Hilbert spaces of both S and S^* are 2-dimensional, and that the state of $S + S^*$ is:

$$|e\rangle = \tfrac{1}{2}|u_1\rangle|v_1\rangle + \tfrac{\sqrt{3}}{2}|u_2\rangle|v_2\rangle$$

Then the reduced state of S is:

$$W = \tfrac{1}{4}P_{u_1} + \tfrac{3}{4}P_{u_2}$$

So the projection operators P_{u_1} and P_{u_2} represent determinate properties of S. That is, either S has the property represented by P_{u_1} and fails to have the property represented by P_{u_2}, or S has the property represented by P_{u_2} and fails to have the property represented by P_{u_1}. But $P_{u_1} \otimes I^*$ and $P_{u_2} \otimes I^*$ are not determinate properties of $S + S^*$, as specified by equation (6.9), because:

$$(P_{u_i} \otimes I^*) \cdot P_e \neq 0$$
$$(P_{u_i} \otimes I^*) \cdot P_e \neq P_e$$

(Note that $\mathrm{tr}(WP_{u_i}) = \mathrm{tr}(P_e \cdot (P_{u_1} \otimes I^*))$, from the definition of W as the reduced state.[107] But $\mathrm{tr}(WP_{u_1}) = \frac{1}{4}$, so $\mathrm{tr}(P_e \cdot (P_{u_i} \otimes I^*)) \neq \mathrm{tr}(P_e) = 1$.) Clifton refers to this as a failure of 'property composition': there are determinate properties of subsystems (for example, P_{u_1} of S and I^* of S^*) that do not compose to yield a determinate property of the composite system $S + S^*$.

As an example of 'property decomposition' failure, Clifton considers a mixed state for $S + S^*$ represented by a reduced density operator like:

$$W_{S+S^*} = \tfrac{1}{8}P_{u_1} \otimes P_{v_1} + \tfrac{3}{8}P_{u_1} \otimes P_{v_2} + \tfrac{1}{6}P_{u_2} \otimes P_{v_1} + \tfrac{1}{3}P_{u_2} \otimes P_{v_2}$$

(supposing $S + S^*$ is a subsystem of a larger system $S + S^* + S^{**}$). The reduced density operator of S is:

$$W = \tfrac{1}{2}I$$

The only determinate properties of S are the trivial properties represented by the zero projection operator and the unit projection operator. But the composite system $S + S^*$ has nontrivial determinate properties represented by $P_{u_i} \otimes P_{v_j}$, for $i = 1,2$ and $j = 1,2$, and hence the determinate property $P_{u_1} \otimes P_{v_1} + P_{u_1} \otimes P_{v_2} = P_{u_1} \otimes (P_{v_1} + P_{v_2}) = P_{u_1} \otimes I^*$. So there are determinate 'non-entangled' properties of the composite system

[107] See the appendix, section A.5.

(for example, $P_{u_1} \otimes I^*$) that do not decompose to yield determinate properties of the subsystems.

As Clifton remarks, if a system has a determinate property only with respect to some perspective, then modal interpretations can hardly claim to provide an acceptable observer-free 'no collapse' interpretation of quantum mechanics in the realist sense. By contrast, in Bohmian mechanics, or in general in any 'no collapse' interpretation based on a determinate sublattice that is selected by a fixed preferred determinate observable, both property composition and property decomposition hold. Note, also, that property composition and decomposition hold in the orthodox interpretation. (If W_{S+S^*} is the density operator of a composite system $S + S^*$ and W the density operator of the component system S, then P is a determinate property of S on the orthodox interpretation if and only if $\text{tr}(WP) = 1$ or 0. Since $\text{tr}(WP) = 1$ or 0 if and only if $\text{tr}(W(P \otimes I^*)) = 1$ or 0, respectively, P is a determinate property of S if and only if $P \otimes I^*$ is a determinate property of $\text{S} + \text{S}^*$.)

Healey's modal interpretation avoids perspectivalism by introducing an elaborate set of conditions for attributing determinate properties to a composite system and its subsystems. Like any modal interpretation designed to avoid the measurement problem, Healey's interpretation rejects the orthodox interpretation principle, that an observable has a determinate value *if and only if* the state is an eigenstate of the observable. But Healey accepts the eigenstate to eigenvalue link, while rejecting the eigenvalue to eigenstate link. More precisely (to avoid terminological confusion), since Healey distinguishes between the 'quantum state' of a system and the 'system representatives,' where the system representatives are the spectral projections of the system's density operator, Healey accepts as part of his interpretation that if the system representative is a projection operator onto an eigenvector of an observable, then that observable has a determinate value given by the eigenvalue corresponding to the eigenvector. Clifton shows that a Kochen–Specker contradiction can be derived from the assumptions that the determinate properties of a system form a partial Boolean algebra, that property composition holds, and that the eigenstate to eigenvalue link holds (in the relevant sense). The proof exploits sixteen of the twenty 1-dimensional projectors in Kernaghan's 20-ray uncolourable configuration in $\mathcal{H}_4$ discussed in section 3.3.

To relate this result to the uniqueness theorem of chapter 4, recall that the orthodox interpretation principle selects the sublattice $\mathcal{L}_e$ of properties as determinate for a quantum system in a state represented by a ray e in a Hilbert space $\mathcal{H}$. This sublattice is maximal, in the sense that no properties can be added to $\mathcal{L}_e$ without generating a lattice, by closure, on which no 2-valued lattice homomorphisms exist, namely the lattice $\mathcal{L}(\mathcal{H})$. Now $\mathcal{L}_e$ is also a partial Boolean algebra. Requiring that the determinate properties of a system form a partial Boolean algebra, and that the eigenstate to eigenvalue link holds, is equivalent to requiring that the partial Boolean algebra of determinate properties of a system in the state e properly includes the partial Boolean algebra $\mathcal{L}_e$. Partial Boolean algebra closure is a weaker constraint than lattice closure, and surely the weakest closure constraint we would want the structure of determinate properties of a system to satisfy. We know that we can't add any properties to the partial Boolean algebra $\mathscr{L}_e$, if we impose the further constraint that the resulting set of properties is not only a partial Boolean algebra but also a lattice. There are no 2-valued homomorphisms on the resulting lattice $\mathscr{L}$, and also no *Boolean* homomorphisms on $\mathscr{L}$ (where a Boolean homomorphism is a 2-valued map on $\mathscr{L}$ that reduces to a 2-valued homomorphism on each Boolean sublattice of $\mathscr{L}$). That is, there are no 2-valued homomorphisms on the structure $\mathscr{L}$ regarded as a partial Boolean algebra. What Clifton's result shows, in effect, is that we can't impose property composition as a further constraint on the resulting set of properties, without generating a structure on which no Boolean homomorphisms exist. No set of determinate properties can include the properties in $\mathscr{L}_e$, even if the set is a partial Boolean algebra and satisfies property composition. So the eigenstate to eigenvalue link is not an artefact of the lattice closure requirement – we can't maintain the link by dropping lattice closure in favour of partial Boolean algebra closure.

It appears, then, that modal interpretations ought to drop the orthodox interpretation principle entirely, as Bohmian mechanics does. In Bohmian mechanics, position in configuration space is always determinate, for every quantum state, so the eigenvalue to eigenstate link fails. But the eigenstate to eigenvalue link also fails. The correlations between measurement outcomes on two separated spin-$\frac{1}{2}$ systems in the singlet state, for example, do not reflect the determinateness of any property of the composite system represented by a projection operator with a range containing the state. Rather, the determinate properties of the two systems are given by their positions and (following the analysis in the previous section) by the projection operators in the determinate sublattice defined by the position observable and the state, and this determinate sublattice does not contain the projection operator onto the state. The correlations between spin measurement outcomes on the two systems arise from dynamical changes in position in configuration space (the preferred determinate observable for the two-particle system) under the local influence of magnetic fields, and not from holistic properties of the composite system, as on Healey's account.

Similarly, in a 'no collapse' interpretation characterized by the uniqueness theorem of chapter 4, the determinate sublattice $\mathscr{L}_{e_{r_1}e_{r_2}\ldots e_{r_k}}$ defined by a quantum state $e(|\psi\rangle)$ and a preferred determinate observable R does not contain the proposition e corresponding to the state (represented by the projection operator onto the state), unless $e(|\psi\rangle)$ is an eigenstate of R (when $\mathscr{L}_{e_{r_1}e_{r_2}\ldots e_{r_k}}$ coincides with the orthodox determinate sublattice $\mathscr{L}_e$). But the Bell–Vink equation of motion for the temporal evolution of the actual value of R, which depends on the quantum state, ensures that a measurement of the proposition e will yield the outcome 1 ('yes') with unit probability in this case.

Clifton (1996b) has shown how to modify the modal interpretation to avoid perspectivalism. The trick involves assigning properties to the most elementary subsystems in a model quantum mechanical universe via the prescription (6.9), and to non-elementary subsystems in such a way as to satisfy property composition and decomposition.

6.3 Proof of the modal recovery theorem[108]

Modal recovery theorem: Consider a quantum system S and its environment E (in principle, the rest of the universe) represented on a Hilbert space $\mathcal{H} \otimes \mathcal{H}_E$. Suppose the state of the composite system $S + E$ is represented by a ray $e \in \mathcal{H} \otimes \mathcal{H}_E$, then

$$\mathcal{D}(e, W \otimes I)|_S = \mathcal{D}_{\text{modal}}(S),$$

where W is the reduced state of S and $|_S$ denotes the restriction of the sublattice to the Hilbert space $\mathcal{H}$ of S.

Proof: The following notation is used in the proof:

W_i: the ith eigenspace of W (in $\mathcal{H}$).
P_{W_i}: the projection operator onto W_i (in $\mathcal{H}$).
$(W \otimes I)_i$: the ith eigenspace of $W \otimes I$ (in $\mathcal{H} \otimes \mathcal{H}_E$).
$P_{(W\otimes I)_i}$: the projection operator onto $(W \otimes I)_i$ (in $\mathcal{H} \otimes \mathcal{H}_E$).
$e_{(W\otimes I)_i}$: the nonzero projection of the ray e onto $(W \otimes I)_i$ (in $\mathcal{H} \otimes \mathcal{H}_E$).
$P_{e(W\otimes I)_i}$: the (nonzero) projection operator onto the ray $e_{(W\otimes I)_i}$ (in $\mathcal{H} \otimes \mathcal{H}_E$).

Note that there is a 1–1 correspondence between the eigenspaces W_i in $\mathcal{H}$ and $(W \otimes I)_i$ in $\mathcal{H} \otimes \mathcal{H}_E$, because w_i is an eigenvalue of W if and only if w_i is an eigenvalue of $W \otimes I$.

The uniqueness theorem in chapter 4 established that

$$\mathcal{D}(e,R) = \mathcal{L}_{e_{r_1}, e_{r_2}, \ldots, e_{r_k}} = \{p\text{: } e_{r_i} \leq p \text{ or } e_{r_i} \leq p^{\perp}, i = 1, \ldots, k\},$$

where the e_{r_i} are the nonzero projections of e onto the k eigenspaces of R. Writing P for the projection operator corresponding to the lattice element p, and $P_{e_{r_i}}$ for the projection operator corresponding to the lattice element e_{r_i}, the determinate sublattice of projection operators can be expressed equivalently as:

$$\{P\text{: } PP_{e_{r_i}} = P_{e_{r_i}} \text{ or } 0, \text{ for all } i\},$$

where the index i ranges over all nonzero projections of e onto the eigenspaces of R.

So we want to show that:

$$\begin{aligned}\mathcal{D}(e, W \otimes I)|_S &\equiv \{P\text{: } (P \otimes I)P_{e(W\otimes I)_i} = P_{e(W\otimes I)_i} \text{ or } 0, \text{ for all } i\} \\ &= \mathcal{D}_{\text{modal}}(S) \\ &\equiv \{P\text{: } PP_{W_i} = P_{W_i} \text{ or } 0, \text{ for all } i\}\end{aligned}$$

The index i in the expression for $\mathcal{D}_{\text{modal}}(S)$ ranges over all projection operators P_{W_i} in $SR(W)$, the set of projection operators defined by the spectral resolution of W (that is, the projection operators onto the eigenspaces corresponding to the nonzero eigenvalues of W). The index i in the expression for $\mathcal{D}(e, W \otimes I)|_S$ ranges over all nonzero

[108] This proof is taken from Bub and Clifton (1996).

projections of e onto the eigenspaces of $W\otimes I$, including in principle the eigenspace corresponding to the zero eigenvalue (the null space of $W\otimes I$), if $W\otimes I$ has a zero eigenvalue. Since W is the reduced state of e for the system S, e lies entirely in the span of the nonzero-eigenvalue eigenspaces of $W\otimes I$, and hence has zero projection onto the zero-eigenvalue eigenspace, if such an eigenspace exists. So the index i in the expression for $\mathscr{D}(e,W\otimes I)|_S$ in fact ranges over the eigenspaces with projection operators in the set $SR(W\otimes I)$, which are in 1–1 correspondence with projection operators in the set $SR(W)$. In the following, the index i ranges over a set of projection operators or subspaces in 1–1 correspondence with the set $SR(W)$.

The proof establishes, first, that $\mathscr{D}_{\text{modal}}(S) \subseteq \mathscr{D}(e,W\otimes I)|_S$, and then that $\mathscr{D}(e,W\otimes I)|_S \subseteq \mathscr{D}_{\text{modal}}(S)$.

$$(i)\ \mathscr{D}_{\text{modal}}(S) \subseteq \mathscr{D}(e,W\otimes I)|_S$$

It suffices to show that the generators of $\mathscr{D}_{\text{modal}}(S)$ are in $\mathscr{D}(e,W\otimes I)|_S$, because $\mathscr{D}(e,W\otimes I)|_S$ is an ortholattice and so must include the lattice $\mathscr{D}_{\text{modal}}(S)$ as a sublattice if it includes the generators of $\mathscr{D}_{\text{modal}}(S)$.

The generators of $\mathscr{D}_{\text{modal}}(S)$ are the projection operators in the set:

$$\{P_{W_i}\text{: for all } i\} \cup \{P\text{: } PP_{W_i} = 0, \text{ for all } i\}$$

To see that $\{P_{W_i}\text{: for all } i\} \subseteq \mathscr{D}(e,W\otimes I)|_S$, note that:

$$P_{(W\otimes I)_i} \in \mathscr{D}(e,W\otimes I), \text{ for all } i$$

because these are just the projection operators onto the nonzero-eigenvalue eigenspaces of the preferred determinate observable $W\otimes I$, which always belong to $\mathscr{D}(e,W\otimes I)$ by assumption. But since

$$P_{(W\otimes I)_i} = P_{W_i}\otimes I, \text{ for all } i$$

it follows that

$$P_{W_i} \in \mathscr{D}(e,W\otimes I)|_S, \text{ for all } i$$

To see that $\{P\text{: } PP_{W_i} = 0, \text{for all } i\} \subseteq \mathscr{D}(e,W\otimes I)|_S$, note that if $PP_{W_i} = 0$ for all $P_{W_i} \in SR(W)$, then:

$$(P\otimes I)(P_{W_i}\otimes I) = (P\otimes I)P_{(W\otimes I)_i} = 0, \text{ for all } i$$

and so:

$$(P\otimes I)P_{e(W\otimes I)_i} = 0, \text{ for all } i$$

But if $(P\otimes I)P_{e(W\otimes I)_i} = 0$ for all nonzero projections of the ray e onto the nonzero-eigenvalue eigenspaces of $W\otimes I$ then, since e has zero projection onto the zero-eigenvalue eigenspace of $W\otimes I$ (if such an eigenspace exists), $(P\otimes I)P_{e(W\otimes I)_i} = 0$ for all nonzero projections of e onto all the eigenspaces of $W\otimes I$, and so, by definition:

$$P \in \mathscr{D}(e,W\otimes I)|_S$$

$$(ii)\ \mathscr{D}(e, W \otimes I)|_S \subseteq \mathscr{D}_{\text{modal}}(S)$$

Suppose the state vector $|e\rangle$ can be represented in *one* of its biorthogonal decompositions with respect to $\mathscr{H} \otimes \mathscr{H}_E$ as:

$$|e\rangle = \sum_{ij} c_{ij}|u_{ij}\rangle|v_{ij}\rangle$$

This vector can be expressed as:

$$|e\rangle = \sum_i \left(\sum_j c_{ij}|u_{ij}\rangle|v_{ij}\rangle \right)$$

where for fixed i, $|c_{ij_1}|^2 = |c_{ij_2}|^2 = \ldots \equiv |c_i|^2$, and the index i ranges over the *distinct* numbers $|c_i|^2$. These are just the nonzero eigenvalues of W that index $SR(W)$, and so:

$$P_{W_i} = \sum_j |u_{ij}\rangle\langle u_{ij}|$$

We now have, for all i:

$$|e_{(W \otimes I)_i}\rangle = P_{(W \otimes I)_i}|e\rangle = P_{W_i} \otimes I|e\rangle = \sum_j c_{ij}|u_{ij}\rangle|v_{ij}\rangle$$

Suppose $P \in \mathscr{D}(e, W \otimes I)|_S$, then:

$$(P \otimes I)P_{e_{(W \otimes I)_i}} = P_{e_{(W \otimes I)_i}} \text{ or } 0, \text{ for all } i$$

It follows that:

$$(P \otimes I)|e_{(W \otimes I)_i}\rangle = |e_{(W \otimes I)_i}\rangle \text{ or } |0\rangle, \text{ for all } i$$

and so:

$$\sum_j c_{ij}P|u_{ij}\rangle|v_{ij}\rangle = \sum_j c_{ij}|u_{ij}\rangle|v_{ij}\rangle \text{ or } 0, \text{ for all } i$$

For a fixed i, this implies that either $P|u_{ij}\rangle = |u_{ij}\rangle$ for all j, or $P|u_{ij}\rangle = |0\rangle$ for all j, because $c_{ij} \neq 0$ for any j. And since $\{|u_{ij}\rangle$, for all $j\}$ is a basis for the eigenspace W_i:

$$PP_{W_i} = P_{W_i} \text{ or } 0, \text{ for all } i$$

That is,

$$P \in \mathscr{D}_{\text{modal}}(S)$$

QED

7
Orthodoxy

Oh, the brave old Duke of York,
He had ten thousand men;
He marched them up to the top of the hill,
And he marched them down again.
And when they were up, they were up,
And when they were down, they were down,
And when they were only half-way up,
They were neither up nor down.
Mother Goose

7.1 The Copenhagen interpretation

The strange thing about the quantum world relative to the classical world is that quantum observables form a noncommutative algebra; equivalently, the idempotent observables or properties represented by projection operators form a non-Boolean algebra. Noncommutativity or non-Booleanity is the feature that underlies the phenomenon of interference in all its forms, from Heisenberg's uncertainty principle for the reciprocal scatter in position and momentum measurements on a particle, to the nonlocality and nonseparability exhibited by systems in 'entangled' quantum states characteristics of EPR-type experiments or measurement interactions.

The Copenhagen interpretation of quantum mechanics originated in discussions between Bohr and Heisenberg on how to understand the physical significance of relatively localizable systems that exhibit interference. Heisenberg describes the genesis of the interpretation as follows (Heisenberg, 1967, pp. 105–6):

> At this time Dirac and Jordan developed the transformation theory to which Born and Jordan in earlier investigations had already laid the foundation, and the completion of this mathematical formalism again confirmed us [i.e., Bohr and Heisenberg] in our belief that there was no more to change in the formal structure of quantum mechanics, and that the remaining problem was to express the connection between the mathematics and experiment in a way free of contradictions. But how this was to be done remained unclear. Our evening discussions quite often lasted till after midnight, and we occasionally parted somewhat discontented, for the difference in the

directions in which we sought the solution seemed often to make the problem more difficult. Still, deeply disquieted after one of these discussions I went for a walk in the Fælledpark, which lies behind the Institute, to breathe the fresh air and calm down before going to bed. On this walk under the stars, the obvious idea occurred to me that one should postulate that nature allowed only experimental situations to occur which could be described within the framework of the formalism of quantum mechanics. This would apparently imply, as one could see from the mathematical formalism, that one could not simultaneously know the position and velocity of a particle. There was no immediate possibility of discussing this idea in detail with Bohr, because just at this time (end of February, 1927) he had left for a skiing holiday in Norway. Bohr was probably also glad to be able to devote himself to a few weeks' completely undisturbed thinking about the interpretation of quantum theory.

Left alone in Copenhagen I too was able to give my thoughts freer play, and I decided to make the above uncertainty the central point in the interpretation. Remembering a discussion I had had long before with a fellow student in Göttingen, I got the idea of investigating the possibility of determining the position of a particle with the aid of a gamma-ray microscope, and in this way soon arrived at an interpretation which I believed to be coherent and free of contradictions. I then wrote a long letter to Pauli, more or less the draft of a paper, and Pauli's answer was decidedly positive and encouraging. When Bohr returned from Norway, I was already able to present him with the first version of a paper along with the letter from Pauli. At first Bohr was rather dissatisfied. He pointed out to me that certain statements in this first version were still incorrectly founded, and as he always insisted on relentless clarity in every detail, these points offended him deeply. Further, he had probably already grown familiar, while he was in Norway, with the concept of complementarity which would make it possible to take the dualism between the wave and particle picture as a suitable starting point for an interpretation. This concept of complementarity fitted well the fundamental philosophical attitude which he had always had, and in which the limitations of our means of expressing ourselves entered as a central philosophical problem. He therefore took objection to the fact that I had not started from the dualism between particles and waves. After several weeks of discussion, which were not devoid of stress, we soon concluded, not least thanks to Oskar Klein's participation, that we really meant the same, and that the uncertainty relations were just a special case of the more general complementarity principle. Thus, I sent my improved paper to the printer and Bohr prepared a detailed publication on complementarity.

The Copenhagen interpretation has been regarded as the 'official' interpretation of quantum mechanics since the Solvay conference of October 1927, when the consensus was that Einstein lost the debate with Bohr. Heisenberg concedes differences between his own position and Bohr's, but concludes that 'we really meant the same.' This reconciliation extends now even to such radical views as Wigner's (1962, p. 290), that 'the quantum description of objects is influenced by impressions entering my consciousness,' and Wheeler's (1983, pp. 184–85) notion of 'observer-participancy,' which are also commonly regarded as consistent with the Copenhagen interpretation:

The dependence of what is observed upon the choice of experimental arrangement made Einstein unhappy. It conflicts with the view that the universe exists 'out there' independent of all acts of observation. In contrast Bohr stressed that we confront here an inescapable new feature of nature, to be welcomed because of the understanding it gives us. In struggling to make clear to Einstein the central point as he saw it, Bohr found himself forced to introduce the word 'phenomenon.' In today's words Bohr's point – and the central point of quantum theory – can be put into a single, simple sentence. 'No elementary phenomenon is a phenomenon until it is a registered (observed) phenomenon.' It is wrong to speak of the 'route' of the photon in the experiment of the beam splitter. It is wrong to attribute a tangibility to the photon in all its travel from the point of entry to its last instant of flight. A phenomenon is not yet a phenomenon until it has been brought to a close by an irreversible act of amplification such as the blackening of a grain of silver bromide emulsion or the triggering of a photodetector. In broader terms, we find that nature at the quantum level is not a machine that goes its inexorable way. Instead what answer we get depends on the question we put, the experiment we arrange, the registering device we choose. We are inescapably involved in bringing about that which appears to be happening.

It is doubtful that Bohr would have endorsed Wigner's dualism or Wheeler's observer-participancy as a friendly amendment to complementarity. In a cautionary remark about misleading terminology, he writes (1949, p. 237):

In this connection I warned especially against phrases, often found in the physical literature, such as 'disturbing of phenomena by observation' or 'creating physical attributes to atomic objects by measurements.' Such phrases, which may serve to remind of the apparent paradoxes in quantum theory, are at the same time apt to cause confusion, since words like 'phenomena' and 'observations,' just as 'attributes' and 'measurements,' are used in a way hardly compatible with common language and practical definition.

The common strand linking these different positions is the rejection of Einstein's realism – the 'ideal of the detached observer,' as Pauli put it. It is the idea that, in some sense (notwithstanding Bohr's discomfort with the terminology), observables 'only have values when you look,' where the notion of 'looking' is understood in a certain way (as involving the specification of a classically describable experimental set-up, or an interaction with a macroscopic measuring instrument that does not involve an ultimate conscious observer, or a measurement process that does involve the activity of a conscious observer, etc.). This claim is justified by citing examples of interference, such as Heisenberg's uncertainty relations, or the double-slit experiment, or beam splitter experiments, or appealing to the irreducible disturbance of the measured systems in quantum mechanical measurement interactions.

Now, it is generally recognized – at least, by all but the most recalcitrant positivists – that the mere fact that measurements disturb what we measure does not preclude the possibility that observables have determinate values, or even that measurements might be exploited to reveal these values in suitably designed measurements contexts. (The 'disturbance' terminology itself suggests the existence of determinate values for observables, prior to measurement, that are 'disturbed' or undergo dynamical change in physical interactions.) And there is no warrant from the theory for interpreting the uncertainty relations for canonically conjugate observables like position and momentum as anything more than a constraint on the possibility of preparing ensembles of systems in which these observables are simultaneously 'sharp' –that is, as anything more than a constraint on the reciprocal distribution of the determinate values of these observables in quantum ensembles.

Even interference phenomena, by themselves, say nothing about whether or not observables have determinate values in the absence of measurements, unless some interpretative principle is introduced. The usual story, in the case of a double-slit photon interference experiment, for example, is that you get the wrong distribution of hits on the screen behind the slits, if you assume that each individual photon goes through one or the other of the two slits, when the photon is in a quantum state that is a linear superposition of a state in which the photon goes through slit 1 and a state in which the photon goes through slit 2. The photon is supposed to exhibit 'wave–particle duality' and 'go through both slits at once' to produce the characteristic interference pattern on the screen, where the photon finally manifests its presence as a particle. In passing through the slits, the photon behaves like a wave, a physical influence spread out over both slits, but in hitting the screen, it behaves like a particle, something localized at a point. Putting it differently, when the state of the photon is the linear superposition, the observable A, corresponding to localization in the plane of the slits to slit 1 or slit 2, has no determinate value.

More precisely, suppose the quantum state of the photon when it passes through the slit system is represented by the linear superposition:

$$|\psi\rangle = \tfrac{1}{\sqrt{2}}|\alpha_1\rangle + \tfrac{1}{\sqrt{2}}|\alpha_2\rangle$$

where $|\alpha_1\rangle$ represents the state of the photon going through slit 1 and $|\alpha_2\rangle$ represents the state of the photon going through slit 2. Then the probability of the photon hitting a certain region b on the screen is:

$$\text{prob}_\psi(b) = \text{tr}(U_t P_\psi U_t^{-1} P_b),$$

where $P_\psi = |\psi\rangle\langle\psi|$, U_t represents the time evolution of the state in the region between the slit system and the screen, and P_b is the projection operator corresponding to b (that is, a projection operator in the spectral measure of an observable representing localization in the plane of the screen). Since $|\psi\rangle\langle\psi| = \tfrac{1}{2}(|\alpha_1\rangle\langle\alpha_1| + |\alpha_2\rangle\langle\alpha_2| + |\alpha_1\rangle\langle\alpha_2| + |\alpha_2\rangle\langle\alpha_1|)$, it follows that:

$$\begin{aligned}\text{prob}_\psi(b) &= \tfrac{1}{2}\text{tr}(U_t|\alpha_1\rangle\langle\alpha_1|U_t^{-1}P_b) + \tfrac{1}{2}\text{tr}(U_t|\alpha_2\rangle\langle\alpha_2|U_t^{-1}P_b)\\ &\quad + \tfrac{1}{2}\text{tr}(U_t|\alpha_1\rangle\langle\alpha_2|U_t^{-1}P_b) + \tfrac{1}{2}\text{tr}(U_t|\alpha_2\rangle\langle\alpha_1|U_t^{-1}P_b)\\ &= \tfrac{1}{2}\text{prob}_{\alpha_1(t)}(b) + \tfrac{1}{2}\text{prob}_{\alpha_2(t)}(b) + \text{interference terms}\end{aligned}$$

If the state of the photon is a mixture of the pure states $|\alpha_1\rangle$ or $|\alpha_2\rangle$ with equal probability represented by the density operator:

$$W = \tfrac{1}{2}P_{\alpha_1} + \tfrac{1}{2}P_{\alpha_2} = \tfrac{1}{2}|\alpha_1\rangle\langle\alpha_1| + \tfrac{1}{2}|\alpha_2\rangle\langle\alpha_2|,$$

then

$$\begin{aligned}\text{prob}_W(b) &= \text{tr}(U_t W U_t^{-1} P_b) \\ &= \tfrac{1}{2}\text{prob}_{\alpha_1(t)}(b) + \tfrac{1}{2}\text{prob}_{\alpha_2(t)}(b)\end{aligned}$$

So, if the initial state of the photon is the mixed state W, the photon goes through one or the other of the two slits, and the distribution on the screen is the sum of the distributions for each slit. If the initial state of the photon is the superposition $|\psi\rangle$, the distribution on the screen is the characteristic interference pattern. Since the sum of the interference terms can be negative for certain regions b on the screen, there are low probability regions on the screen when both slits are open that correspond to high probability regions when only slit 1 or only slit 2 is open. If we assume that a photon in the state $|\psi\rangle$ either goes through slit 1 or through slit 2, exclusively, it would appear that the second slit being open can prevent a photon from hitting a certain region on the screen that it would have hit if that slit had been closed. The interference pattern appears to be incompatible with the assumption that the photon either goes through slit 1 or through slit 2, exclusively; that is, incompatible with the assumption that the observable A, corresponding to localization in the plane of the slits to slit 1 or slit 2, has a determinate value.

The loophole in the argument is the link between attributing a determinate value to the observable A and attributing a specific quantum state to the photon. This is the orthodox (Dirac–von Neumann) interpretation principle; specifically, the eigenvalue to eigenstate link. If we reject this principle, then we can attribute a determinate value to the observable A, associated with the photon going through slit 1 or through slit 2, exclusively, without assigning the state $|\alpha_1\rangle$ or the state $|\alpha_2\rangle$ to the photon. And this, of course, is precisely what observer-free 'no collapse' interpretations like Bohmian mechanics accomplish.

Similar remarks apply to beam splitter experiments. In the schematic representation of a Mach–Zehnder interferometer in figure 7.1, a beam of particles enters the interferometer from a source and is split into two beams by the first beam splitter. The two beams move towards mirrors (indicated by the solid angled bars) that reflect the beams towards a second beam splitter. The positions of the mirrors and beam splitters are arranged so that the two beams emerging from the second beam splitter in the direction of a detector, indicated by D_1, interfere constructively, while the two beams emerging in the direction of the detector D_2 interfere destructively. So all the particles entering the interferometer are detected at D_1, and no particles are detected at D_2. If a barrier is placed in one of the paths, say the 'up' path between the first beam splitter and the mirror, then no interference occurs: half the particles entering the beam splitter are scattered or absorbed by the barrier, and of the remaining half, one quarter are detected at D_1 and one quarter at D_2.

Interference experiments of this sort are commonly performed with photons or neutrons, and have even been done with sources that produce single particle states, so that the particles pass through the interferometer one at a time. The standard argument is that a particle cannot be moving along a definite path in the interferometer, either

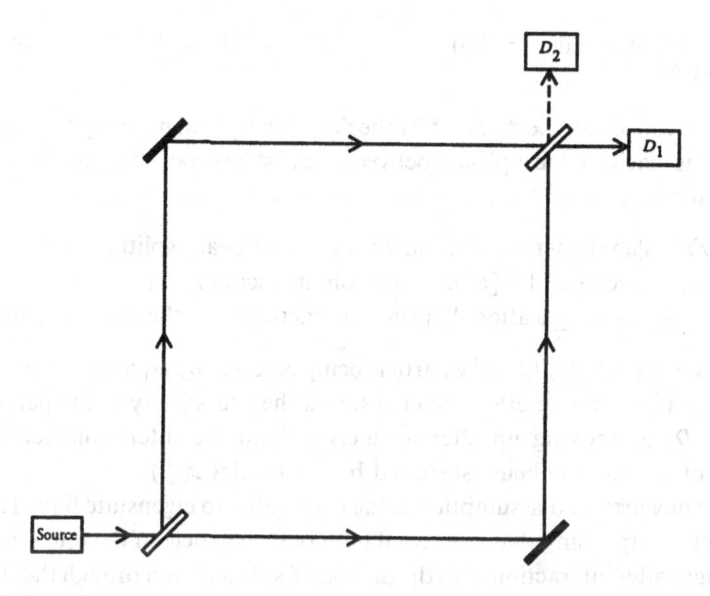

Figure 7.1. Schematic representation of a beam splitter interference experiment.

first to the right and then up towards the second beam splitter, or first up and then to the right towards the second beam splitter, because if this were the case, the particle would be detected at D_1 or at D_2, depending on its path. In many repetitions of the experiment, we would expect to see both detectors registering the particles at the same rate. The fact that one of these event-types, the detection at D_2, is impossible, is taken as demonstrating the indefiniteness or indeterminateness of the particle's path in the interferometer.

The action of a beam splitter on a single particle (assuming a transmission coefficient of $\frac{1}{2}$) can be represented by the state transitions:

$$|r\rangle \rightarrow \tfrac{1}{\sqrt{2}}(|r\rangle + \mathrm{i}|u\rangle)$$
$$|u\rangle \rightarrow \tfrac{1}{\sqrt{2}}(|u\rangle + \mathrm{i}|r\rangle),$$

where $|r\rangle$ represents the state of a particle moving to the right in the diagram, and $|u\rangle$ represents the state of a particle moving up in the diagram. Since the two mirrors induce the transitions:

$$|r\rangle \rightarrow \mathrm{i}|u\rangle$$
$$|u\rangle \rightarrow \mathrm{i}|r\rangle$$

it follows that a particle entering the interferometer in the state $|r\rangle$ will emerge in the same state:

$$|r\rangle \rightarrow \tfrac{1}{\sqrt{2}}(|r\rangle + \mathrm{i}|u\rangle) \text{ [after interaction with the first beam splitter]}$$
$$\rightarrow \tfrac{1}{\sqrt{2}}(\mathrm{i}|u\rangle - |r\rangle) \text{ [after reflection by the mirrors]}$$

$\rightarrow \frac{1}{2}[(\mathrm{i}|u\rangle - |r\rangle) - (|r\rangle + \mathrm{i}|u\rangle)] = -|r\rangle$ [after interaction with the second beam splitter][109]

So the particle will always be detected by the detector D_1, and never by the detector D_2.

If a barrier is placed in the 'up' path between the first beam splitter and the mirror, the transitions are:

$|r\rangle \rightarrow \frac{1}{\sqrt{2}}(|r\rangle + \mathrm{i}|u\rangle)$ [after interaction with the first beam splitter]
$\rightarrow \frac{1}{\sqrt{2}}(\mathrm{i}|u\rangle + \mathrm{i}|\text{scattered}\rangle)$ [after reflection and scattering]
$\rightarrow \frac{1}{2}(\mathrm{i}|u\rangle - |r\rangle) + \frac{\mathrm{i}}{\sqrt{2}}|\text{scattered}\rangle$ [after interaction with the second beam splitter]

It follows that the probability of the particle being detected by D_1 as moving to the right after it emerges from the interferometer is $\frac{1}{4}$, and the probability of the particle being detected by D_2 as moving up after it emerges from the interferometer is $\frac{1}{4}$. (The probability of the particle being scattered by the barrier is $\frac{1}{2}$.)

Again, the unwarranted assumption is the eigenvalue to eigenstate link. There is no inconsistency in supposing that the actual path of the particle in the interferometer is, say, to the right after interaction with the first beam splitter, even though the state of the particle is $\frac{1}{\sqrt{2}}(|r\rangle + \mathrm{i}|u\rangle)$.

Interference *per se* presents no obstacle to the simultaneous determinateness of noncommuting observables. The justification for assuming constraints on the simultaneous determinateness of quantum mechanical observables comes, rather, from the 'no go' theorems, and it is only these constraints that are relevant to the problem of constructing a viable interpretation of quantum mechanics.

I have throughout distinguished complementarity from what I refer to as the orthodox (Dirac–von Neumann) interpretation. I have already shown (in sections 4.1 and 5.1) how the orthodox prescription for attributing determinate values to the observables of a quantum system can be accommodated by the framework for 'no collapse' interpretations provided by the uniqueness theorem. We can regard this as a particular way of cashing out the Copenhagen idea that 'observables only have values when you look,' on the basis that observables acquire determinate values in quantum measurements that prepare the corresponding eigenstates. Complementarity, too, can be understood as a type of 'no collapse' interpretation. From the perspective of the uniqueness theorem, Bohr's complementarity and Einstein's realism (in the 'beable' sense, stripped of Einstein's stringent separability and locality requirements, as in Bohmian mechanics) appear as two quite different proposals for selecting the preferred determinate observable – either fixed, once and for all, as the realist would require, or settled pragmatically by what we choose to observe. (So complementarity is not an *observer-free* 'no collapse' interpretation.)

For Bohr, a quantum 'phenomenon' is an individual process that occurs under conditions defined by a specific, classically describable experimental arrangement, and an observable can be said to have a determinate value only in the context of an

[109] The vectors $|r\rangle$ and $-|r\rangle$ span the same ray and so represent the same quantum state.

experiment suitable for measuring the observable. The experimental arrangements suitable for locating an atomic object in space and time, and for a determination of momentum–energy values, are mutually exclusive. We can choose to investigate either of these 'complementary' phenomena at the expense of the other, so there is no unique description of the object in terms of determinate properties.

Summing up a discussion on causality and complementarity, Bohr writes (1948, p. 317):

Recapitulating, the impossibility of subdividing the individual quantum effects and of separating a behaviour of the objects from their interaction with the measuring instruments serving to define the conditions under which the phenomena appear implies an ambiguity in assigning conventional attributes to atomic objects which calls for a reconsideration of our attitude towards the problem of physical explanation. In this novel situation, even the old question of an ultimate determinacy of natural phenomena has lost its conceptual basis, and it is against this background that the viewpoint of complementarity presents itself as a rational generalization of the very ideal of causality.

Pauli characterizes Bohr's position this way (1948, p. 307–8):

While the means of observation (experimental arrangements and apparatus, records such as spots on photographic plates) have still to be described in the usual 'common language supplemented with the terminology of classical physics,' the atomic 'objects' used in the theoretical interpretation of the 'phenomena' cannot any longer be described 'in a unique way by conventional physical attributes.' Those 'ambiguous' objects used in the description of nature have an obviously symbolic character.

In terms of the uniqueness theorem, we can understand the complementarity interpretation as the proposal to take the classically describable experimental arrangement (suitable for either a space–time or a momentum–energy determination) as defining the preferred determinate observable in what Bohr calls a quantum 'phenomenon.' So the preferred determinate observable for a model quantum mechanical universe is not fixed, but is defined by the classically described 'means of observation.' On this view, the determinate sublattice of a quantum universe depends partly on what we choose to measure, not on objective features of the system itself. To echo Pauli, the properties we attribute to a quantum object in a measurement are 'ambiguous,' or merely 'symbolic.' The complementarity interpretation, unlike a beable interpretation, which selects a fixed preferred determinate observable, is not a realist interpretation.

It is instructive to consider the application of the complementarity interpretation to the Einstein–Podolsky–Rosen experiment. The problem for the complementarity interpretation posed by the Einstein–Podolsky–Rosen argument is this: According to quantum mechanics, there are cases where a *local* measurement suffices to assign determinate values or ranges of values to observables of a *distant* system. For example, in the EPR experiment a measurement of the position (or momentum) of a subsystem S_2 in a spatially separated two-particle composite system $S_1 + S_2$ suffices to determine the position (or momentum) of the subsystem S_1. On the complementarity interpretation, an observable has a determinate value only in the context of a classically describable experimental arrangement suitable for measuring the observable. Prior to the EPR experiment, this was understood to mean that an S_1-observable can come to have a determinate value only in the context of (and in virtue of) an experimental arrangement that is localized in the vicinity of S_1. But the EPR experiment seems to require that a local experimental arrangement at S_2 can somehow make observables of the distant system S_1 determinate – observables that were indeterminate according to the complementarity interpretation before the measurement at S_2. Because they rejected the possibility that S_1-observables, indeterminate in the experimental arrangement used to prepare the state $|\psi\rangle$ of the composite system, can become determinate in virtue of what we subsequently choose to measure on S_2, Einstein, Podolsky, and Rosen concluded that there are S_1-observables that are already determinate in the state $|\psi\rangle$, prior to the measurement at S_2. In their terminology, there are 'elements of reality' at S_1 corresponding to observables that are not assigned any determinate value when the composite system $S_1 + S_2$ is in the state $|\psi\rangle$. Hence, they argued, the quantum state $|\psi\rangle$ is an incomplete description of the system $S_1 + S_2$.

The central concept of the complementarity interpretation is Bohr's concept of a quantum 'phenomenon' (Bohr, 1939, p. 20):

> The essential lesson of the analysis of measurements in quantum theory is thus the emphasis on the necessity, in the account of the phenomena, of taking the whole experimental arrangement into consideration, in complete conformity with the fact that all unambiguous interpretation of the quantum mechanical formalism involves the fixation of the external conditions, defining the initial state of the atomic system concerned and the character of the possible predictions as regards subsequent observable properties of that system. Any measurement in quantum theory can in fact only refer either to a fixation of the initial state or to the test of such predictions, and it is first the combination of measurements of both kinds which constitutes a well-defined phenomenon.

So a quantum phenomenon is defined by two measurements. The first measurement, under external conditions defined by a suitable experimental arrangement, prepares the state; that is, the outcome of the measurement is a quantum state. I shall refer to these conditions as C_{state} and the measurement as M_{state}. The second measurement, under (in general) altered external conditions defined by an altered experimental arrangement, tests a prediction; that is, the outcome of the measurement is a value or range of values of an observable, equivalently a proposition. I shall refer to these conditions as C_{test} and the measurement as M_{test}. The conditions C_{test} will differ from the conditions C_{state} if the observable involved in the test (an observable that is determinate in the context of the conditions C_{test}) does not commute with the observable represented by the 1-dimensional projector onto the ray spanned by the state (an observable that is determinate in the context of the conditions C_{state}).

I take 'the fixation of the external conditions, defining . . . the character of the possible predictions as regards subsequent observable properties of that system' to refer to the selection, by the conditions C_{test}, of a family of determinate properties in the lattice of properties of the system. Other properties are not applicable to the system as manifested in the phenomenon, and the corresponding propositions are neither true nor false – they are not 'possible predictions.' Since the phenomenon is defined by the first measurement M_{state} that prepares the state as well as the subsequent measurement M_{test}, this family of properties must be a determinate sublattice defined by the state and a preferred observable associated with the conditions C_{test}.

For Bohr, the complementary relationship between the different determinate sublattices in the non-Boolean property structure of a quantum mechanical system is grounded in the relation between classical and quantum mechanics (Bohr, 1939, pp. 23–5):

> In the system to which the quantum mechanical formalism is applied, it is of course possible to include any intermediate auxiliary agency employed in the measuring processes. . . . The only significant point is that in each case some ultimate measuring instruments, like the scales and clocks which determine the frame of space–time coordination – on which, in the last resort, even the definitions of momentum and energy quantities rest – must always be described entirely on classical lines, and consequently kept outside the system subject to quantum mechanical treatment.
>
> The unaccustomed features of the situation with which we are confronted in quantum theory necessitate the greatest caution as regards all questions of terminology. . . . It is certainly far more in accordance with the structure and interpretation of the quantum mechanical symbolism, as well as with elementary physical principles, to reserve the word 'phenomenon' for the comprehension of the effects observed under given experimental conditions.
>
> These conditions, which include the account of the properties and manipulation of all measuring instruments essentially concerned, constitute in fact the only basis for the definition of the concepts by which the phenomenon is described. It is just in this sense that phenomena defined by different concepts, corresponding to mutually exclusive experimental arrangements, can unambiguously be regarded as complementary aspects of the whole obtainable evidence concerning the objects under investigation.

The heart of Bohr's response to the EPR argument is contained in a footnote.[110] The structure of the paper is this: There is an initial acknowledgement of the difficulty raised by EPR. Then the crucial footnote is introduced (Bohr, 1935, pp. 696, 697). This is followed by a review of complementarity, after which Bohr argues that the criterion of physical reality proposed by EPR is ambiguous for quantum systems. Finally, Bohr returns to the footnote as an illustration of this ambiguity, linking the analysis in the footnote to the composite system considered by EPR in a further footnote that comments on the original footnote (Bohr, 1935, p. 699). No part of the paper other than

[110] For a contrasting analysis, see Beller and Fine (1994). They conclude (p. 29) that the EPR argument 'shifted complementarity away from physical analysis and into the realm of philosophical counsel,' forcing Bohr towards positivism.

the two footnotes and the brief remarks preceding the second footnote specifically addresses the EPR argument.

What Bohr says in the first footnote is this (Bohr, 1935; pp. 696, 697):

The deductions contained in the article cited may in this respect be considered as an immediate consequence of the transformation theorems of quantum mechanics, which perhaps more than any other feature of the formalism contribute to secure its mathematical completeness and its rational correspondence with classical mechanics. In fact, it is always possible in the description of a mechanical system, consisting of two partial systems (1) and (2), interacting or not, to replace any two pairs of canonically conjugate variables (q_1p_1), (q_2p_2) pertaining to systems (1) and (2), respectively, and satisfying the usual commutation rules

$$[q_1p_1] = [q_2p_2] = ih/2\pi$$
$$[q_1q_2] = [p_1p_2] = [q_1p_2] = [q_2p_1] = 0$$

by two pairs of new conjugate variables (Q_1P_1), (Q_2P_2) related to the first variables by a simple orthogonal transformation, corresponding to a rotation of angle θ in the planes (q_1q_2), (p_1p_2)

$$q_1 = Q_1 \cos\theta - Q_2 \sin\theta \qquad p_1 = P_1 \cos\theta - P_2 \sin\theta$$
$$q_2 = Q_1 \sin\theta + Q_2 \cos\theta \qquad q_1 = P_1 \sin\theta + P_2 \cos\theta.$$

Since these variables will satisfy analogous commutation rules, in particular

$$[Q_1P_1] = ih/2\pi \qquad [Q_1P_2] = 0,$$

it follows that in the description of the state of the combined system definite numerical values may not be assigned to both Q_1 and P_1, but that we may clearly assign such values to both Q_1 and P_2. In that case it further results from the expressions of these variables in terms of (q_1p_1) and (q_2p_2), namely,

$$Q_1 = q_1 \cos\theta + q_2 \sin\theta, \qquad P_2 = -\, p_1 \sin\theta + p_2 \cos\theta,$$

that a subsequent measurement of either q_2 or p_2 will allow us to predict the value of q_1 or p_1 respectively.

The remarks preceding the second footnote contain the substance of Bohr's analysis of the EPR argument (Bohr, 1935, p. 699):

The particular quantum-mechanical state of two free particles, for which [EPR] give an explicit mathematical expression, may be reproduced, at least in principle, by a simple experimental arrangement, comprising a rigid diaphragm with two parallel slits, which are very narrow compared with their separation, and through each of which one particle with given initial momentum passes independently of the other. If the momentum of this diaphragm is measured accurately before as well as after the passing of the particles, we shall in fact know the sum of the components perpendicular to the slits of the momenta of the two escaping particles, as well as the difference of their initial positional coordinates in the same direction; while of course the conjugate quantities, i.e., the difference of the components of their momenta, and the sum of their positional coordinates, are entirely unknown.* In this arrangement, it is therefore clear that a subsequent single measurement either of the position or of the momentum of one of the particles will automatically determine the position or momentum, respectively, of the other particle with any desired accuracy; at least if the wave-length corresponding to the free motion of each particle is sufficiently short compared with the width of the slits.

The footnote (*) – Bohr's second footnote addressing the EPR argument – points out that this set-up corresponds (apart from a trivial normalizing factor) to the case discussed in the first footnote, if (q_1, p_1), (q_2, p_2) represent the position and momentum coordinates of the two particles, respectively, and $\theta = -\frac{\pi}{4}$. For then $Q_1 = \frac{1}{\sqrt{2}}(q_1 - q_2)$ and $P_2 = \frac{1}{\sqrt{2}}(p_1 + p_2)$. The EPR correlated state $|\psi\rangle$ is prepared by this experimental arrangement with the value 0 of P_2 in the limit of infinitely narrow slits.

In the terminology introduced above, C_{state} refers to the conditions required to prepare an eigenstate of Q_1 and P_2 while C_{test} refers to the conditions defined by the experimental arrangement suitable for the measurement M_{test} of q_2 (or p_2). The state preparation and the measurement of q_2 (or p_2) defines the phenomenon. Bohr's point is that the determinate value for q_1 (or p_1) is part of the phenomenon. That is, q_1 (or p_1) inherits determinate status from the state and the preferred determinate observable q_2 (or p_2) – preferred in virtue of the classically described experimental arrangement at S_2. We can see how this comes about if we take the phenomenon as defining a determinate sublattice, for the determinate sublattice defined by the EPR state and the preferred observable q_2 (or p_2) does contain q_1-propositions (or q_2-propositions).

Referring to the EPR example, Bohr can then claim (1935, p. 699):

From our point of view we now see that the wording of the above-mentioned criterion of physical reality proposed by Einstein, Podolsky and Rosen contains an ambiguity as regards the meaning of the expression 'without in any way disturbing a system.' Of course there is in a case like that just considered no question of a mechanical disturbance of the system under investigation during the last critical stage of the measuring procedure. But even at this stage there is essentially the question of *an influence on the very conditions which define the possible types of predictions regarding the future behaviour of the system.* Since these conditions constitute an inherent element of the description of any phenomenon to which the term 'physical reality' can be properly attached, we see that the argumentation of the mentioned authors does not justify their conclusion that quantum-mechanical description is essentially incomplete.

The 'system under investigation' here is the subsystem S_1. The 'last critical stage of the measuring procedure' is the measurement M_{test} under the conditions C_{test}. Clearly M_{test} does not disturb S_1 in a mechanical sense. But M_{test} does change *something* at S_1. After M_{state}, which mechanically affects $S_1 + S_2$, and M_{test}, which mechanically affects only S_2, certain propositions become determinately true or false of $S_1 + S_2$, hence (it turns out) of S_1, which were not determinately true or false (according to the complementarity interpretation) prior to the phenomenon defined by M_{state} and M_{test} – so M_{test} (through C_{test}) affects 'the very conditions which define the possible types of predictions regarding the future behavior of [S_1].' This is not a change of truth values (as would be effected by a mechanical disturbance) but *a change in the set of propositions that have truth values.*

To see how this comes about in terms of the relevant determinate sublattice, consider the argument for two separated spin-$\frac{1}{2}$ systems in the singlet spin state. Suppose the initial measurement M_{state} prepares the singlet state, while the conditions C_{test} for the subsequent measurement M_{test} selects spin in the z-direction of S_2 as the preferred observable. The singlet state, $|\psi\rangle$, could, in principle, be prepared by selecting the appropriate outcome of a measurement of a complete commuting set of spin observables, $\{S^2, S_z\}$, where S represents the total spin and S_z the component of spin in the z-direction on the composite system $S_1 + S_2$ (ignoring spatial motion). The corresponding complete orthonormal set of eigenvectors in the 4-dimensional spin factor space of $S_1 + S_2$ includes the three vectors of the triplet state:

$$|z+\rangle|z+\rangle$$
$$\tfrac{1}{\sqrt{2}}(|z+\rangle|z-\rangle + |z-\rangle|z+\rangle)$$
$$|z-\rangle|z-\rangle$$

and the vector of the singlet state:

$$|\psi\rangle = \tfrac{1}{\sqrt{2}}(|z+\rangle|z-\rangle - |z-\rangle|z+\rangle),$$

where $|z+\rangle$ and $|z-\rangle$ are abbreviations for the two eigenstates $|\sigma_z = +1\rangle$ and $|\sigma_z = -1\rangle$, respectively, of the Pauli operator $\sigma_z = 2S_z$, with eigenvalues ± 1 in units of $\hbar = 1$. The singlet state is invariant under a change in direction from the z-direction to any other direction. That is, $\frac{1}{\sqrt{2}}(|z+\rangle|z-\rangle - |z-\rangle|z+\rangle) = \frac{1}{\sqrt{2}}(|x+\rangle|x-\rangle - |x-\rangle|x+\rangle)$, etc., so the direction of the spin component S_z is arbitrary. Physically, it is more natural to think of $|\psi\rangle$ as resulting from a preparation procedure involving a scattering or decay process. For example, $|\psi\rangle$ could be obtained by a spin-zero particle decaying into two spin-$\frac{1}{2}$ particles (Bohm, 1951, p. 614), or antiprotons could be brought to rest in liquid hydrogen and the cases selected in which the anti-proton–proton pairs formed decay into two kaons (d'Espagnat, 1976, p. 80), or the singlet state could be prepared by s-wave scattering of a beam of protons with hydrogen gas (Rae, 1986, p. 207).

The determinate sublattice $\mathscr{D}(e(|\psi\rangle), \sigma_z(S_2))$, for the preferred determinate observable $\sigma_z(S_2)$ selected by C_{test}, is obtained by projecting the state $e(|\psi\rangle)$ onto the two eigenplanes associated with the eigenvalues ± 1 of $\sigma_z(S_2)$ in the 4-dimensional spin factor space of $S_1 + S_2$. These are the planes associated with the projection operators

$$P_+^2 = P_{++} \vee P_{-+} = P_{++} + P_{-+}$$
$$P_-^2 = P_{+-} \vee P_{--} = P_{+-} + P_{--},$$

where P_{++} is the 1-dimensional projector onto the ray spanned by the vector $|z+\rangle|z+\rangle$, etc. The projection of the state $e(|\psi\rangle)$ onto P_+^2 is the ray e_+ spanned by the vector $|z-\rangle|z+\rangle$. The projection of $e(|\psi\rangle)$ onto P_-^2 is the ray e_- spanned by the vector $|z+\rangle|z-\rangle$. The determinate sublattice $\mathscr{D}(e(|\psi\rangle), \sigma_z(S_2)) = \mathscr{L}_{e_+e_-}$ is generated from the rays e_+, e_-, and all the rays in the plane orthogonal to the plane $e_+ \vee e_-$ spanned by e_+ and e_-.

Since the rays spanned by the vectors $|z+\rangle|z+\rangle$ and $|z-\rangle|z-\rangle$ lie in this plane,

$\mathscr{L}_{e_+e_-}$ contains the Boolean subalgebra generated by the rays spanned by the vectors $|z+\rangle|z+\rangle$, $|z+\rangle|z-\rangle$, $|z-\rangle|z+\rangle$, $|z-\rangle|z-\rangle$ (the Boolean subalgebra associated with the complete commuting set of spin observables in the spin factor space of $S_1 + S_2$: $\sigma_z(S_1), \sigma_z(S_2)$). So both the observables $\sigma_z(S_1)$ and $\sigma_z(S_2)$ are determinate in the phenomenon defined by the experimental conditions yielding the singlet state $|\psi\rangle$ and the experimental conditions associated with the measurement of $\sigma_z(S_2)$. Note that in this case $\sigma_z(S_2)$, and functions of $\sigma_z(S_2)$, are the only S_2-observables (apart from the unit) that are determinate (restricting attention to the spin factor space). Similarly, $\sigma_z(S_1)$, and functions of $\sigma_z(S_1)$, are the only S_1-observables (apart from the unit) that are determinate. All other observables associated with propositions in the sublattice $\mathscr{L}_{e_+e_-}$ are global observables of the system $S_1 + S_2$.

I have distinguished Bohr's complementarity interpretation from the Dirac–von Neumann interpretation in terms of the determinate sublattice selected: $\mathscr{D}(e,I) = \mathscr{L}_e$ in the case of the orthodox interpretation, and $\mathscr{D}(e,R) = \mathscr{L}_{e_{r_1}e_{r_2}\ldots e_{r_k}}$ in the case of complementarity, where e denotes the ray representing the quantum state, I is the unit observable, and R is some preferred observable that is taken as determinate on the basis of external conditions imposed on the system by a classically described experimental arrangement. The sublattice $\mathscr{L}_e$ contains all the propositions represented by subspaces that either include e (and hence are assigned probability 1 by e), or are orthogonal to e (and hence are assigned probability 0 by e). So $\mathscr{L}_e$ is generated from the ray e and every ray in the subspace orthogonal to e by the lattice operations. The sublattice $\mathscr{L}_{e_{r_1}e_{r_2}\ldots e_{r_k}}$ is generated from the rays e_{r_i}, the nonzero projections of e onto the k eigenspaces of R, and every ray in the subspace $(e_{r_1} \vee e_{r_2} \vee \ldots \vee e_{r_k})^{\perp}$ orthogonal to the subspace spanned by the e_{r_i}, for $i = 1, \ldots, k$.

These two sublattices can be related in the following way: Consider the elements in $\mathscr{L}_e$ that are compatible with R – the elements represented by projection operators that commute with R (or, equivalently, commute with the projection operators in the Boolean algebra of projections defined by the spectral measure of R, the projection operators representing R-propositions). Call the sublattice generated by lattice closure from these elements in $\mathscr{L}_e$ and the R-propositions, $\mathscr{L}_{e/R}$. I shall show that $\mathscr{L}_{e/R}$ is a sublattice of $\mathscr{L}_{e_{r_1}e_{r_2}\ldots e_{r_k}}$. In fact, $\mathscr{L}_{e_{r_1}e_{r_2}\ldots e_{r_k}}$ differs from $\mathscr{L}_{e/R}$ only in including certain elements that are assigned measure zero by every probability measure on $\mathscr{L}_{e_{r_1}e_{r_2}\ldots e_{r_k}}$.

To see this, first note that $e_{r_1} \vee e_{r_2} \vee \ldots \vee e_{r_k} \in \mathscr{L}_e$, and hence $(e_{r_1} \vee e_{r_2} \vee \ldots \vee e_{r_k})^{\perp} \in \mathscr{L}_e$ (because $e \in e_{r_1} \vee e_{r_2} \vee \ldots \vee e_{r_k}$). Since the e_{r_i} are orthogonal rays, $e_{r_1} \vee e_{r_2} \vee \ldots \vee e_{r_k} = e_{r_1} + e_{r_2} + \ldots + e_{r_k}$, and so $e_{r_1} \vee e_{r_2} \vee \ldots \vee e_{r_k}$ is compatible with R (because each e_{r_i} is compatible with R).[111] It follows that

$$(e_{r_1} \vee e_{r_2} \vee \ldots \vee e_{r_k}) \wedge r_i = (e_{r_1} + e_{r_2} + \ldots + e_{r_k}) \cdot r_i = e_{r_i}$$

belongs to $\mathscr{L}_{e/R}$, for all $i = 1, \ldots, k$, and since $e_{r_i} \in \mathscr{L}_{e/R}$ and $r_i \in \mathscr{L}_{e/R}$, $r_i - e_{r_i} \in \mathscr{L}_{e/R}$. Also, any elements in $\mathscr{L}_e$ representing subspaces included in r_i are compatible with R and

[111] I use the same symbol for the subspace and corresponding projection operator here and in the following section.

hence belong to $\mathscr{L}_{e/R}$. In particular, all the rays in $r_i - e_{r_i}$, for $i = 1, \ldots, k$, which are orthogonal to e and hence in $\mathscr{L}_e$, belong to $\mathscr{L}_{e/R}$.

The rays e_{r_i}, $i = 1, \ldots, k$, and all the rays in the subspaces $r_i - e_{r_i}$, represent all the atoms in $\mathscr{L}_{e/R}$. Apart from the rays in the k subspaces $r_i - e_{r_i}$ and spans of these rays, there are no elements representing subspaces of $(e_{r_1} \vee e_{r_2} \vee \ldots \vee e_{r_k})^{\perp}$ that belong to $\mathscr{L}_{e/R}$. Any such element would have to be skew to a subspace r_i, for some i (that is, neither included in r_i, nor orthogonal to r_i), and so could not be compatible with every r_i (and every element in $\mathscr{L}_{e/R}$ is either an element in $\mathscr{L}_e$ that is compatible with every r_i, for $i = 1, \ldots, k$, or can be generated from such an element by taking the lattice closure with the r_i, in which case it is compatible with every r_i but does not belong to $\mathscr{L}_e$). Similarly, apart from the k rays e_{r_i} and spans of these rays, there are no elements representing proper subspaces of $e_{r_1} \vee e_{r_2} \vee \ldots \vee e_{r_k}$ that belong to $\mathscr{L}_{e/R}$. Again, any such element would have to be incompatible with some of the r_i. So $\mathscr{L}_{e/R}$ is generated from the k rays e_{r_i} and all the rays in the k subspaces $r_i - e_{r_i}$ by lattice closure.

Now, the difference between the sublattice $\mathscr{L}_{e/R}$ and the sublattice $\mathscr{L}_{e_{r_1}e_{r_2}\ldots e_{r_k}}$ is just this: $\mathscr{L}_{e_{r_1}e_{r_2}\ldots e_{r_k}}$ is generated by the k rays e_{r_i} and *all* the rays in the subspace $(e_{r_1} \vee e_{r_2} \vee \ldots \vee e_{r_k})^{\perp}$. So $\mathscr{L}_{e_{r_1}e_{r_2}\ldots e_{r_k}}$ includes additional elements in $(e_{r_1} \vee e_{r_2} \vee \ldots \vee e_{r_k})^{\perp}$, not in $\mathscr{L}_{e/R}$, representing rays spanned by vectors that are linear superpositions of vectors from different subspaces r_i (that is, elements that are incompatible with some of the r_i). These are rays in $(e_{r_1} \vee e_{r_2} \vee \ldots \vee e_{r_k})^{\perp}$ that do not lie in any of the subspaces $r_i - e_{r_i}$, for $i = 1, \ldots, k$. Since they lie in the subspace $(e_{r_1} \vee e_{r_2} \vee \ldots \vee e_{r_k})^{\perp}$, they are assigned probability zero by the state e and by every probability measure on $\mathscr{L}_{e_{r_1}e_{r_2}\ldots e_{r_k}}$, just as the rays in each $r_i - e_{r_i}$ are assigned probability zero by the state e and by every probability measure on $\mathscr{L}_{e_{r_1}e_{r_2}\ldots e_{r_k}}$. So the sublattice $\mathscr{L}_{e_{r_1}e_{r_2}\ldots e_{r_k}}$ represents a kind of completion of the sublattice $\mathscr{L}_{e/R}$ by adding 'null' rays in this sense (and any further elements generated by lattice closure from the added elements). From the uniqueness theorem of chapter 4, we know that $\mathscr{L}_{e_{r_1}e_{r_2}\ldots e_{r_k}}$ is maximal, so we can't add any of the other 'null' rays that are in $e^{\perp}$ but not in $(e_{r_1} \vee e_{r_2} \vee \ldots \vee e_{r_k})^{\perp}$ to $\mathscr{L}_{e/R}$.

From this perspective, Bohr's complementarity interpretation can be regarded as a 'minimal revision' of the orthodox Dirac–von Neumann interpretation without the projection postulate, constrained by the determinateness of the measured observable R and the requirement of maximizing the set of propositions that (i) are determinate in the (unprojected) quantum state (according to the orthodox interpretation) and (ii) can be maintained as determinate, consistently with the determinateness of R. In other words, $\mathscr{L}_{e_{r_1}e_{r_2}\ldots e_{r_k}}$ contains all the propositions in $\mathscr{L}_e$ that it makes sense to talk about consistently with R-propositions: the lattice $\mathscr{L}_{e/R}$, derived from $\mathscr{L}_e$ and R by compatibility constraints, extended by propositions generated by lattice closure from a maximal set of atomic propositions (rays) assigned probability zero by the state e.

The state e assigns a probability of 0 or 1 to each proposition in $\mathscr{L}_e$, and this map defines a 2-valued lattice homomorphism on $\mathscr{L}_e$ that is also a 2-valued homomorphism on each Boolean sublattice of $\mathscr{L}_e$, a Boolean homomorphism or classical truth value assignment to the propositions in $\mathscr{L}_e$. Any proposition in $\mathscr{L}_e$ that survives the minimal revision with respect to R and is retained as an element of $\mathscr{L}_{e_{r_1}e_{r_2}\ldots e_{r_k}}$, is assigned the same truth value, 0 or 1, by *every* 2-valued homomorphism on $\mathscr{L}_{e_{r_1}e_{r_2}\ldots e_{r_k}}$, and this is the same as the truth value it is assigned as an element of $\mathscr{L}_e$ by the state e as a 2-valued homomorphism on $\mathscr{L}_e$. The proposition $e_{r_1} \vee e_{r_2} \vee \ldots \vee e_{r_k} \in \mathscr{L}_e$, which is assigned the value 1 by e, is assigned the value 1 as an element of $\mathscr{L}_{e_{r_1}e_{r_2}\ldots e_{r_k}}$ by every 2-valued homomorphism on $\mathscr{L}_{e_{r_1}e_{r_2}\ldots e_{r_k}}$ (since the 2-valued homomorphisms on $\mathscr{L}_{e_{r_1}e_{r_2}\ldots e_{r_k}}$ are in 1–1 correspondence with the atoms e_{r_i}, for $i = 1, \ldots, k$). The propositions that are represented by spans of the 'null' rays in $(e_{r_1} \vee e_{r_2} \vee \ldots \vee e_{r_k})^{\perp}$ are assigned the value 0 as elements of $\mathscr{L}_e$ by the state e, and the value 0 by every 2-valued homomorphism on $\mathscr{L}_{e_{r_1}e_{r_2}\ldots e_{r_k}}$.

In the following section, I illustrate these features of the lattices $\mathscr{L}_e$, $\mathscr{L}_{e/R}$, and $\mathscr{L}_{e_{r_1}e_{r_2}\ldots e_{r_k}}$ with some elementary constructions in $\mathscr{H}_3$ and $\mathscr{H}_4$.

7.2 Some formal constructions

Bohr's complementarity interpretation and the orthodox (Dirac–von Neumann) interpretation select different determinate sublattices for a system in a given (pure) quantum state e: a lattice $\mathscr{L}_{e_{r_1}e_{r_2}\ldots e_{r_k}} = \{p: e_{r_i} \leq p \text{ or } e_{r_i} \leq p^{\perp}, i = 1, \ldots, k\}$, generated by a preferred determinate observable R defined by a classically describable experimental arrangement, in the case of complementarity, and the lattice $\mathscr{L}_e = \{p: e \leq p \text{ or } e \leq p^{\perp}\}$ in the case of the orthodox interpretation. As I showed in the previous section, $\mathscr{L}_{e_{r_1}e_{r_2}\ldots e_{r_k}}$ includes the lattice $\mathscr{L}_{e/R}$, where $\mathscr{L}_{e/R}$ is generated by lattice closure from the projection operators in the spectral measure of R and the elements in $\mathscr{L}_e$ compatible with R. Here I illustrate this relationship between the lattices in terms of some explicit constructions, showing how $\mathscr{L}_{e_{r_1}e_{r_2}\ldots e_{r_k}}$ can be obtained from $\mathscr{L}_{e/R}$ by adding a maximal set of 'null' rays assigned probability zero by the state e that are not already in $\mathscr{L}_{e/R}$.

Consider the sublattice $\mathscr{L}_{e/R}$ in $\mathscr{H}_3$, where e is not an eigenstate of R. Suppose the preferred observable R is a nonmaximal observable with two distinct eigenspaces, r_+, a plane, and r_-, a 1-dimensional subspace. Let e_{r_+} and e_{r_-} denote the rays that are the projections of the state e onto the planes r_+ and r_-, respectively, f denote the ray orthogonal to e_{r_+} in the plane r_+, and g denote the ray orthogonal to e in the plane spanned by e and e_{r_+} (shown in figure 7.2). Since r_- is 1-dimensional, e_{r_-} coincides with r_-.

The lattice $\mathscr{L}_e$ is generated by the ray e and every ray in the plane $e^{\perp}$ orthogonal to e. This lattice is also a partial Boolean algebra, where each Boolean sublattice in $\mathscr{L}_e$ is either an 8-element maximal Boolean sublattice generated by the ray e and two orthogonal rays in the plane $e^{\perp}$ as atoms, or a sublattice of one of these maximal Boolean sublattices. One of these maximal Boolean sublattices is generated by the rays, e, f, and g as atoms. Call this sublattice $\mathscr{B}$.

It is easy to see that (i) any two Boolean sublattices in $\mathscr{L}_e$ intersect in a Boolean sublattice in $\mathscr{L}_e$, (ii) all the Boolean sublattices in $\mathscr{L}_e$ have the same minimum and maximum elements (corresponding to the null space and the whole Hilbert space,

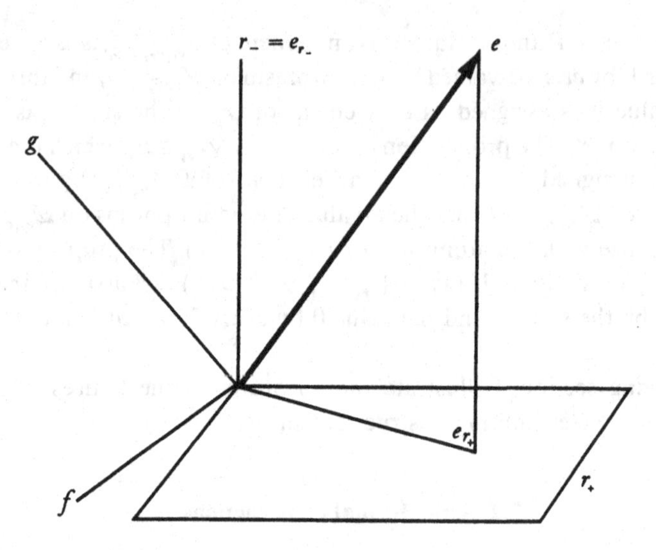

Figure 7.2. Construction of $\mathscr{L}_{e/R}$ in $\mathscr{H}_3$.

respectively), (iii) for any element or pair of elements in $\mathscr{L}_e$ that belong to the intersection of two Boolean sublattices in $\mathscr{L}_e$, the Boolean complements, meets, and joins taken with respect to either of the two Boolean sublattices coincide (and correspond, respectively, to the lattice orthocomplements, infima, and suprema, respectively, in $\mathscr{L}_e$), and (iv) for any n elements in $\mathscr{L}_e$, such that for every pair of the n elements there is a Boolean sublattice in $\mathscr{L}_e$ containing the pair, there exists a Boolean sublattice in $\mathscr{L}_e$ containing all n elements. These conditions characterize the 'pasting' required for a set of Boolean lattices to form a partial Boolean algebra.

Notice that every nonmaximal Boolean sublattice in $\mathscr{L}_e$ is a proper sublattice of a maximal Boolean sublattice in $\mathscr{L}_e$. So $\mathscr{L}_e$ is completely specified by the set of its maximal Boolean sublattices. This is also characteristic of $\mathscr{L}_e$ in the general n-dimensional case.

To construct $\mathscr{L}_{e/R}$, we must first, in effect, canvas all the maximal Boolean sublattices in $\mathscr{L}_e$ and discard any elements that are incompatible with R. Consider the maximal Boolean sublattice $\mathscr{B}$, generated by the rays e, f, and g as atoms. The element f is a ray in the plane r_+ and so is compatible with R. Neither e nor g is compatible with R, but $e \vee g = e_{r_+} \vee e_{r_-} = f^\perp$ is compatible with R. So we discard the rays e and g and the planes $e \vee f$ and $g \vee f$, and retain the ray f and the plane $e \vee g$ (and, of course, the elements 0 and 1).

The second step in the construction of $\mathscr{L}_{e/R}$ is the generation of new elements by lattice closure. Taking the infimum of $e \vee g$ with r_+ yields the ray e_{r_+}:

$$(e \vee g) \wedge r_+ = (e_{r_+} \vee e_{r_-}) \wedge r_+ = e_{r_+}$$

Similarly, taking the infimum with r_- yields the ray r_-. So we generate the 8-element

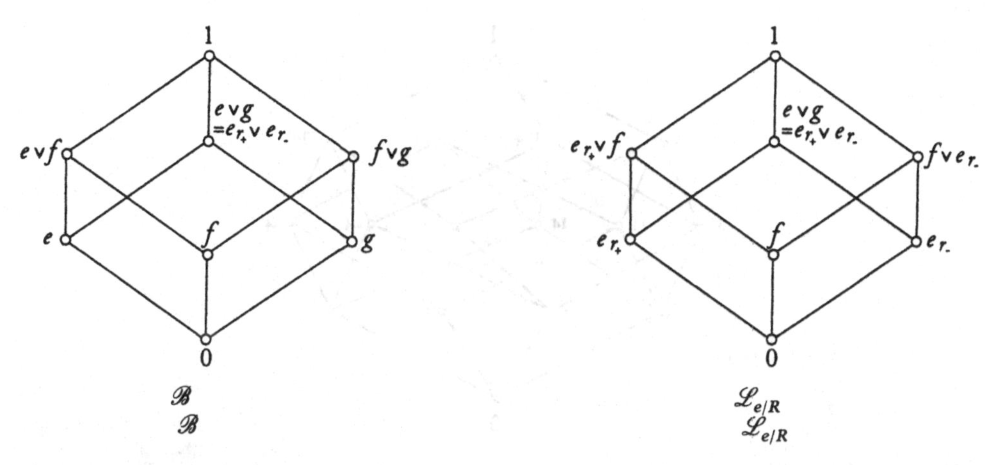

Figure 7.3. The Boolean algebras $\mathscr{B}$ and $\mathscr{L}_{e/R}$.

maximal Boolean sublattice with the atoms f, e_{r_+} and e_{r_-} from $\mathscr{B}$ and r_+ and r_-, which is evidently just the lattice $\mathscr{L}_{e_{r_+}e_{r_-}}$, since $(e_{r_+} \vee e_{r_-})^{\perp} = f$.

Now, in this case, none of the other maximal Boolean sublattices in $\mathscr{L}_e$ contain elements that are compatible with R (apart from 0 and 1), because all these elements are rays in the plane $e^{\perp}$ that are skew to the planes r_+ and r_-, or the plane $e^{\perp}$ itself, which is skew to e_{r_+} and e_{r_-}. So $\mathscr{L}_{e/R}$ reduces to a maximal Boolean subalgebra of $\mathscr{L}(\mathscr{H}_3)$. In this 3-dimensional case, then, where the subspace orthogonal to the plane $e_{r_+} \vee e_{r_-}$ is 1-dimensional, $\mathscr{L}_{e/R} = \mathscr{L}_{e_{r_+}e_{r_-}}$, as shown in figure 7.3.

Note that the elements of $\mathscr{L}_e$ that survive the transition to $\mathscr{L}_{e/R}$ (the elements, f, $e \vee g = e_{r_+} + e_{r_-}$, and 1) retain their truth value in $\mathscr{L}_{e/R}$ whatever the value of R: f is assigned the value 0 and $e \vee g$ is assigned the value 1 by e in $\mathscr{L}_e$, and these are the values assigned to these elements by either of the two 2-valued homomorphisms on $\mathscr{L}_{e/R}$ associated with the two possible values of R.

Consider now a 4-dimensional example. Let $e = e(|\psi\rangle)$ represent the ray spanned by the singlet state

$$|\psi\rangle = \tfrac{1}{\sqrt{2}}(|z+\rangle|z-\rangle - |z-\rangle|z+\rangle),$$

where $|z+\rangle$ and $|z-\rangle$ represent the 'up' and 'down' eigenstates of spin in the z-direction of two separated spin-$\frac{1}{2}$ systems, S_1 and S_2. The lattice $\mathscr{L}_e$ is generated by the ray e and every ray in the 3-dimensional subspace orthogonal to e in $\mathscr{H}_4 = \mathscr{H}_2(S_1) \otimes \mathscr{H}_2(S_2)$. Again, $\mathscr{L}_e$ can be regarded as a partial Boolean algebra, where the maximal Boolean sublattices in $\mathscr{L}_e$ are generated by the ray e and some selection of three mutually orthogonal rays in $e^{\perp}$. Take $R = \sigma_z(S_2)$ as the preferred determinate observable. To construct $\mathscr{L}_{e/R}$, we canvas all the maximal Boolean sublattices in $\mathscr{L}_e$ and discard elements incompatible with R. One of these maximal Boolean sublattices is generated by the ray e and the rays corresponding to the three vectors of the triplet state:

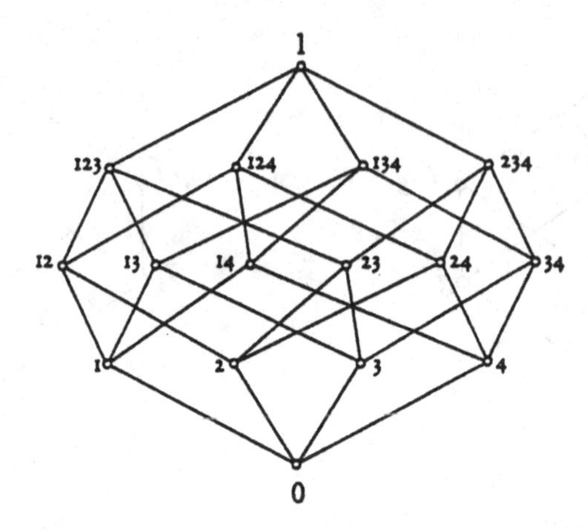

Figure 7.4. Coordinatization of $\mathscr{B}_4$.

$$|z+\rangle|z+\rangle$$
$$\tfrac{1}{\sqrt{2}}(|z+\rangle|z-\rangle+|z-\rangle|z+\rangle)$$
$$|z-\rangle|z-\rangle$$

This is the maximal Boolean sublattice associated with the complete commuting set of observables $\{S^2, S_z\}$. Call this Boolean sublattice $\mathscr{B}_z$. It is a 16-element Boolean lattice generated by four atoms, the rays spanned by the singlet state and the three vectors of the triplet state.

For notational convenience here, I denote the atoms of a general 4-atom Boolean lattice by 1, 2, 3, 4, and the minimum and maximum elements by O and I. The remaining elements can then be denoted by $12(=1\vee 2)$, $13(=1\vee 3)$, $14(=1\vee 4)$, $23(=2\vee 3)$, $24(=2\vee 4)$, $34(=3\vee 4)$, $123(=1\vee 2\vee 3=4^{\perp})$, $124(=1\vee 2\vee 4=3^{\perp})$, $134(=1\vee 3\vee 4=2^{\perp})$, $234(=2\vee 3\vee 4=1^{\perp})$. So the Hasse diagram of a 4-atom Boolean lattice is as shown in figure 7.4, where the order relations can be inferred from the coordinates of the positions in the diagram: 1, 2, 3, 4 represent atoms; 123, 124, 134, 234 represent the co-atoms $4^{\perp}, 3^{\perp}, 2^{\perp}, 1^{\perp}$, respectively; $1\leq 12$, $2\leq 12$, etc., and $12\leq 123$, $13\leq 123$, $23\leq 123$, etc.

The Boolean sublattice $\mathscr{B}_z$ can now be coordinatized as follows (where the entries in the square brackets denote the corresponding projection operators):

O: the null subspace, [0]
1: the ray spanned by $|z+\rangle|z+\rangle$, $[P_{++}]$
2: the ray spanned by $|z-\rangle|z-\rangle$, $[P_{--}]$
3: the ray spanned by $\tfrac{1}{\sqrt{2}}(|z+\rangle|z-\rangle+|z-\rangle|z+\rangle)$, $[\mathrm{P}_t]$
4: the ray spanned by $\tfrac{1}{\sqrt{2}}(|z+\rangle|z-\rangle-|z-\rangle|z+\rangle)$, $[\mathrm{P}_s]$
12: $1\vee 2$, the plane spanned by $|z+\rangle|z+\rangle$ and $|z-\rangle|z-\rangle$, $[P_{++}+P_{--}]$
13: $1\vee 3$, $[P_{++}+P_t]$
14: $1\vee 4$, $[P_{++}+P_s]$

23: $2 \vee 3$, $[P_{--} + P_t]$
24: $2 \vee 4$, $[P_{--} + P_s]$
34: $3 \vee 4$, $[P_s + P_t = P_{+-} + P_{-+}]$
123: $1 \vee 2 \vee 3$, the 3-dimensional subspace spanned by $|z+\rangle|z+\rangle$, $|z-\rangle|z-\rangle$, and $\frac{1}{\sqrt{2}}(|z+\rangle|z-\rangle + |z-\rangle|z+\rangle)$, or equivalently, the subspace orthogonal to the singlet state $\frac{1}{\sqrt{2}}(|z+\rangle|z-\rangle - |z-\rangle|z+\rangle)$, $[P^{\perp}{}_s = P_{++} + P_{--} + P_t]$
124: $1 \vee 2 \vee 4$, $[P_t^{\perp} = P_{++} + P_{--} + P_s]$
134: $1 \vee 3 \vee 4$, $[P_{--}^{\perp} = P_{++} + P_s + P_t = P_{++} + P_{+-} + P_{-+}]$
234: $2 \vee 3 \vee 4$, $[P_{++}^{\perp} = P_{+-} + P_{-+} + P_{--}]$
I: the whole space, $[I]$

The preferred determinate observable $R = \sigma_z(S_2)$ has two eigenplanes in the Hilbert space $\mathcal{H}_4 = \mathcal{H}_2(S_1) \otimes \mathcal{H}_2(S_2)$: the eigenplane r_+ (associated with spin 'up' for S_2) spanned by the rays $|z+\rangle|z+\rangle$ and $|z-\rangle|z+\rangle$, and the eigenplane r_- (associated with spin 'down' for S_2) spanned by the rays $|z+\rangle|z-\rangle$ and $|z-\rangle|z-\rangle$. The Boolean sublattice of R-propositions is the 4-element Boolean lattice generated by these two planes as atoms.

The elements in $\mathcal{B}_z$ that are compatible with R are:

$$O, 1, 2, 12, 34, 134 = 2^{\perp}, 234 = 1^{\perp}, I$$

This is immediately obvious if we consider these elements as projection operators: O, P_{++}, P_{--}, $(P_{++} + P_{--})$, $(P_{+-} + P_{-+})$, $(P_{++} + P_{+-} + P_{-+})$, $(P_{+-} + P_{-+} + P_{--})$, I all commute with the projection operators: $P_+^2 = P_{++} + P_{-+}$, $P_-^2 = P_{+-} + P_{--}$ onto the R-eigenplanes r_+ and r_-, respectively. The remaining elements are:

3: P_t
4: P_s
13: $P_{++} + P_t$
14: $P_{++} + P_s$
23: $P_{--} + P_t$
24: $P_{--} + P_s$
123: $P_s^{\perp} = P_{++} + P_{--} + P_t$
124: $P_t^{\perp} = P_{++} + P_{--} + P_s$

and none of these elements commute with P_+^2 or P_-^2, because they represent rays, planes, or 3-dimensional subspaces skew to the planes r_+ and r_-. (Note the P_t and P_s – the projection operators onto $\frac{1}{\sqrt{2}}(|z+\rangle|z-\rangle + |z-\rangle|z+\rangle)$ and $\frac{1}{\sqrt{2}}(|z+\rangle|z-\rangle - |z-\rangle|z+\rangle)$, respectively – do not commute with P_{+-} or P_{-+}.)

The eight elements in $\mathcal{B}_z$ compatible with R, together with the four elements in $\mathcal{B}(R)$ (the elements O, $P_{++} + P_{-+}$, $P_{+-} + P_{--}$, I) generate a new 16-element maximal Boolean sublattice $\mathcal{B}_{z/R} \subset \mathcal{L}$ by lattice closure. The elements of $\mathcal{B}_{z/R}$ are:

O: (the null operator)
1: P_{++}
2: P_{+-}
3: P_{-+}
4: P_{--}
12: $P_{++} + P_{+-}$
13: $P_{++} + P_{-+}$
14: $P_{++} + P_{--}$
23: $P_{+-} + P_{-+}$
24: $P_{+-} + P_{--}$
34: $P_{-+} + P_{--}$
123: $P_{++} + P_{+-} + P_{-+}$
124: $P_{++} + P_{+-} + P_{--}$
134: $P_{++} + P_{-+} + P_{--}$
234: $P_{+-} + P_{-+} + P_{--}$
I: I (the unit operator)

The elements labelled O, 1, 4, 14, 23, 123, 234, I are derived from $\mathscr{B}_z$. The elements labelled 13, 24 are derived from $\mathscr{B}(R)$. The remaining elements are generated as suprema or infima or orthocomplements of these elements (that is, as spans or intersections or orthogonal complements of the subspaces that are the ranges of the projection operators). The Boolean lattice $\mathscr{B}_{z/R}$ is just the maximal Boolean sublattice generated by the atoms corresponding to the rays spanned by the eigenvectors $|z+\rangle|z+\rangle, |z+\rangle|z-\rangle, |z-\rangle|z+\rangle, |z-\rangle|z-\rangle$ of the complete commuting set of observables $\{\sigma_z(S_1), \sigma_z(S_2)\}$ of $S_1 + S_2$ (the maximal observable $\sigma_z(S_1)\&\sigma_z(S_2)$ in the notation of section 4.2).

Since the singlet state is invariant under a change of spin direction:

$$|\psi\rangle = \tfrac{1}{\sqrt{2}}(|z+\rangle|z-\rangle - |z-\rangle|z+\rangle)$$
$$= \tfrac{1}{\sqrt{2}}(|x+\rangle|x-\rangle - |x-\rangle|x+\rangle), \text{ for any other direction } x \neq z,$$

the 3-dimensional subspace orthogonal to $e(|\psi\rangle)$ contains the rays spanned by the three vectors of the triplet state for the direction x:

$$|x+\rangle|x+\rangle$$
$$\tfrac{1}{\sqrt{2}}(|x+\rangle|x-\rangle + |x-\rangle|x+\rangle)$$
$$|x-\rangle|x-\rangle$$

These three rays and the singlet state e generate a 16-element maximal Boolean sublattice $\mathscr{B}_x$ that coincides with the sublattice $\mathscr{B}_z$ on the minimum and maximum elements, on the atom corresponding to e, and on the orthocomplement of this atom (the element corresponding to the 3-dimensional subspace orthogonal to the singlet state). There are no elements in $\mathscr{B}_x$ compatible with the elements in $\mathscr{B}(R)$, apart from the elements O and I, so $\mathscr{B}_{x/R} = \mathscr{B}(R)$ for all $x \neq z$.

This is perhaps obvious if we consider the corresponding sets of observables. In the complete commuting set of observables $\{\sigma^2, \sigma_z\}$ associated with $\mathscr{B}_z$, only the observable σ_z is compatible with $\sigma_z(S_2)$. But in the complete commuting set $\{\sigma^2, \sigma_x\}$, there are no observables compatible with $\sigma_z(S_2)$.

There are other maximal Boolean sublattices in $\mathscr{L}_e$ that are not sublattices generated by the singlet and triplet spin states for some spin direction (for example, the sublattice generated by the ray spanned by the singlet state, the ray spanned by the orthogonal state $\frac{1}{\sqrt{2}}(|z+\rangle|z-\rangle + |z+\rangle|z+\rangle)$, and any two rays in the plane spanned by $|z+\rangle|z+\rangle$ and $|z-\rangle|z-\rangle$ other than the rays spanned by these vectors). It is easy to see that none of these sublattices contains elements compatible with the elements in $\mathscr{B}(R)$ (apart from the elements O and I), so they make no contribution to $\mathscr{L}_{e/R}$. It follows that $\mathscr{L}_{e/R} = \mathscr{B}_{z/R}$.

The projection of the ray spanned by the singlet state onto the subspace r_+ associated with spin 'up' for S_2 (the subspace that is the range of the projection operator $P_{++} + P_{-+}$) is just e_{r_+}, the ray spanned by the vector $|z-\rangle|z+\rangle$. Similarly, the projection of the ray spanned by the singlet state onto the subspace r_- associated with spin 'down' for S_2 (the subspace that is the range of the projection operator $P_{+-} + P_{--}$) is just e_{r_-}, the ray spanned by the vector $|z+\rangle|z-\rangle$. The lattice $\mathscr{L}_{e_{r_+}e_{r_-}}$ is generated by the rays e_{r_+}, e_{r_-}, and all the rays in the 2-dimensional subspace $(e_{r_+} \vee e_{r_-})^\perp$. The maximal Boolean sublattices in $\mathscr{L}_{e_{r_+}e_{r_-}}$ are all 4-atom Boolean lattices generated by the atoms e_{r_+}, e_{r_-}, and any two atoms corresponding to an orthogonal pair of rays in $(e_{r_+} \vee e_{r_-})^\perp$. One such Boolean sublattice is $\mathscr{B}_{z/R} = \mathscr{L}_{e/R}$, so in this case $\mathscr{L}_{e/R} \subset \mathscr{L}_{e_{r_+}e_{r_-}}$.

Again, the elements in $\mathscr{L}_e$ that survive the transition to $\mathscr{L}_{e_{r_+}e_{r_-}}$ retain their truth values, for either of the 2-valued homomorphisms on $\mathscr{L}_{e_{r_+}e_{r_-}}$ associated with the two possible values of $R = \sigma_z(S_2)$. Since $\mathscr{B}_z$ is the only maximal Boolean sublattice in $\mathscr{L}_e$ that contributes elements to $\mathscr{B}_{z/R} = \mathscr{L}_{e/R}$ that are not in $\mathscr{B}(R)$, and $\mathscr{L}_{e_{r_+}e_{r_-}}$ differs from $\mathscr{B}_{z/R}$ by the addition of 'null' elements only, it suffices to consider the transition from $\mathscr{B}_z$ to $\mathscr{B}_{z/R}$. In fact, if the elements that survive the transition from $\mathscr{B}_z$ to $\mathscr{B}_{z/R}$ are required to preserve their truth value, the elements true in $\mathscr{B}_z$ and the elements true in $\mathscr{B}(R)$ together determine the elements true in $\mathscr{B}_{z/R}$.

The elements true in $\mathscr{B}_z$ are contained in the 8-element ultrafilter generated by the singlet state, consisting of the elements 4, 14, 24, 34, 124, 134, 234, I (the elements represented by the projection operators: P_s, $P_{++} + P_s$, $P_{--} + P_s$, $P_{+-} + P_{-+}$, $P_{++} + P_{--} + P_s$, $P_{++} + P_{+-} + P_{-+}$, $P_{+-} + P_{-+} + P_{--}$, I). The elements true in $\mathscr{B}(R)$ are contained in the ultrafilter $P_{++} + P_{-+}$, I, corresponding to the value $+1$ for $\sigma_z(S_2)$, or the ultrafilter $P_{+-} + P_{--}$, I, corresponding to the value -1 for $\sigma_z(S_2)$. The only elements in the $\mathscr{B}_z$ ultrafilter that transfer to $\mathscr{B}_{z/R}$ are:

$P_{+-} + P_{-+}$
$P_{++} + P_{+-} + P_{-+}$
$P_{+-} + P_{-+} + P_{--}$
I

In $\mathscr{B}_{z/R}$ these are the elements at positions 23 ($P_{+-} + P_{-+}$), 123 ($P_{++} + P_{+-} + P_{-+}$), 234 ($P_{+-} + P_{-+} + P_{--}$), and I. A truth value assignment to $\mathscr{B}_{z/R}$ is a 2-valued homomorphism on $\mathscr{B}_{z/R}$. Since the element 23 is mapped onto 1, and either the element 13 ($P_{++} + P_{-+}$), or the element 24 ($P_{+-} + P_{--}$), is also mapped onto 1, it follows that either the element 2 is mapped onto 1 (the ray spanned by the vector $|z+\rangle|z-\rangle$) or the element 3 is mapped onto 1 (the ray spanned by the vector $|z-\rangle|z+\rangle$).

8

The new orthodoxy

> It is easy to imagine a state vector for the whole universe, quietly pursuing its linear evolution through all of time and containing somehow all possible worlds. But the usual interpretive axioms of quantum mechanics come into play only when the system interacts with something else, is 'observed'. For the universe there *is* nothing else, and quantum mechanics in its traditional form has simply nothing to say. It gives no way of, indeed no meaning in, picking out from the wave of possibility the single unique thread of history.
>
> *Bell and Nauenberg (1966, p. 284)*

8.1 Decoherence

For most physicists, the measurement problem of quantum mechanics would hardly rate as even a 'small cloud' on the horizon. The standard view is that Bohr had it more or less right, and that anyone willing to waste a little time on the subject could easily straighten out the sort of muddle philosophers might get themselves into. There seems to be a growing consensus that a modern, definitive version of the Copenhagen interpretation has emerged, in terms of which the Bohr–Einstein debate can be seen as a rather old-fashioned way of dealing with issues that are now much more clearly understood.

This 'new orthodoxy' weaves together several strands: the physical phenomenon of environment-induced decoherence (Joos and Zeh, 1985; Zurek, 1981, 1982, 1991, 1993a,b), elements of Everett's 'relative state' formulation of quantum mechanics (Everett, 1957, 1973), popularized as the 'many worlds' interpretation, and the notion of 'consistent histories' developed by Griffiths (1984, 1987) and extended in different ways by Omnès (1990, 1992, 1994), Gell-Mann and Hartle (1990, 1991a,b), and others. Omnès (1994, p. xiii) refers to 'the interpretation of quantum mechanics, not an interpretation,' and characterizes the view as 'simply a modernized version of the interpretation first proposed by Bohr in the early days of quantum mechanics' (1994, p. 498).

The idea behind environment-induced decoherence is that the sort of environmental 'monitoring' discussed in sections 5.4 and 6.2 leads almost instantaneously to the destruction of interference between the component states of a superposition of macroscopically distinguished states, such as distinct pointer-reading states of a measuring instrument.

Consider the simple example discussed in section 5.4. A spin-$\frac{1}{2}$ system S, initially in the state $\frac{1}{\sqrt{2}}|z+\rangle + \frac{1}{\sqrt{2}}|z-\rangle$, where $|z+\rangle$ and $|z-\rangle$ are the eigenstates of σ_z, interacts with a second system M, whose states are also represented on a 2-dimensional Hilbert space, so that after a certain time the state of $S + M$ takes the form:

$$|\psi\rangle = \tfrac{1}{\sqrt{2}}|z+\rangle|r_{z+}\rangle + \tfrac{1}{\sqrt{2}}|z-\rangle|r_{z-}\rangle \tag{8.1}$$

The entangled state $|\psi\rangle$ is a pure state, a superposition of product states, that differs from the corresponding mixture defined by the density operator:

$$W = \tfrac{1}{2}\, P_{z+} \otimes P_{r_{z+}} + \tfrac{1}{2}\, P_{z-} \otimes P_{r_{z-}} \tag{8.2}$$

The density operator associated with the state $|\psi\rangle$ is the projection operator P_ψ, where:

$$\begin{aligned} P_\psi &= \tfrac{1}{2}\, P_{z+} \otimes P_{r_{z+}} + \tfrac{1}{2}\, P_{z-} \otimes P_{r_{z-}} + \tfrac{1}{2}|z+\rangle\langle z-| \otimes |r_{z+}\rangle\langle r_{z-}| \\ &\quad + \tfrac{1}{2}|z-\rangle\langle z+| \otimes |r_{z-}\rangle\langle r_{z+}| \\ &= W + \text{interference terms} \end{aligned}$$

So the difference between the pure state $|\psi\rangle$ and the mixture W is that the contributions to the probability of any proposition P by the component states $|z+\rangle|r_{z+}\rangle$ and $|z-\rangle|r_{z-}\rangle$ are simply additive in the mixture W:

$$\begin{aligned} \text{prob}_W(P) &= \text{tr}(WP) \\ &= \tfrac{1}{2}\, \text{tr}(P_{z+} \otimes P_{r_{z+}} \cdot P) + \tfrac{1}{2}\, \text{tr}(P_{z-} \otimes P_{r_{z-}} \cdot P) \end{aligned}$$

while they produce interference effects in the pure state $|\psi\rangle$:

$$\text{prob}_\psi(P) = \text{tr}(P_\psi P) = \text{prob}_W(P) + \text{interference terms}$$

Classically, of course, we would expect that the probability assigned to P by an equal-weight distribution of two stochastic states is just the sum of the probabilities assigned to P by the two states separately, with equal weights.

As Furry (1936) observed for the similar entangled state arising in the spin version of the Einstein–Podolsky–Rosen experiment (the only difference being a '$-$' instead of a '$+$' in the superposition of the component states), the mixture W and the pure state $|\psi\rangle$ will yield the same statistics for any S-observable alone, or any M-observable alone, or even any $(S + M)$-observable of the form $V_S \otimes V_M$, where V_S is any observable that commutes with σ_z or V_M is any observable that commutes with R_z.

This becomes obvious if we note that:

$$\text{tr}(|z+\rangle\langle z-| \otimes |r_{z+}\rangle\langle r_{z-}| \cdot V_S \otimes V_M) = \text{tr}(|z-\rangle\langle z+| \otimes |r_{z-}\rangle\langle r_{z+}| \cdot V_S \otimes V_M) = 0$$

if $|z+\rangle$ and $|z-\rangle$ are eigenstates of V_S and $|r_{z+}\rangle$ and $|r_{z-}\rangle$ are eigenstates of V_M. For observables $V_S \otimes I_M$ or $I_S \otimes V_M$, this also follows immediately from the fact that W and P_ψ have the same reduced density operators for the subsystems S and M:[112]

$$W_S = \tfrac{1}{2} P_{z+} + \tfrac{1}{2} P_{z-}$$
$$W_M = \tfrac{1}{2} P_{r_{z+}} + \tfrac{1}{2} P_{r_{z-}}$$

and so:

$$\mathrm{Exp}_W(V_S \otimes I_M) = \mathrm{tr}(W_S V_S) = \mathrm{Exp}_\psi(V_S \otimes I_M)$$
$$\mathrm{Exp}_W(I_S \otimes V_M) = \mathrm{tr}(W_M V_M) = \mathrm{Exp}_\psi(I_S \otimes V_M)$$

So the pure state $|\psi\rangle$ and the mixture W can only be distinguished experimentally by measuring an observable of the composite system $S + M$ of the form $V_S \otimes V_M$, where V_S does not commute with σ_z and V_M does not commute with R, or an observable of $S + M$ that does not reduce to a tensor product of an S-observable and an M-observable.

Consider, for example, the observable $\sigma_x - R_x$ (more precisely, $\sigma_x \otimes I_M - I_S \otimes R_x$),[113] where the eigenstates of R_x are

$$|r_{x+}\rangle = \tfrac{1}{\sqrt{2}}|r_{z+}\rangle + \tfrac{1}{\sqrt{2}}|r_{z-}\rangle$$
$$|r_{x-}\rangle = -\tfrac{1}{\sqrt{2}}|r_{z+}\rangle + \tfrac{1}{\sqrt{2}}|r_{z-}\rangle$$

Then:

$$|\psi\rangle = \tfrac{1}{\sqrt{2}}|z+\rangle|r_{z+}\rangle + \tfrac{1}{\sqrt{2}}|z-\rangle|r_{z-}\rangle$$
$$= \tfrac{1}{\sqrt{2}}|x+\rangle|r_{x+}\rangle + \tfrac{1}{\sqrt{2}}|x-\rangle|r_{x-}\rangle$$

So $|\psi\rangle$ is an eigenstate of $\sigma_x - R_x$ (as well as $\sigma_z - R_z$), with the eigenvalue 0, and a measurement of $\sigma_x - R_x$ on the system $S + M$ in the state $|\psi\rangle$ will yield the value 0 with probability 1.

To compute the probability of the eigenvalue 0 for a measurement of $\sigma_x - R_x$ on $S + M$ in the mixture W, notice that the 0-eigenspace of $\sigma_x - R_x$ in the 4-dimensional Hilbert space of $S + M$ is a 2-dimensional eigenspace spanned by the vectors $|x+\rangle|r_{x+}\rangle$ and $|x-\rangle|r_{x-}\rangle$. Neither of the component states in the mixture W, the states $|z+\rangle|r_{z+}\rangle$ and $|z-\rangle|r_{z-}\rangle$, lie in this subspace. Since:

$$|z+\rangle|r_{z+}\rangle = (\tfrac{1}{\sqrt{2}}|x+\rangle - \tfrac{1}{\sqrt{2}}|x-\rangle)(\tfrac{1}{\sqrt{2}}|r_{x+}\rangle - \tfrac{1}{\sqrt{2}}|r_{x-}\rangle)$$
$$= \tfrac{1}{2}|x+\rangle|r_{x+}\rangle - \tfrac{1}{2}|x+\rangle|r_{x-}\rangle - \tfrac{1}{2}|x-\rangle|r_{x+}\rangle + \tfrac{1}{2}|x-\rangle|r_{x-}\rangle$$

the projection of $|z+\rangle|r_{r+}\rangle$ onto the 0-eigenspace of $\sigma_x - R_x$ is:

$$\tfrac{1}{2}|x+\rangle|r_{x+}\rangle + \tfrac{1}{2}|x-\rangle|r_{x-}\rangle$$

and the square of the length of this projection is $\frac{1}{4} + \frac{1}{4} = \frac{1}{2}$. Similarly, the square of the length of the projection of $|z-\rangle|r_{z-}\rangle$ onto the 0-eigenspace of σ_x is $\frac{1}{2}$. So the

[112] See the appendix, section A.5.

[113] Readers familiar with David Albert's *Quantum Mechanics and Experience* (1992) will recognize this observable as Albert's 'Zip – Colour' observable (more precisely: $\sigma_x - R_x = -$(Zip – Colour)).

probability of getting the 0-eigenvalue in a measurement of $\sigma_x - R_x$ on the mixture W, where the states $|z+\rangle|r_{z+}\rangle$ and $|z-\rangle|r_{z-}\rangle$ occur with equal weights is $\frac{1}{2}\cdot\frac{1}{2} + \frac{1}{2}\cdot\frac{1}{2} = \frac{1}{2}$.

Since $|\psi\rangle$ is an eigenstate of $\sigma_z - R_z$ with eigenvalue 0, and $|z+\rangle|r_{z+}\rangle$ and $|z-\rangle|r_{z-}\rangle$ are also eigenstates of $\sigma_z - R_z$ with the eigenvalue 0, the probability of getting the eigenvalue 0 in a measurement of $\sigma_z - R_z$ in the mixture W is 1, as for the pure state $|\psi\rangle$. So there are observables of $S + M$ that do not reduce to tensor products of S-observables and M-observables for which the pure state $|\psi\rangle$ and the mixture W yield the same statistics.

This example demonstrates two features of entangled states of composite systems: Firstly, certain entangled states of $S + M$ can define a correlation between σ_z-values and the values of an observable R_z of M, and at the same time a similar correlation between σ_x-values and the values of an observable R_x of M, where R_x does not commute with R_z. So, regarded as a quantum measuring device, M simultaneously 'measures' the two noncommuting observables, σ_x and σ_z. Secondly, the difference between the pure state $|\psi\rangle$ and the corresponding mixture W, for the representation of $|\psi\rangle$ as a superposition of tensor products of eigenstates of σ_z and R_z, only becomes apparent in the measurement statistics of a certain class of observables of the composite system. In particular, $|\psi\rangle$ and W generate the same statistics for observables of S alone or observables of M alone.

If M interacts with a third system E (the 'environment'), whose states are also represented on a 2-dimensional Hilbert space, resulting in the transition, after a certain time, to a pure state of the form:

$$|\psi'\rangle = \tfrac{1}{\sqrt{2}}|z+\rangle|r_{z+}\rangle|\varepsilon_{z+}\rangle + \tfrac{1}{\sqrt{2}}|z-\rangle|r_{z-}\rangle|\varepsilon_{z-}\rangle, \qquad (8.3)$$

where $|\varepsilon_{z+}\rangle$ and $|\varepsilon_{z-}\rangle$ are two orthogonal states that span the Hilbert space of E, then the state $|\psi'\rangle$ is no longer an eigenstate of $\sigma_x - R_x$ (more precisely, an eigenstate of $\sigma_x \otimes I_M \otimes I_E - I_S \otimes R_x \otimes I_E$). We can't distinguish the pure state $|\psi'\rangle$ from the corresponding mixture:

$$W' = \tfrac{1}{2}\, P_{z+} \otimes P_{r_{z+}} \otimes P_{\varepsilon_{z+}} + \tfrac{1}{2}\, P_{z-} \otimes P_{r_{z-}} \otimes P_{\varepsilon_{z-}} \qquad (8.4)$$

by the measurement statistics of observables of $S + M$ alone, because $|\psi'\rangle$ and W' yield the same reduced state for $S + M$: the mixed state represented by the density operator W.

To distinguish $|\psi'\rangle$ and W', we would have to measure an appropriate observable of the composite system $S + M + E$ that is not representable as a tensor product of the form $V_S \otimes V_M \otimes V_E$, or an observable of the form $V_S \otimes V_M \otimes V_E$ where the eigenstates of V_S and V_M differ from the eigenstates of σ_z and R_z, respectively, and the eigenstates of V_E differ from the vectors $|\varepsilon_{z+}\rangle$ and $|\varepsilon_{z-}\rangle$.

If the system E is a composite system consisting of n subsystems, each with its own 2-dimensional Hilbert space, and the interaction (after a certain time) preserves the form of the original biorthogonal decomposition of $|\psi\rangle$ for the eigenstates of σ_z and R_z as an n-orthogonal decomposition, then as n increases, it becomes more and more difficult to distinguish between the statistics generated by the pure state representing the n-orthogonal decomposition, and the statistics generated by the corresponding mixture. The n-orthogonal pure state and the corresponding mixture will differ only with respect to a certain class of observables of the *entire* composite system of $n + 2$ systems.

As I pointed out in section 5.4, this is essentially the example considered by Zurek (1981) to illustrate the origin of environment-induced decoherence, the only difference being inessential phase differences in the pure states. In Zurek's example, the system S is a spin-$\frac{1}{2}$ particle in a reversible Stern–Gerlach apparatus interacting with a bistable atom M on one of the trajectories in the apparatus. The environment E consists of a second bistable atom that interacts with the atom M via an interaction Hamiltonian that commutes with R_z, not with R_x, or n bistable atoms interacting similarly. Such an interaction preserves the original correlations in the state $|\psi\rangle$ between σ_z and R_z, but not between σ_x and R_x (or any other observables of S and M).

Consider again the state (8.3):

$$|\psi'\rangle = \tfrac{1}{\sqrt{2}}|z+\rangle|r_{z+}\rangle|\varepsilon_{z+}\rangle + \tfrac{1}{\sqrt{2}}|z-\rangle|r_{z-}\rangle|\varepsilon_{z-}\rangle$$

Taking the partial trace over the Hilbert space of the environment system E yields the reduced state of $S + M$ as the mixture (8.2):

$$W = \tfrac{1}{2}\, P_{z+} \otimes P_{r_{z+}} + \tfrac{1}{2}\, P_{z-} \otimes P_{r_{z-}}$$

During the interaction between M and E, when the states $|\varepsilon_{z+}(t)\rangle$ and $|\varepsilon_{z-}(t)\rangle$ are not orthogonal, the reduced density operator of $S + M$ takes the form:

$$W(t) = W + \text{interference terms},$$

where the off-diagonal interference terms (in the density matrix) are time dependent and decay to zero.

So it seems that the interaction between the measuring instrument and the environment leads to a destruction of interference for the composite system consisting of the measured system S and the measuring instrument M, with respect to the particular pointer basis selected by the instrument–environment interaction (or, more precisely, with respect to the basis defined by tensor products of eigenstates of this pointer observable and eigenstates of the S-observable correlated with pointer eigenstates by the original measurement interaction). By performing what can be regarded as a nondemolition or non-disturbing measurement on a particular observable of the instrument M, the environment leads to decoherence or loss of interference between certain product states of $S + M$ in the pure state arising from what is sometimes referred to as the initial 'pre-measurement' interaction. For the sort of macrosystems we are able to use as measuring instruments, the environment E typically consists of an enormous number of subsystems, and decoherence turns out to occur almost instantaneously. So the environment, it is said, induces a 'superselection rule,' effectively forbidding certain superpositions of states.[114]

For the single-particle environment considered above, as the state of $S + M + E$ evolves in time, $S + M$ becomes periodically disentangled from the environment E. The state of $S + M + E$ takes the form of a product state for $S + M$ and E, so that the state of $S + M$ alone (the reduced state of $S + M$) becomes a pure state, with the reappearance of interference. For a composite environment consisting of many subsystems, the time scale for the reappearance of interference becomes longer as the number of subsystems in the environment increases.

I have sketched the basic idea behind environmental decoherence using an elementary model of two-state systems. Current discussions in the literature by Zurek and others develop considerably more sophisticated models of the environment, but the underlying mechanism is similar. The essential feature of all these models is that the specific sorts of interactions that take place in our world, between the systems we are able to use as measuring instruments and the environment of these systems, select particular observables as suitable measurement pointers. These are the observables that become correlated with observables of the systems we measure in quantum mechanical measurement interactions, for correlations that remain stable over time while the systems functioning as measuring instruments undergo an effective 'monitoring' by the environment.

A similar model is discussed in Zurek (1982). Omnès (1994, p. 279) calls this the 'simplest model.' A two-state system S interacts with an environment E consisting of n two-state systems via the interaction Hamiltonian:

$$H_{\text{int}} = \sum_j g_j R \otimes R_j \otimes \prod_{j' \neq j} I_{j'}$$

The eigenstates of the observable R of S are $|+\rangle$ and $|-\rangle$, and the eigenstates of the observable R_j of the jth environmental system coupled with R in the interaction, for $j = 1, \ldots, n$, are $|+\rangle_j$ and $|-\rangle_j$. The g_j are coupling constants. Solving the Schrödinger equation for this Hamiltonian (assuming a zero Hamiltonian for $S + E$ before the interaction) yields the transition from an initial state:

$$|\psi(0)\rangle = (a|+\rangle + b|-\rangle) \prod_{j=1}^{n} \otimes (\alpha_j|+\rangle_j + \beta_j|-\rangle_j)$$

to the state at time t:

$$|\psi(t)\rangle = a|+\rangle \prod_{j=1}^{n} \otimes (\alpha_j \exp(\mathrm{i}g_j t)|+\rangle_j + \beta_j \exp(-\mathrm{i}g_j t)|-\rangle_j)$$
$$+ b|-\rangle \prod_{j=1}^{n} \otimes (\alpha_j \exp(-\mathrm{i}g_j t)|+\rangle_j + \beta_j \exp(\mathrm{i}g_j t)|-\rangle_j)$$

[114] A superselection rule partitions the Hilbert space into orthogonal superselection sectors or 'coherent' subspaces. The superposition principle applies to vectors within a given coherent subspace only. Hence the term 'decoherence.' An observable whose subspaces are the coherent subspaces commutes with every observable. So a quantum system with a superselection rule is characterized by a noncommutative algebra of observables with a 'nontrivial center': there are observables other than the unit that commute with all observables.

Taking the partial trace of the density operator $|\psi(t)\rangle\langle\psi(t)|$ over the environment yields the reduced state of S as the mixture:

$$W(t) = |a|^2|+\rangle\langle+| + |b|^2|-\rangle\langle-| + z(t)ab^*|+\rangle\langle-| + z^*(t)a^*b|-\rangle\langle+|$$

where

$$z(t) = \prod_{j=1}^{n}[\cos 2g_j t + i(|\alpha_j|^2 - |\beta_j|^2)\sin 2g_j t]$$

When $t = 0$, $z(t) = 1$, as we expect, and W represents a pure state with interference terms. When $t \neq 0$, $z(t)$ is a product of n complex numbers, each of which has absolute value < 1, unless $2g_j t$ is a multiple of π, or $|\alpha_j|^2 = 1$ or 0 (in which case $|\beta_j|^2 = 1 - |\alpha_j|^2 = 0$ or 1). If we assume that the initial states of the environmental systems and the coupling constants g_j are distributed randomly, then $z(t)$ will have a very small absolute value if n is sufficiently large.

Since $z(t)$ is multiply periodic, it will return arbitrarily close to its initial value of 1, so environmental decoherence – the spontaneous dynamical diagonalization of the reduced density matrix of the system in a specific basis – will persist for a period of time, until there is a reappearance of interference. If n is sufficiently large, the 'recurrence time' can be shown to be very long.

More sophisticated models treat the environment as a collection of harmonic oscillators or as a quantum field. Zurek (1991) considers a particle initially in a state represented by a wave function that is a coherent superposition of two narrow Gaussian functions with widths δ that are separated by a distance $\Delta \gg \delta$. The density matrix in the position representation has four peaks: two peaks on the diagonal, corresponding to the ranges of position of width δ where the Gaussians take values that are appreciably different from zero, and two peaks off the diagonal representing the interference terms. A master equation is proposed to describe the evolution of the density matrix in the position representation, as the particle interacts with its environment. The master equation involves three terms: a term characterizing the usual unitary Schrödinger evolution of a closed system; a dissipation term characterizing a loss of energy and a decrease in the average momentum; and a decoherence term that has the effect of rapidly reducing the off-diagonal peaks, with negligible effect on the diagonal peaks. For a system at room temperature with a mass of 1 gram and an initial separation between the Gaussians of 1 centimetre, Zurek (1991, p. 41) estimates the ratio of the decoherence time to the relaxation time as 10^{-40}. Even if the relaxation time is of the order of the age of the universe, $\sim 10^{17}$ seconds, interference between the peaks is destroyed in $\sim 10^{-23}$ seconds.

This analysis is similar to an earlier calculation by Joos and Zeh (1985) for a dust grain interacting with an air molecule.[115] If the state of the dust grain is $|q\rangle$, where q represents the position of the center of mass, the collision with an air molecule initially in the state $|\varepsilon_0\rangle$ is described by the transition:

$$|q\rangle|\varepsilon_0\rangle \rightarrow |q\rangle|\varepsilon_q\rangle,$$

assuming the mass of the dust grain is infinite relative to that of the air molecule, so that its position is unchanged in the interaction. If the state of the dust grain is a superposition $|\alpha\rangle = \int dq|q\rangle\langle q|\alpha\rangle = \int dq\phi(q)|q\rangle$, the collision with the air molecule is described by the transition:

$$|\alpha\rangle|\varepsilon_0\rangle \rightarrow |\Psi\rangle = \int dq\phi(q)|q\rangle|\varepsilon_q\rangle$$

The density operator of the composite two-particle system after the interaction is:

$$|\Psi\rangle\langle\Psi| = \iint dq dq'\phi(q)\phi^*(q')|q\rangle\langle q'| \otimes |\varepsilon_{q'}\rangle\langle\varepsilon_q|$$

The reduced density operator of the dust particle alone is:

$$W = \iint dq dq'\phi(q)\phi^*(q')\langle\varepsilon_{q'}|\varepsilon_q\rangle|q\rangle\langle q'|$$

with the corresponding density matrix:

$$\begin{aligned} W(q'',q''') &= \langle q''|W|q'''\rangle \\ &= \iint dq dq'\phi(q)\phi^*(q')\delta(q - q'')\delta(q' - q''')|\varepsilon_{q'}\rangle\langle\varepsilon_q| \\ &= \phi(q'')\phi^*(q''')\langle\varepsilon_{q'''}|\varepsilon_{q''}\rangle \end{aligned}$$

In the absence of the collision, the q,q'-term in the density matrix of the dust particle would have been $\phi(q)\phi^*(q')$. So the effect of the collision is to multiply this term by the factor $\langle\varepsilon_q|\varepsilon_{q'}\rangle$, which is small if $|q' - q|$ is sufficiently large, since $|\varepsilon_q\rangle$ and $|\varepsilon_{q'}\rangle$ are two states of the air molecule scattered from scattering centers at different positions. In fact, it turns out that $\langle\varepsilon_q|\varepsilon_{q'}\rangle \approx 0$ if $|q' - q| \gg \lambda$, where λ is the wavelength of the particle. Joos and Zeh show that if there are many such scattering processes for randomly distributed scattered air molecules, then the off-diagonal elements of W decay exponentially:

$$W(q, q', t) = W(q,q')\exp[-\Lambda(q - q')^2 t]$$

Joos (1986) calculates the 'localization rate' Λ as 10^{19} cm^2 s^{-1} for a dust particle with radius 10^{-5} cm, even in the best available laboratory vacuum! So the off-diagonal elements in the reduced density matrix of the dust particle in the position representation decay almost instantaneously to zero.

[115] For an insightful critique, see d'Espagnat (1995, pp. 257–67). I follow d'Espagnat's simplified discussion.

So this is apparently what environmental decoherence does for us: When the evolution of an open system – a system that interacts with its environment – is governed by an interaction Hamiltonian that commutes with a certain observable R of the system, the reduced density matrix of the system is rapidly diagonalized in the R-basis (or, more generally, with respect to the R-eigenspaces, if R is degenerate in the Hilbert space of the system). In particular, since any macroscopic measuring instrument is an open system, environmental decoherence – a dynamical effect of the interaction between a measuring instrument and its environment – singles out a preferred pointer basis for the instrument.

The question is: Does this solve the measurement problem?

Consider, first, in what sense a preferred basis is selected by the instrument–environment interaction. In a model quantum mechanical universe consisting of a system S, a measuring instrument M, and an environment E, an initial pure state of $S + M + E$ evolves, after a measurement interaction between S and M and a further interaction between M and E, to a new pure state. The reduced density matrix for $S + M$ is diagonal in the pointer basis (that is, the product basis of pointer eigenstates and S-eigenstates correlated with pointer states by the measurement) only if the environmental states coupled with the pointer states in the pure state of $S + M + E$ are mutually orthogonal. So environmental decoherence is really a dynamical process in which the pure state of the universe evolves to an n-orthogonal form with respect to the pointer basis of the instrument.

For example, suppose in a measurement of σ_z on a spin-$\frac{1}{2}$ system S, the composite system $S + M + E$ ends up in a pure state of the form (8.3):

$$|\psi'\rangle = \tfrac{1}{\sqrt{2}}|z+\rangle|r_{z+}\rangle|\varepsilon_{z+}\rangle + \tfrac{1}{\sqrt{2}}|z-\rangle|r_{z-}\rangle|\varepsilon_{z-}\rangle,$$

where the environmental states $|\varepsilon_{z+}\rangle$ and $|\varepsilon_{z-}\rangle$ are orthogonal. We can think of $|r_{z+}\rangle$ and $|r_{z-}\rangle$ as two states in different multi-dimensional eigenspaces of a pointer observable R_z of M, if M is a system associated with a multi-dimensional Hilbert space (rather than a bistable atom). It is because the specific nature of the interaction between M and E preserves the form of the biorthogonal decomposition (8.1) arising from the interaction between S and M:

$$|\psi\rangle = \tfrac{1}{\sqrt{2}}|z+\rangle|r_{z+}\rangle + \tfrac{1}{\sqrt{2}}|z-\rangle|r_{z-}\rangle$$

as the triorthogonal decomposition (8.3) or, more precisely, as an n-orthogonal decomposition, since the environmental states $|\varepsilon_{z+}\rangle$ and $|\varepsilon_{z-}\rangle$ are product states, that the reduced density matrix of $S + M$ is diagonal with respect to the eigenspaces of the pointer observable R. What the dynamical decoherence models show is that the environmental states $|\varepsilon_{z+}(t)\rangle$ and $|\varepsilon_{z-}(t)\rangle$ become very nearly orthogonal almost instantaneously.

In the discussion of environmental monitoring in section 5.4, I showed that the pointer observable is selected by the instrument–environmental interaction, because the form of the resulting state as a triorthogonal or n-orthogonal decomposition in the pointer basis is *unique*. This uniqueness result holds as a general tridecompositional or n-decompositional theorem for *any* distinct (noncollinear) environmental states (not necessarily orthogonal states), if the states of S (and M) are even linearly independent (and not necessarily the eigenstates of any observable). So the pointer observable is selected by the form of the instrument–environment interaction, which leads to a tridecomposed pure state for $S + M + E$ after an ideal or non-ideal measurement interaction between S and M, and does not depend on 'tracing out' the environment E to yield a reduced density matrix for $S + M$ alone that is diagonal in the basis defined by the pointer and the measured S-observable. In other words, even if the off-diagonal terms in the reduced density matrix of $S + M$ do not decay to zero, the pointer is selected as preferred by the uniqueness of the form of the state of $S + M + E$ induced by environmental monitoring, after an ideal or non-ideal measurement.

The decay to zero of the off-diagonal terms in the reduced density matrix of $S + M$ serves quite a different purpose: decoherence, or the destruction of interference between different pointer-reading states. Decoherentists suppose, firstly, that a necessary condition for a solution to the measurement problem is the destruction of interference between different pointer-reading states. As I pointed out in section 7.1, this supposition stands and falls with the orthodox (Dirac–von Neumann) interpretation principle (the 'eigenvalue–eigenstate link'), and the whole point of the theorem of chapter 4 is to show that this principle is not forced on us by any structural features of quantum mechanics. Secondly, decoherentists suppose that this is also a sufficient condition for a solution to the problem.

To see that this is not the case, let's accept that the environmental states $|\varepsilon_{z+}\rangle$ and $|\varepsilon_{z-}\rangle$ in the decomposition of the state $|\psi'\rangle$ given by equation (8.3) with respect to the pointer eigenstates $|r_{z+}\rangle$ and $|r_{z-}\rangle$ are strictly orthogonal. The reduced state of $S + M$ is then the mixture (8.2):

$$W = \tfrac{1}{2}\, P_{z+} \otimes P_{r_{z+}} + \tfrac{1}{2}\, P_{z-} \otimes P_{r_{z-}}$$

There would be no measurement problem if we could now say that the pointer observable R of M has a determinate value, either r_{z+} or r_{z-}, and that the observable σ_z of S has the correlated value, $+1$ or -1. That is, the measurement problem would be resolved if we could interpret the mixture (8.2) as representing the occurrence of a particular event, either the event associated with the pair of eigenvalues r_{z+} for R and $+1$ for σ_z, or the event associated with the pair of eigenvalues r_{z-} for R and -1 for σ_z, with the coefficients representing a measure of our ignorance as to the actual event, expressed here in terms of equal probabilities for the two events. But the fact that the reduced density operator of $S + M$ takes this form does not justify any such claim.

The reduced density operator of $S + M$ is derived by taking the partial trace of the pure state of $S + M + E$ over the system E. Zurek (1991, p. 39) describes this as 'ignoring (tracing over) the uncontrolled (and unmeasured) degrees of freedom' of the environment. The suggestion is that this is similar to the procedure of deriving a probability of $\frac{1}{2}$ for 'heads' and $\frac{1}{2}$ for 'tails' in a coin toss experiment by averaging over the uncontrolled and unmeasured degrees of freedom of the environment of the coin. But the two procedures are not at all analogous. When we 'ignore' the environment to claim that the probability of getting 'heads' on a particular toss of the coin is $\frac{1}{2}$, we can also claim that we *do* in fact get *either* 'heads' *or* 'tails' on each particular toss, and whether we get 'heads' or 'tails' on a particular toss depends on the precise values of certain environmental parameters, which we do not attempt to control or measure. But in the quantum mechanical case, we can't claim that taking full account of the environment on each particular occasion would fix the value of the pointer R as either r_{z+} or r_{z-}. Taking full account of the environment will, of course, give us back the pure state of $S + M + E$ from which we derived the mixture W. So, given its origin, the mixture (8.2) is actually *inconsistent* with the occurrence of either of the events associated with the pointer readings r_{z+} and r_{z-}.

Zurek does not take this position, but opts instead for what he calls an 'existential interpretation' of quantum mechanics. Recognizing that environmental decoherence does not justify the claim that one and only one event associated with the pointer states in the diagonal representation of the reduced density matrix actually occurs, Zurek proposes that all such events occur in some Everettian relative sense. The move here is to finesse the issue by appealing to Everett's 'relative state' interpretation of quantum mechanics.

Here is how Zurek introduces the existential interpretation (1993a, pp. 88–9):

In the quantum setting the observer must be demoted from an all-powerful external experimenter dealing *from without* with one more physical system (the universe) to a subsystem of that universe, with all of the limitations arising from such a confinement *to within* the physical entity he or she is supposed to monitor. Correlation – between the memory of the observer and the outcomes (records) of the past observations – emerges as a central concept. . . .

Modifications of the observer's state as a result of quantum events may be drastic (as would be the case for Schrödinger's cat) or subtle (as for Wigner's friend). Observers may or may not be conscious of them. Only states that can continue to define both the observers and the state of their knowledge for prolonged periods (at least as long as the characteristic information processing time scale of the observer's own nervous system – which, for us is more than a millisecond, orders of magnitude longer than a macroscopic open system typically takes to decohere) will correspond to perceptions. Memory is the stable existence of records – correlations with the state of the relevant branch of the universe. The requirement of stable existence and the recognition of ultimate interdependence between the identities of the observers (determined in part by the physical states of their memories) and their perceptions define the *existential interpretation* of quantum mechanics.

The role of decoherence is to cause negative selection and thus define the stable alternatives – states of the observer's identity – that can exist in spite of immersion in the environment. . . . Events happen because the environment helps define a set of stable options that is rather small compared with the set of possibilities available in principle in the Hilbert space. Each time the system of interest (or the memory of an apparatus, computer or nervous system) is forced into a superposition that violates environment-induced superselection rules, it will decohere on a time scale that is nearly instantaneous when the options are macroscopically distinguishable. This onset of decoherence is the apparent 'collapse of the wavepacket.' Thereafter each of the alternatives becomes a 'matter of fact' to the observer who has recorded it: It will evolve on its own, with negligible chances of interference with the other alternatives, but with the correlation of the records with all the relevant states of the measured observables intact.

In spite of the Everett-like framework of this discussion, the picture that emerges in the end – when described from the point of view of an observer – is very much in accord with the views of Bohr. A macroscopic observer will have recording and measuring devices that will behave classically. Any quantum measurement will lead to an almost instantaneous reduction of the wavepacket, so that the resulting mixture can safely be regarded as corresponding to just one unknown measurement outcome. According to the existential interpretation, what is perceived is not a 'complete wavefunction of the universe' but a few characteristics of its specific branch consistent with all of the records the state of the observer happens to include. The freedom to partition the global state vector into nearly arbitrary sets of branches (present in the original work of Everett) has been constrained by the requirement that the effectively classical states should be able to persist on dynamical time scales, that is, for much longer than the decoherence time.... Such an observer will remember events, perceive specific 'matters of fact' and agree about them with other observers.

In a subsequent elaboration (1993b, p. 311), Zurek summarizes the existential interpretation as follows:

The interpretation that emerges from these considerations is obviously consistent with Everett's 'Relative State' point of view. It is supplemented by a process – decoherence – which arises when the division of the Universe into separate entities is recognized. Its key consequence – emergence of the preferred set of states which can exist for time long compared to the decoherence timescale for a state randomly chosen from the Hilbert space of the system of interest – is responsible for the selection of the individual branches. As reported by an observer, whose memory and state becomes modified each time a 'splitting' involving him regarded as the physical system takes place, the apparent collapses of the state vector occur in complete accord with Bohr's 'Copenhagen Interpretation.' The role of the decoherence is to establish a boundary between quantum and classical. This boundary is in principle movable, but in practice largely immobilized by the irreversibility of the process of decoherence... which is in turn closely tied to the number of the degrees of freedom coupled to a macroscopic body. The equivalence between 'macroscopic' and 'classical' is then validated by the decoherence considerations, but only as a consequence of the practical impossibility of keeping objects which are macroscopic perfectly isolated.

To see what Zurek has in mind, we will have to consider what to make of Everett's interpretation, which is the subject of the following section.

8.2 Many worlds

In two lengthy papers (1957, 1973), Everett set out to show that there was a way of understanding quantum mechanics in which the linear Schrödinger dynamics and the orthodox (Dirac–von Neumann) interpretation principle (the 'eigenvalue–eigenstate link') could both be maintained consistently, without requiring a collapse postulate for an account of measurements. Exactly how to understand this proposal is the subject of an ongoing debate. I shall consider a suggestion developed in some detail by Albert (1992), and by Barrett (1992, 1994, 1996a), as the 'bare theory.'[116]

Everett (1957, p. 459) puts it this way:

> We thus arrive at the following picture: Throughout all of a sequence of observation processes there is only one physical system representing the observer, yet there is no single unique *state* of the observer (which follows from the representations of interacting systems). Nevertheless, there is a representation in terms of a *superposition*, each element of which contains a definite observer state and a corresponding system state. Thus with each succeeding observation (or interaction), the observer state 'branches' into a number of different states. Each branch represents a different outcome of the measurement and the *corresponding* eigenstate for the object–system state. All branches exist simultaneously in the superposition after any given sequence of observations.

A footnote added in the proof offers the following amplification (1957, pp. 459–60):

> In reply to a preprint of this article some correspondents have raised the question of the 'transition from possible to actual,' arguing that in 'reality' there is – as our experience testifies – no such splitting of observer states, so that only one branch can ever actually exist. Since this point may occur to other readers the following is offered in explanation.
>
> The whole issue of the transition from 'possible' to 'actual' is taken care of in the theory in a very simple way – there is no such transition, nor is such a transition necessary for the theory to be in accord with our experience. From the viewpoint of the theory *all* elements of a superposition (all 'branches') are 'actual,' none any more 'real' than the rest. It is unnecessary or suppose that all but one are somehow destroyed, since all the separate elements of a superposition individually obey the wave equation with complete indifference to the presence or absence ('actuality' or not) of any other elements. This total lack of effect of one branch on another also implies that no observer will ever be aware of any 'splitting' process.
>
> Arguments that the world picture presented by this theory is contradicted by experience, because we are unaware of any branching process, are like the criticism of the Copernican theory that the mobility of the earth as a real physical fact is incompatible with the common sense interpretation of nature because we feel no such motion. In both cases the argument fails when it is shown that the theory itself predicts that our experience will be what it in fact is. (In the Copernican case the addition of Newtonian physics was required to be able to show that the earth's inhabitants would be unaware of any motion of the earth.)

The popular way of understanding the 'relative state' interpretation of quantum mechanics is as a 'many worlds' theory in a literal sense. Albert (1992, p. 115) refers to

[116] Albert sees this theory as an attractive way to elaborate Everett's ideas as a 'one world' theory that preserves the standard way of thinking about superpositions (that is, a theory based on the 'eigenvalue–eigenstate link' as the criterion for value-definiteness or determinateness). While Albert cites Lockwood (1992), who follows Deutsch (1985a,b), Geroch (1984) has also proposed an interpretation of Everett that is very much along the same lines. Note that these 'bare' interpretations of Everett differ from Bell's (1987) 'one world' interpretation of Everett. Bell's interpretation – as he puts it (1987, p. 133), 'the pilot-wave theory without the trajectories' is a preferred-observable interpretation of the type considered in chapter 5, with position in configuration space as the preferred, always determinate observable.

this as the 'vulgar' reading of Everett. In the case of Schrödinger's cat, for example, the state of the radioactive atom + cat:

$$|\psi\rangle = c_1|\text{atom undecayed}\rangle|\text{cat alive}\rangle + c_2|\text{atom decayed}\rangle|\text{cat dead}\rangle$$

is supposed to describe two worlds, a world in which the atom has not decayed and the cat is alive, and a world in which the atom has decayed and the cat is dead. Here the states $|\text{cat alive}\rangle$ and $|\text{cat dead}\rangle$ can be understood as two states in orthogonal subspaces $\mathscr{K}_{\text{alive}}$ and $\mathscr{K}_{\text{dead}}$ that partition the Hilbert space of the cat.

If $c_1 = c_2 = \frac{1}{\sqrt{2}}$, this biorthogonal decomposition of $|\psi\rangle$ is not unique and $|\psi\rangle$ can be expressed as a different biorthogonal decomposition, say:

$$|\psi\rangle = \tfrac{1}{\sqrt{2}}|a_+\rangle|r_+\rangle + \tfrac{1}{\sqrt{2}}|a_-\rangle|r_-\rangle,$$

where $|a_+\rangle$ and $|a_-\rangle$ are two states of the atom defined by:

$$|a_+\rangle = \tfrac{1}{\sqrt{2}}\,|\text{atom undecayed}\rangle + \tfrac{1}{\sqrt{2}}\,|\text{atom decayed}\rangle$$
$$|a_-\rangle = \tfrac{1}{\sqrt{2}}\,|\text{atom undecayed}\rangle - \tfrac{1}{\sqrt{2}}\,|\text{atom decayed}\rangle$$

and $|r_+\rangle$ and $|r_-\rangle$ are two states of the cat defined by:

$$|r_+\rangle = \tfrac{1}{\sqrt{2}}\,|\text{cat alive}\rangle + \tfrac{1}{\sqrt{2}}\,|\text{cat dead}\rangle$$
$$|r_-\rangle = \tfrac{1}{\sqrt{2}}\,|\text{cat alive}\rangle - \tfrac{1}{\sqrt{2}}\,|\text{cat dead}\rangle$$

But now the state $|\psi\rangle$ apparently describes a different 'splitting,' into a world in which some observable A of the atom takes the value a_+ and some observable R of the cat takes the value r_+, and a world in which A takes the value a_- and R takes the value r_-. In neither of these worlds is the cat either alive or dead, since the eigenstates $|r_+\rangle$ and $|r_-\rangle$ of R do not belong to the subspace $\mathscr{K}_{\text{alive}}$ or to the subspace $\mathscr{K}_{\text{dead}}$.

In fact, for any coefficients c_1 and c_2, nothing rules out partitioning the Hilbert space of the atom + cat in some arbitrary way, so that $|\psi\rangle$ can be expressed as:

$$|\psi\rangle = \sum_{i=1}^{n} d_i|\xi_i\rangle|\zeta_i\rangle,$$

which would describe n worlds, in each of which two entangled subsystems that together make up the atom + cat have certain values for observables X, with eigenstates $|\xi_i\rangle$, and Z, with eigenstates $|\zeta_i\rangle$.

So, if there is literally going to be a 'splitting' of worlds whenever a measurement occurs, or whenever the quantum state takes a certain form, this ambiguity will have to be resolved. Some of Zurek's remarks suggest that he sees decoherence as ensuring a unique 'splitting,' by selecting a preferred basis: 'The freedom to partition the global state vector into nearly arbitrary sets of branches (present in the original work of Everett) has been constrained by the requirement that the effectively classical states should be able to persist on dynamical time scales, that is, for much longer than the decoherence time.' But the emphasis on 'correlation – between the memory of the observer and the outcomes (records) of the past observations' as 'a central concept' indicates that the existential interpretation draws on a reading of Everett that is much closer to the 'bare theory.'[117]

Consider an ideal measurement of the spin of an electron S along some direction by a measuring instrument M. If the initial state of S is $c_1|+\rangle_S + c_2|-\rangle_S$, after a certain time the state of $S + M$ will take the form:

$$c_+|+\rangle_S|r_+\rangle_M + c_-|-\rangle_S|r_-\rangle_M,$$

where $|r_+\rangle_M$ and $|r_-\rangle_M$ are two pointer-reading states of M that become correlated with the two spin states of S by the measurement interaction. An observer – call her Eve – who acquires a belief about the spin of S by looking at M (again, ideally, assuming the universal validity of the linear dynamics of quantum mechanics) will generate the state:

$$c_+|+\rangle_S|r_+\rangle_M|\text{believes `}+\text{'}\rangle_{\text{Eve}} + c_-|-\rangle_S|r_-\rangle_M|\text{believes `}-\text{'}\rangle_{\text{Eve}} \quad (8.5)$$

Here the two states of Eve in the superposition are states lying in distinct eigenspaces of some observable of Eve's brain that records the information obtained by looking at the pointer of M.

Albert's reading of the 'relative state' interpretation is to see Everett as resolving the tension between the linearity of the unitary Schrödinger dynamics and the orthodox interpretation principle by posing (and answering) the following question (Albert, 1992, p. 116):

Suppose that there is only one world, and suppose that there is only one full story about that world that's true, and suppose that the linear quantum-mechanical equations of motion are the true and complete equations of motion of the world, and suppose that the standard way of thinking about what it means to be in a superposition is the right way of thinking about what it means to be in a superposition, and consider the question of what it would *feel* like to *be* in a state like the one in [(8.5)] (that is: the question of what it would feel like to be *the experimenter* in a state like the one in [(8.5)]).

Albert reads Everett as showing how observers will always *come to believe* that they have definite beliefs about the systems with which they interact, by exploiting only the 'bare theory': the universal validity of the Schrödinger dynamics, and the 'standard way of thinking' about superpositions – the orthodox (Dirac–von Neumann) interpretation principle or 'eigenvalue–eigenstate link.'

The trick is to consider what has come to be known as a 'Wigner's friend' argument. Wigner (1962) considered what quantum mechanics without a collapse postulate would say about a situation in which a friend reported to him the observed outcome of a measurement. In this case, suppose Eve's partner Adam wants to find out whether Eve has a determinate belief about the spin of the electron while she is in state (8.5). Adam can't simply ask Eve what her spin-belief is, because the linearity of the Schrödinger equation (applied to the process of reflection or self-monitoring by Eve in

[117] The discussion of the bare theory that follows is taken from Bub, Clifton, and Monton (1996).

virtue of which she comes up with an answer) will put Eve in a superposition of responding to Adam in two different ways, and 'standard thinking' applied to *this* superposition leaves the issue unresolved (because the answer is indeterminate). But Adam can ask Eve whether she has a definite belief, just one of either '+' or '−' (and not which particular belief she has), and since the answer to that question is 'yes' in both the states $|\text{believes '+'}\rangle_{\text{Eve}}$ and $|\text{believes '−'}\rangle_{\text{Eve}}$, by linearity Eve will answer 'yes' to the question in the state (8.5).

So according to the bare theory, Eve will respond that she has a definite belief about the spin of the electron even when she in fact has no definite belief (according to the orthodox interpretation principle applied to the state (8.5)): Eve is 'apparently going to be radically deceived even about what *her own occurrent mental state* is' (Albert, 1992, p. 118). A similar analysis shows that observers will take immediately repeated measurements to yield the same determinate result, that different observers measuring the same observable will take themselves to agree on the outcome, and that observers will take the statistics of measurement results to confirm the statistical predictions of quantum mechanics. See Barrett (1994) for a detailed discussion.

Consider, now, how Eve's answering Adam's question about the determinateness or definiteness of her belief about the spin of the electron would have to be modelled on the bare theory in order to justify the conclusion that she answers 'yes' to the question. We can think of Eve's physical state as an element of the tensor product of the Hilbert space that represents the features of her brain that register beliefs about electron spin, with the Hilbert space associated with the parts of Eve's brain involved in reflecting on what her beliefs are and responding to questions about them. For Eve to answer 'yes' to the question when she believes the spin is '+' and when she believes the spin is '−,' the interaction (between the two parts of her brain) must induce a unitary evolution of the following sort:

$$\left.\begin{aligned} &|\text{believes '+'}\rangle_{\text{Eve}}|\text{ready to answer}\rangle_{\text{Eve}} \rightarrow |\text{believes '+'}\rangle_{\text{Eve}}|\text{yes}\rangle_{\text{Eve}} \\ &|\text{believes '−'}\rangle_{\text{Eve}}|\text{ready to answer}\rangle_{\text{Eve}} \rightarrow |\text{believes '−'}\rangle_{\text{Eve}}|\text{yes}\rangle_{\text{Eve}} \end{aligned}\right\} \qquad (8.6)$$

Linearity will then ensure that when Eve responds to the question in the state (8.5), the state of $S + M +$ Eve will end up as:

$$c_+|+\rangle_S|r_+\rangle_M|\text{believes '+'}\rangle_{\text{Eve}}|\text{yes}\rangle_{\text{Eve}} + c_-|-\rangle_S|r_-\rangle_M|\text{believes '−'}\rangle_{\text{Eve}}|\text{yes}\rangle_{\text{Eve}}, \qquad (8.7)$$

which is an eigenstate of Eve responding 'yes.'

This way of modelling Eve's response implicitly appeals to her competency to ascertain and report her mental state when it is an eigenstate of some definite belief about the spin of the electron S. But the following parallel line of argument is equally compelling, and equally natural in the context of the bare theory.

Suppose, instead, that Eve is competent to report her mental state when it is *not* an eigenstate of some definite belief about S-spin. If:

$$|?\rangle_{\text{Eve}} = \tfrac{1}{\sqrt{2}}|\text{believes `+'}\rangle_{\text{Eve}} + \tfrac{1}{\sqrt{2}}\,|\text{believes `}-\text{'}\rangle_{\text{Eve}}$$
$$|¿\rangle_{\text{Eve}} = \tfrac{1}{\sqrt{2}}|\text{believes `+'}\rangle_{\text{Eve}} - \tfrac{1}{\sqrt{2}}\,|\text{believes `}-\text{'}\rangle_{\text{Eve}}$$

then Eve should respond 'no' to the question. 'Do you have a definite S-spin belief?' in each of these non-eigenstates of S-spin belief. By linearity, Eve will now respond 'no' for any brain state in the plane, in particular the states:

$$|\text{believes `+'}\rangle_{\text{Eve}} = \tfrac{1}{\sqrt{2}}\,|?\rangle_{\text{Eve}} + \tfrac{1}{\sqrt{2}}\,|¿\rangle_{\text{Eve}}$$
$$|\text{believes `}-\text{'}\rangle_{\text{Eve}} = \tfrac{1}{\sqrt{2}}\,|?\rangle_{\text{Eve}} - \tfrac{1}{\sqrt{2}}\,|¿\rangle_{\text{Eve}}$$

If Eve correctly responds 'no' to the question about the definiteness of her S-spin belief in the states $|?\rangle_{\text{Eve}}$ and $|¿\rangle_{\text{Eve}}$:

$$\left.\begin{array}{l} |?\rangle_{\text{Eve}}|\text{ready to answer}\rangle_{\text{Eve}} \rightarrow |?\rangle_{\text{Eve}}|\text{no}\rangle_{\text{Eve}} \\ |¿\rangle_{\text{Eve}}|\text{ready to answer}\rangle_{\text{Eve}} \rightarrow |¿\rangle_{\text{Eve}}|\text{no}\rangle_{\text{Eve}} \end{array}\right\} \quad (8.8)$$

by linearity she will incorrectly respond 'no' to the question in the states $|\text{believes `+'}\rangle_{\text{Eve}}$ and $|\text{believes `}-\text{'}\rangle_{\text{Eve}}$. So when Eve responds to the question in the state (8.5), the state of $S + M +$ Eve will end up as:

$$c_+|+\rangle_S|r_+\rangle_M|\text{believes `+'}\rangle_{\text{Eve}}|\text{no}\rangle_{\text{Eve}} + c_-|-\rangle_S|r_-\rangle_M|\text{believes `}-\text{'}\rangle_{\text{Eve}}|\text{no}\rangle_{\text{Eve}}, \quad (8.9)$$

which is an eigenstate of Eve responding 'no' to the question.

So there are two completely parallel arguments for how to model Eve's response that yield utterly incompatible results, and incompatible stories about the way in which Eve is deluded. The bare theory *per se* does not tell us which is the correct model, because it contains no principle that stipulates in what ways observers should be competent to introspect about or report their beliefs, and in what ways they should not. In other words, the bare theory contains no prescription that says that when Eve reflects about her own beliefs, Eve's brain is hard-wired so that the unitary evolution (8.6) gets turned on inside her brain, as opposed to the unitary evolution (8.8). The bare theorist cannot treat (8.6) as a kind of calibration condition that defines what we mean by Eve's response acting as a good measurement of the structure of her own beliefs. For the evolution which induces her to respond 'no' in both the $|?\rangle_{\text{Eve}}$ and $|¿\rangle_{\text{Eve}}$ states is based on an equally good calibration condition, given that under the bare theory the 'no' response would be the correct response for *those* states.

Now one might argue that it is sufficient that the bare theory should be able to model (if not strictly deduce) the fact that observers in situations like Eve never take their beliefs to be indefinite after observing the instrument pointer. But since, on the bare theory, there is nothing special about the set of belief eigenspaces in the Hilbert space representing Eve's brain over any other observable's eigenspace, the bare theorist can only justify the particular model that works by saying that *it just so happens* in our world that the brain is hard-wired to give us the kind of beliefs we take ourselves to have about our beliefs. In this regard, the bare theory's strategy for dealing with the problem of what observers take themselves to believe, on the basis of their measuring instruments, is on a par with a preferred observable interpretation like Bohm's causal interpretation. A Bohmian would say that 'belief' is the kind of physical observable for which it always makes sense to speak about having one belief or another, because in purely physical terms that amounts to saying that the particles in observers' brains take up one position configuration over another. The difference between the unitary evolution that would make Eve respond 'yes' and the one that would make her respond 'no' is just the difference in choice of basis: either we decide that she is in fact competent to report the definiteness of her belief in both of two orthogonal belief eigenstates, or we decide that she is in fact competent to report her lack of definite belief in both of two orthogonal non-eigenstates of belief that span the same plane. The comparable choice on a 'no collapse' interpretation based on a preferred determinate observable is between Eve's belief observable as determinate versus, say, the observable with $|?\rangle_{\mathrm{Eve}}$ and $|¿\rangle_{\mathrm{Eve}}$ as eigenstates. The formal structure of quantum mechanics that needs to be distinguished by the bare theory to model observers' reports is exactly the same structure that a 'no collapse' preferred observable interpretation exploits: a preferred basis in the Hilbert space spanned by $|+\rangle_{\mathrm{Eve}}$ and $|-\rangle_{\mathrm{Eve}}$ picked out by those two states.

So, despite its name, the bare theory can't get by with less than nonstandard 'no collapse' interpretations based on a preferred determinate observable to explain why observers take themselves to have definite beliefs about the outcomes of measurements, even when the state of the observer is entangled with states of the measuring instrument and measured system in a superposition, why different observers measuring the same observable will take themselves to agree on the outcome, and so on. But the bare theory is actually *worse off* than a preferred observable interpretation, because it does not succeed in explaining all that needs to be explained.

In order to make the case that Eve is deluded about the definiteness of her S-spin beliefs, the bare theory requires a distinction between Eve's brain as a register of beliefs about the spin of the electron S, and Eve's brain as a register of reflections about the status of those beliefs. Call these two registers the 'primary' and 'secondary' belief registers. What the bare theorist shows is that the two registers disagree. When the state of $S + M +$ Eve is (8.5), then Eve doesn't believe '$+$' and she doesn't believe '$-$,' but the secondary register records 'yes' when Eve reflects on whether she has exactly one of the two beliefs in the set {believes '$+$', believes '$-$'}, on the assumption that the act of reflection is modelled by the unitary evolution (8.6).

Now, while the observation that Eve is deluded when answering 'yes' to the question about the definiteness of her primary belief is correct (on the above analysis), one can also consider what the bare theory will predict if Eve reflects about what particular primary belief it is that she believes to be definite. Evidently, assuming she can ascertain her specific beliefs in the eigenstate cases, Eve would then get into a superposition of believing that she believes '$+$' and believing that she believes '$-$.' But then, according to the bare theory (applied, as always, to the state of $S + M +$ Eve), Eve will not end up believing that she believes '+,' and she will not end up believing that she believes '−,' because these secondary beliefs are indeterminate in the superposition, just as she doesn't believe '+' and doesn't believe '−' in the superposition (8.5), because these primary beliefs are indeterminate in (8.5).

The point is that a 'no collapse' interpretation of quantum mechanics should be able to account for the fact that we take ourselves to have *specific* beliefs about the outcomes of measurements in states like (8.5), and not just that we have definite beliefs about measurement outcomes in some noncommittal sense. This failure of the bare theory is minimized by emphasizing what the bare theory *can* explain. If we introduce a tertiary belief register that records Eve's reflections about her secondary beliefs about her primary beliefs about the spin of the electron, then Eve can be modelled to answer 'yes' if she is asked whether her reflection about what her S-spin belief is agrees with her actual S-spin belief. The analysis is similar to the bare theory's explanation of why it seems that immediately repeated measurements yield the same determinate result, or why there is the impression of agreement between different observers measuring the same observable.

Specifically:

$$|\text{believes '+'}\rangle_{\text{Eve}(1)}|\text{believes primary belief is '+'}\rangle_{\text{Eve}(2)}|\text{ready to answer}\rangle_{\text{Eve}(3)}$$
$$\rightarrow |\text{believes '+'}\rangle_{\text{Eve}(1)}|\text{believes primary belief is '+'}\rangle_{\text{Eve}(2)}|\text{yes}\rangle_{\text{Eve}(3)}$$
$$|\text{believes '−'}\rangle_{\text{Eve}(1)}|\text{believes primary belief is '−'}\rangle_{\text{Eve}(2)}|\text{ready to answer}\rangle_{\text{Eve}(3)}$$
$$\rightarrow |\text{believes '−'}\rangle_{\text{Eve}(1)}|\text{believes primary belief is '−'}\rangle_{\text{Eve}(2)}|\text{yes}\rangle_{\text{Eve}(3)}$$

(where I have indexed the three registers of Eve), and so linearity will ensure that:

$$c_+|+\rangle_S|r_+\rangle_M|\text{believes '+'}\rangle_{\text{Eve}(1)}|\text{believes primary belief is '+'}\rangle_{\text{Eve}(2)}|\text{ready to answer}\rangle_{\text{Eve}(3)}$$
$$+\, c_-|-\rangle_S|r_-\rangle_M|\text{believes '−'}\rangle_{\text{Eve}(1)}|\text{believes primary belief is '−'}\rangle_{\text{Eve}(2)}|\text{ready to answer}\rangle_{\text{Eve}(3)}$$
$$\rightarrow$$
$$c_+|+\rangle_S|r_+\rangle_M|\text{believes '+'}\rangle_{\text{Eve}(1)}|\text{believes primary belief is '+'}\rangle_{\text{Eve}(2)}|\text{yes}\rangle_{\text{Eve}(3)}$$
$$+\, c_-|-\rangle_S|r_-\rangle_M|\text{believes '−'}\rangle_{\text{Eve}(1)}|\text{believes primary belief is '−'}\rangle_{\text{Eve}(2)}|\text{yes}\rangle_{\text{Eve}(3)}$$

Granted, the bare theory can explain Eve's impression that her primary and secondary beliefs are in accord, but that does nothing to give her any sense of what specific S-spin the agreement is about. Mere agreement here is not enough. A similar comment applies to the bare theory's explanation of intersubjective agreement between observers about what their measurement outcomes are. What cannot be explained is why sometimes the observers will take that agreement to be established because both their instruments registered '+,' say.

The same criticism undermines the bare theory's account of why observers take themselves to have confirmed the statistical predictions of quantum mechanics through their measurements. In the limit of an infinite sequence of measurements, the state of the system + measuring instrument + observer's brain (with multiple registers) will get into a quantum state that lies in an eigenspace of the relative frequency operator, with an eigenvalue equal to the quantum mechanically predicted frequency for the measured observable. So observers will take themselves to have obtained sequences that agree, in their long-run frequencies, with quantum statistical predictions. But observers seem to believe more than this: they take themselves to have obtained particular sequences, with particular characteristics. Merely getting into an eigenspace of the relative frequency operator will not ensure *that*, since there are many distinct sequences that can yield the same long-run frequencies.

The conclusion seems inescapable that the bare theory is too bare to explain why we take ourselves to have *specific* beliefs about measurement outcomes or outcome sequences in generic experimental situations. And this follows even if we allow the bare theory the luxury of a preferred basis.

So the objection to environmental decoherence as a solution to the measurement problem is not avoided by appealing to Everett. 'Tracing out' the environment, it was claimed, yields a reduced density matrix for $S + M$ that is effectively diagonal in the pointer basis: the off-diagonal terms decay almost instantaneously to zero. Environmental 'monitoring' generates the preferred basis, and the destruction of interference between alternative events associated with this basis generates what is formally a classical probability space. But, as I argued above, the destruction of interference between events associated with different measurement outcomes is neither necessary nor sufficient for a solution to the measurement problem. The fact that the reduced density matrix of $S + M$ obtained by 'ignoring' the environment takes the form of a mixture with respect to events associated with the pointer basis not only fails to account for the occurrence of *just one* of these events, but is actually *inconsistent* with such an occurrence (considering the origin of this mixture, and assuming the orthodox interpretation principle). Similarly, what's right about Everett appears to be the insight that memory correlations can be exploited to explain certain features of our experience that, at first sight, seem to be in conflict with the linear dynamics of quantum mechanics and the orthodox way of interpreting quantum states. But what is not explained by this manoeuvre is, again, the specificity or individuality of our experience. In this case, though, there is no inconsistency. In the state (8.7), Eve does believe she has exactly one of two S-spin beliefs, even though she doesn't believe that she believes '+,' and she doesn't believe that she believes '−.' But we can't conclude that Eve believes that she doesn't believe '+,' and that she believes that she doesn't believe '−.' To draw either of the latter conclusions – which *would* involve an inconsistency in her secondary beliefs – Eve would have to get into an eigenstate of belief with respect to the record in her secondary register concerning a particular disbelief about S-spin, and this she cannot do in the above scenario.

Oddly, when the problem is acknowledged by decoherentists, it is disarmed by being declared 'formidable,' or even beyond the reach of theory. The suggestion is that it transcends in profundity the debate about the measurement problem and the interpretation of quantum mechanics. Omnès remarks (1994, p. 492):

Whereas quantum mechanics can only envision all the various data that can occur in the experiment on the same footing, reality shows a unique real datum: an actual fact. When one repeats a measurement many times under the same conditions, one finds that the data are distributed randomly, their probabilities agreeing with the theory, but the status of actual facts in the theory remains nevertheless an open and troublesome question. Where does this uniqueness and even this existence of facts come from? This is undoubtedly the main remaining problem of quantum mechanics and the most formidable one.

In his ninth thesis, one of twenty-one theses summarizing his interpretation of quantum mechanics, Omnès declares (1994, p. 507):

The theory is unable to give an account of the existence of facts, as opposed by their uniqueness to the multiplicity of possible phenomena. This impossibility could mean that quantum mechanics has reached an ultimate limit in the agreement between a mathematical theory and physical reality. It might also be the underlying reason for the probabilistic character of the theory.

But in fact, quantum mechanics, like classical mechanics, *is* able to account for the existence of facts, and the theorem of chapter 4 shows how to construct 'no collapse' interpretations that do just that. What has to be given up is the orthodox (Dirac–von Neumann) interpretation principle, the 'eigenvalue–eigenstate link,' in favour of an interpretation that takes an appropriate preferred observable as having a determinate value independently of the quantum state.

8.3 Consistent histories

A further strand in the decoherence solution to the measurement problem is the notion of 'consistent histories' introduced by Griffiths.

A *history* is a sequence of events of a closed system occurring at different times, where each event is a property of the system represented by a projection operator:

$$h = \{P_1(t_1), P_2(t_2), \ldots, P_n(t_n)\}$$

Histories are assigned probabilities by quantum states. For example, the quantum state represented by the density operator W assigns the two-event history $h_0 = \{P_1(t_1), P_2(t_2)\}$ the probability:

$$\text{prob}_W(h_0) = \text{tr}(WP_1(t_1)) \cdot \text{tr}(W'P_2(t_2)),$$

where the operator W' represents the normalized projection of W onto the subspace associated with $P_1(t_1)$, as given by the Lüders rule (see section 1.4):

$$W' = \frac{P_1(t_1)WP_1(t_1)}{\text{tr}(P_1(t_1)WP_1(t_1))} = \frac{P_1(t_1)WP_1(t_1)}{\text{tr}(WP_1(t_1))}$$

So:

$$\begin{aligned}\text{prob}_W(h_0) &= \text{tr}(P_1(t_1)WP_1(t_1)P_2(t_2)) \\ &= \text{tr}(P_2(t_2)P_1(t_1)WP_1(t_1)P_2(t_2)),\end{aligned}$$

using the idempotence of the projection operators and the cyclic invariance of the trace.

In general, then:

$$\text{prob}_W(h) = \text{tr}(P_n(t_n) \ldots P_1(t_1) W P_1(t_1) \ldots P_n(t_n)) \tag{8.10}$$

If $W = P_\psi$, the probability takes the form:

$$\begin{aligned}\text{prob}_\psi(h) &= \text{tr}(P_1(t_1) \ldots P_n(t_n) P_n(t_n) \ldots P_1(t_1) P_\psi) \\ &= \langle \psi | (P_1(t_1) \ldots P_n(t_n) P_n(t_n) \ldots P_1(t_1) | \psi \rangle,\end{aligned}$$

which is just the norm of $P_1(t_1) \ldots P_n(t_n) | \psi \rangle$.

The probabilities of histories defined in this way are positive numbers, and the probabilities of two complementary histories sum to 1, where the complement of the history h is the history $h' = \{I - P_1(t_1), \ldots, I - P_n(t_n)\}$. So a family of histories will form a classical probability space if the probabilities of disjoint histories are additive. This will be the case for a 'consistent family' of histories satisfying a certain consistency condition.

As a simple example, consider a two-slit experiment and the three histories:

$$\begin{aligned}h_1 &= \{P_A(t_1), P(t_2)\} \\ h_2 &= \{P_B(t_1), P(t_2)\} \\ h &= \{P_A(t_1) + P_B(t_1), P(t_2)\},\end{aligned}$$

where $P_A(t_1)$ and $P_B(t_1)$ are two projection operators in the spectral measure of the observable representing localization in the plane of the slits, for position ranges associated with the slits A and B, and P is a projection operator representing localization to a region d on the detecting screen or photographic plate. The two histories h_1 and h_2 represent disjoint sequences of events: localization to the slit A at time t_1 and to the region d at time t_2, and localization to the slit B at t_1 and to d at t_2, respectively. The history h represents the sequence of events: localization to either slit A or slit B at t_1 and to d at t_2.

As we might expect, the probabilities of these histories are not additive, and the family of histories $\{h_1, h_2, h\}$ is not consistent, but the example serves to illustrate how the consistency conditions are generated. We have:

$$\begin{aligned}prob(h) &= \text{tr}(P(t_2)[P_A(t_1) + P_B(t_2)]W[P_A(t_1) + P_B(t_2)]) \\ &= \text{tr}(P(t_2)P_A(t_1)WP_A(t_1)) + \text{tr}(P(t_2)P_B(t_1)WP_B(t_1)) \\ &\quad + \text{tr}(P(t_2)P_A(t_1)WP_B(t_1)) + \text{tr}(P(t_2)P_B(t_1)WP_A(t_1)) \\ &= prob(h_1) + p(h_2) + \text{tr}(P(t_2)P_A(t_1)WP_B(t_1)) + \text{tr}(P(t_2)P_B(t_1)WP_A(t_1))\end{aligned}$$

The terms:

$$\text{tr}(P(t_2)P_A(t_1)WP_B(t_1)) + \text{tr}(P(t_2)P_B(t_1)WP_A(t_1))$$

are the usual interference terms in the two-slit experiment for an initial state W, and the condition for the histories $\{h_1, h_2, h\}$ to form a consistent family is just that these interference terms should vanish.

This condition can be reformulated by noting that:

$$[\text{tr}(P(t_2)P_A(t_1)WP_B(t_1))]^* = \text{tr}([WP_B(t_1)]^\dagger[(P(t_2)P_A(t_1)]^\dagger)$$
$$= \text{tr}(P_B(t_1)WP_A(t_1)(P(t_2))$$
$$= \text{tr}(P(t_2)P_B(t_1)WP_A(t_1))$$

where '*' denotes the complex conjugate and '†' denotes the adjoint of an operator. (Recall that projection operators and density operators are self-adjoint.) So the consistency condition becomes:

$$\text{Re}[\text{tr}(P(t_2)P_A(t_1)WP_B(t_1))] = 0 \tag{8.11}$$

The concept of consistent histories provides an elegant way of conforming to the restrictions of the Copenhagen interpretation concerning what we can say about quantum systems in various experimental situations, without invoking the primacy of classical concepts, or a 'cut' between the observer and what is being observed. In fact, the appeal of the consistent histories approach to quantum mechanics is just that it purports to provide the basis for an 'observer-free' interpretation that can be applied to the universe as a closed system, with obvious advantages for quantum cosmology. The way this is done is to reformulate the consistency condition (which is a 'no interference' condition) as a condition about *decohering* histories.

Gell-Mann and Hartle (1991a, p. 435) define a *decoherence functional* on a family of histories as a complex functional, $D(h', h)$, where h and h' are any two histories in the family. Consider a family of histories:

$$\{P_1^{\alpha_1}(t_1), P_2^{\alpha_2}(t_2), \ldots, P_n^{\alpha_n}(t_n)\}, \text{ for all } \alpha_1, \alpha_2, \ldots, \alpha_n$$

defined for a sequence of times $t_1, t_2, \ldots, t_n$, where at each time, t_i, the projection operator $P_i^{\alpha_i}(t_i)$ belongs to a set of orthogonal projection operators $\{P_i^1(t_i), P_i^2(t_i), \ldots\}$ that sum to the unit operator (representing a mutually exclusive and collectively exhaustive set of alternatives at t_i):

$$\sum_\alpha P_i^\alpha(t_i) = I$$
$$P_i^\alpha(t_i)P_i^\beta(t_i) = \delta_{\alpha\beta}P_i^\alpha(t_i)$$

A history, denoted by $[P^\alpha]$, is a particular sequence of alternatives:

$$[P^\alpha] = \{P_1^{\alpha_1}(t_1), P_2^{\alpha_2}(t_2), \ldots, P_n^{\alpha_n}(t_n)\},$$

where the subscript i labels the set of alternatives at t_i (the set of projection operators in a 'decomposition of the unit'), and the superscript α_i labels the particular alternative in the ith set.

A history is *completely fine-grained* if it is specified by a set of 1-dimensional projectors onto the eigenstates of a complete set of observables at all times. A history h' is a *coarse-graining* of history h if the set of projectors in h' are sums of projectors in h. (A completely coarse-grained history is the history consisting of the unit operator.)

The decoherence functional is defined as:

$$D_W([P^{\alpha'}], [P^{\alpha}]) = \text{tr}(P^{\alpha'_n}_n(t_n) \ldots P^{\alpha'_1}_1(t_1) W P^{\alpha_1}_1(t_1) \ldots P^{\alpha_n}_n(t_n)) \tag{8.12}$$

A family of alternative coarse-grained histories is said to *decohere* when the off-diagonal elements in the decoherence functional are sufficiently small. This generalizes Griffiths's consistency condition by requiring the approximate vanishing of the trace, rather than the vanishing of the real part of the trace. Note that the probabilities of the histories $[P^{\alpha}]$ are given by the diagonal elements of D_W, so for a decoherent family of histories:

$$\begin{aligned}\text{prob}_w([P^{\alpha}]) &\approx D_W([P^{\alpha}],[P^{\alpha}]) \\ &= \text{tr}(P^{\alpha_n}_n(t_n) \ldots P^{\alpha_1}_1(t_1) W P^{\alpha_1}_1(t_1) \ldots (P^{\alpha_n}_n(t_n))\end{aligned}$$

Of course, the probabilities in a family of alternative decohering histories will only be approximately classical. A decoherent family for which there is no finer-grained family that decoheres is said to be maximal. Some maximal families of decoherent histories will be more 'classical' than others, and these are said to define *quasiclassical domains.*

The connection with environmental decoherence arises in the following way: The time sequences of alternative approximate localizations of a macroscopic system interacting with its environment can be represented as a suitably *coarse-grained* family of histories that decohere in virtue of environmental monitoring. Coarse-graining here corresponds to 'tracing out' the environment.

Gell-Mann and Hartle (1991a, p. 430) characterize the decoherent histories formulation of quantum mechanics as 'an attempt at extension, clarification, and completion of the Everett interpretation.' The picture that emerges is of a universe with quasiclassical domains, or possibly even one unique quasiclassical domain, generated by decoherence through coarse-graining, in virtue of the initial state of the universe and the dynamics induced by the Hamiltonian, exploited by 'information gathering and utilizing systems' for the purpose of prediction (1991a, pp. 453–4):

Our picture is of a universe that, as a consequence of a particular initial condition and of the underlying Hamiltonian, exhibits at least one quasiclassical domain made up of suitably defined maximal sets of alternative histories with as much classicity as possible. . . .

Both singly and collectively we are examples of the general class of complex adaptive systems. When they are considered within quantum mechanics as portions of the universe, making observations, we refer to such complex adaptive systems as information gathering and utilizing systems (IGUSes). . . . Probabilities of interest to the IGUS include those for correlations between its memory and the external world. . . .

To carry on in this way, an IGUS uses probabilities for histories referring both to the future and the past. An IGUS uses decohering sets of alternative histories and therefore performs further coarse graining on a quasiclassical domain. Naturally, its coarse graining is very much coarser than that of the quasiclassical domain since it utilizes only a few of the variables in the universe.

The reason such systems as IGUSes exist, functioning in such a fashion, is to be sought in their evolution within the universe. It seems likely that they evolved to make predictions because it is adaptive to do so. The reason, therefore, for their focus on decohering variables is that these are the *only* variables for which predictions can be made. The reason for their focus on the histories of a quasiclassical domain is that these present enough regularity over time to permit the generation of models (schemata) with significant predictive power.

If there is essentially only one quasiclassical domain, then naturally the IGUS utilizes further coarse grainings of it. If there are many essentially inequivalent quasiclassical domains, then we could adopt a subjective point of view, as in some traditional discussions of quantum mechanics, and say that the IGUS 'chooses' its coarse graining of histories and, therefore, 'chooses' a particular quasiclassical domain, or a subset of such domains, for further coarse graining. It would be better, however, to say that the IGUS evolves to exploit a particular quasiclassical domain or set of such domains.

What the new orthodoxy achieves is an observer-free formulation of the Copenhagen interpretation. One might say that it provides a way of generating some of the Boolean virtues of classicality purely quantum mechanically, through the particular quasiclassical domain or set of domains exploited by IGUSs. But even with a unique quasiclassical domain, we end up with many alternatives, whether they are referred to as 'branches' or as 'histories,' where no one alternative can be regarded as any more actual than another. So the measures introduced as probabilities, in spite of their formal quasiclassicality, have no cash value as probabilities in the usual sense. There are still two quasiclassical histories for Schrödinger's cat, one with a dead cat and one with a live cat, whatever the probabilities, and while it can be argued that Eve will always take her impressions concerning the outcome of her spin measurements to agree with her own self-monitoring of those impressions, what that agreement is specifically about, either + or − for the spin, remains indeterminate.

The only way to get definiteness or determinateness into the story is to put it in somewhere. One way to do this, for a unique quasiclassical domain, would be to reject the orthodox interpretation principle and take the quasiclassical domain as generating a suitable preferred observable R in the sense of chapters 4 and 5. But then we have a very different picture. In such a quantum world, specific events correlated with the determinate values of R actually *happen* to IGUSs like Schrödinger's cat and Eve, as those values evolve stochastically in time according to the equation of motion of section 5.2. So probabilities cash out in the usual way as measures over a set of alternative histories, only one of which is actual, and we preserve the core features of Einstein's realism.

Coda

The previous chapters have all been variations on a central theme: the question of how to make sense of the probabilities generated by quantum states, or how to interpret superpositions so as to avoid the measurement problem.

On the orthodox interpretation, a system in a quantum state:

$$|\psi\rangle = c_1|a_1\rangle + c_2|a_2\rangle,$$

where $|a_1\rangle$ and $|a_2\rangle$ are eigenstates of some observable A, has no determinate A-property: neither the property corresponding to the value a_1 of A, nor the property corresponding to the value a_2 of A. It is said to be 'meaningless' to ascribe an A-property to the system in the superposition $|\psi\rangle$: the system has an A-property if and only if the state is an eigenstate of A. In the superposition $|\psi\rangle$, the probability 'amplitudes' c_1 and c_2 are associated with wave aspects of the system, which manifest themselves in the probabilities of finding a determinate value for A, either a_1 with probability $|c_1|^2$ or a_2 with probability $|c_2|^2$, if we make an A-measurement on the system. But if a measurement is a quantum mechanical interaction that conforms to the linear dynamics of the theory, the system and measuring instrument won't end up in a state that is an eigenstate of A for the eigenvalue a_1 with a relative frequency of $|c_1|^2$, or a state that is an eigenstate of A for the eigenvalue a_2 with a relative frequency of $|c_2|^2$.[118]

The notion of 'wave–particle duality' – that a system acts like a wave in some experimental contexts and like a particle in other experimental contexts – should have been rejected, once and for all, with Schrödinger's proof of the equivalence of wave mechanics and matrix mechanics, and von Neumann's unification of the two theories in the Hilbert space formulation of quantum mechanics. The transition from the classical phase space description of the states and dynamical variables of a physical system to the Hilbert space description of quantum mechanics involves the transition from a commutative algebra of dynamical variables to a noncommutative algebra or, equivalently, the transition from a Boolean to a non-Boolean algebraic structure for the properties of physical systems. From this perspective, quantum mechanics is not about physical systems that exhibit a peculiar and elusive ontology, but rather about

[118] In fact, as Fine (1970), Shimony (1974), Brown (1986), and others have shown under very general assumptions about measurement, it is *impossible* for the system and measuring instrument to end up in a mixed state represented by the density operator $W = |c_1|^2 P_{a_1} + |c_2|^2 P_{a_2}$.

physical systems with a non-Boolean property structure. The problem is then how to make sense of a quantum world in which the properties of systems 'fit together' in a non-Boolean way.

As it turns out, the property structure of a quantum world is strongly non-Boolean, in the sense that it cannot be embedded into a Boolean algebra (except in the case of a property structure associated with a 2-dimensional Hilbert space), so we can't simply understand the quantum mechanical description as an incomplete classical description, in the sense that the description can be extended to a Boolean classical description by introducing additional dynamical variables. Rather, we need to find a way of introducing notions of actuality, possibility, and probability into the non-Boolean structure itself.

Here we must be guided by some underlying metaphysical principles: what we ought to aim for is a description that preserves as much as possible of Einstein's realist intuitions, subject to the constraints of the 'no go' hidden variable theorems, which limit the applicability of these intuitions in a quantum world in various ways. The uniqueness theorem of chapter 4 characterizes the broad features of such a description.

The Boolean property structure of a classical world defines a fixed space of possibilities, given by the set of 2-valued homomorphisms on the Boolean algebra. These 2-valued homomorphisms define property states – alternative 'lists' of properties that 'fit together,' one of which is the actual property state. The property states correspond to atoms in the Boolean property structure, or 1-point subsets in phase space, the Stone space of the Boolean algebra. Dynamical change is tracked by the evolution (via Hamilton's equations of motion) of the actual property state over time. So the property state coincides with the dynamical state in a classical world.

In a quantum world, the properties that define the space of possibilities at a given time form a proper sublattice of the non-Boolean property structure. This 'determinate' sublattice is itself non-Boolean and non-embeddable into a Boolean lattice, although there do exist 2-valued homomorphisms that select property states on the sublattice – in fact, sufficiently many so that the probabilities specified by the quantum state for the properties in the sublattice can be represented as measures over these 2-valued homomorphisms or property states in the standard way. The determinate sublattices are maximal, in the sense that they contain the maximal subcollections of properties that can 'fit together' in a quantum world, or the maximal subcollections of propositions that can be determinately true or false, given the constraints of the theorem. Each determinate sublattice is uniquely defined by a (pure) quantum state – an atom in the non-Boolean property structure – and some preferred observable that is stipulated as determinate. Dynamical change in a quantum world is tracked by the evolution of the quantum state over time, via Schrödinger's equation of motion. So the quantum state functions as a dynamical state in tracking the temporal evolution of what is possible and what is probable, through the temporal evolution of the determinate sublattice associated with the quantum state, with respect to a preferred determinate observable. The actual property state evolves stochastically in time, via the

Bell–Vink equation of motion, so as to conform to the dynamically evolving sublattice of possible properties.

The picture that emerges from the analysis underlying the uniqueness theorem is this: The actual properties in a classical world are selected by a 2-valued homomorphism from a set of alternatives defined by a *fixed* Boolean algebra – the Boolean algebra of subsets of the phase space of the classical world – representing the same collection of possible properties at all times. The actual properties in a quantum world at time t are selected by a 2-valued homomorphism from a set of alternatives defined by a *dynamically evolving* non-Boolean sublattice in the lattice of all subspaces of the Hilbert space of the quantum world. So the actual properties in a classical world evolve in a fixed Boolean possibility space, while the actual properties in a quantum world evolve in a dynamically changing non-Boolean possibility space. Classically, only the actual properties are time-indexed; quantum mechanically, both the actual properties and the possible properties are time-indexed.

Modulo the usual philosophical worries about modality, there is nothing inherently strange about the notions of possibility or actuality in quantum mechanics. Once we have a precise handle on what is possible and what is actual, and how change proceeds, the rest is calculation. So measurement comes out as a process internal to the theory on the basis of the dynamics alone, without requiring any privileged status for observers. To be sure, the choice of the preferred determinate observable is not arbitrary, but depends on contingent dynamical features of our quantum world, if we want reliable instruments with pointers that remain stable after measurement. But the fact that we find ourselves in a quantum world where measurement is possible – dynamical change of a certain sort that characterizes the transfer of information between systems – will surely involve the same sort of explanation as the fact that we find ourselves in a world where we are able to exist at all as carbon-based life forms.

The measurement problem arises on the orthodox interpretation through the identification of the dynamical state of quantum mechanics with the property state. This is the eigenvalue–eigenstate link: the assumption that an observable of a system has a value, or the system has a property, if and only if the state of the system is an eigenstate of the observable, or an eigenstate of the projection operator representing the property. The uniqueness theorem of chapter 4 shows that both the eigenvalue to eigenstate link and the eigenstate to eigenvalue link must be dropped to avoid the measurement problem, and this cannot be avoided by weakening the lattice closure requirement on the determinate sublattice to partial Boolean algebra closure, as I pointed out in section 6.2. Orthodox attempts to resolve the measurement problem by exploiting features of the instrument–environment interaction must fail, because decoherence – in the sense of loss of interference –cannot yield determinateness, and the fundamental problem of the orthodox interpretation is the inconsistency of the eigenvalue–eigenstate link with any intelligible notion of determinateness that makes sense of measurement in a quantum world.

Einstein rejected Bohm's causal interpretation as 'too cheap for me' (see section 2.1).

Perhaps this was because, on the usual formulation of Bohm's theory, dynamical change in the particle positions occurs through the medium of a guiding field that propagates in configuration space. One might well see this as a quick and dirty way of getting a working causal theory out of the interference properties of wave functions. But if we see Bohm's theory as an observer-free 'no collapse' interpretation in the sense of the uniqueness theorem, then the theory appears as one possible way of 'completing' quantum mechanics in Einstein's sense, while preserving as much as can be preserved of Einstein's realist principles.

The really essential thing about a quantum world is the irreducible indeterminism associated with noncommutativity or non-Booleanity, and the really puzzling feature of irreducible indeterminism is how to make sense of irreducible probabilities. Von Neumann saw these probabilities as 'logical.' In an address on 'Unsolved Problems in Mathematics' to the International Mathematical Congress, Amsterdam, Sept. 2, 1954,[119] von Neumann put it this way:

If you take a classical mechanism of logics, and if you exclude all those traits of logics which are difficult and where all the deep questions of the foundations come in, so if you limit yourself to logics referred to a finite set, it is perfectly clear that logics in that range is equivalent to the theory of all sub-sets of that finite set, and that probability means that you have attributed weights to single points, that you can attribute a probability to each event, which means essentially that the logical treatment corresponds to set theory in that domain and that a probabilistic treatment corresponds to introducing measures. I am, of course, taking both things now in the completely trivialized finite case.

But it is quite possible to extend this to the usual infinite sets. And one also has this parallelism that logics corresponds to set theory and probability theory corresponds to measure theory and that given a system of logics, so given a system of sets, if all is right, you can introduce measures, you can introduce probability and you can always do it in very many different ways.

In the quantum mechanical machinery the situation is quite different. Namely instead of the sets use the linear sub-sets of a suitable space, say of a Hilbert space. The set theoretical situation of logics is replaced by the machinery of projective geometry, which in itself is quite simple.

However all quantum mechanical probabilities are defined by inner products of vectors. Essentially if a state of a system is given by one [unit] vector, the transition probability in another state is the inner product of the two which is the square of the cosine of the angle between them. In other words, probability corresponds precisely to introducing the angles geometrically. Furthermore, there is only one way to introduce it. The more so because in the quantum mechanical machinery the negation of a statement, so the negation of a statement which is represented by a linear set of vectors, corresponds to the orthogonal complement of this linear space.

And therefore, as soon as you have introduced into the projective geometry the ordinary machinery of logics, you must have introduced the concept of orthogonality. This actually is rigorously true and any axiomatic elaboration of the subject bears it out. So in order to have logics you need in this set a projective geometry with a concept of orthogonality in it. In order to have probability all you need is a concept of all angles, I mean angles other than 90°. Now it is

[119] Typescript, von Neumann Archives, Library of Congress, Washington, DC; pp. 21, 22.

perfectly quite true that in a geometry, as soon as you can define the right angle, you can define all angles. Another way to put it is that, if you take the case of an orthogonal space, those mappings of this space on itself, which leave orthogonality intact, leave all angles intact, in other words, in those systems which can be used as models of the logical background for quantum theory, it is true that as soon as all the ordinary concepts of logics are fixed under some isomorphic transformation, all of probability theory is already fixed.

So when you replace the set-theoretical (Boolean) logic or possibility structure of a classical world with the 'machinery of projective geometry,' probabilities emerge as a new structural feature of the way properties can 'fit together' in a quantum world. The problem is how to understand these probabilities on the projective geometry, the non-Boolean lattice $\mathscr{L}$ of subspaces of a Hilbert space. The theory describes their dynamical evolution, through the dynamical evolution of quantum states, so it appears that what is physically meaningful in a quantum world is just the lattice $\mathscr{L}$ of possibilities, and the dynamically evolving probabilities defined on $\mathscr{L}$. These probabilities cannot be reduced to measures over 2-valued homomorphisms on $\mathscr{L}$, or measures over Boolean homomorphisms on $\mathscr{L}$ – there are no 2-valued homomorphisms or Boolean homomorphisms on $\mathscr{L}$ in the general case – hence the apparently irreducible indeterminism.

So far so good. But it is impossible to make sense of probabilities over a domain of possibilities without introducing the notion of actuality, or of truth. At some point we want something to actually happen in a quantum world. Von Neumann's proposal was to identify actuality with unit probability. This is not something that is forced on us by the projective structure itself, but a quite separate assumption. What the projective structure mandates with respect to actuality is given by the uniqueness theorem of chapter 4. Von Neumann's proposal is then seen as the decision to choose the preferred determinate observable as the unit observable, and it is just this choice that leads to the measurement problem.

To avoid the measurement problem, we have to look to features of the projective structure itself. Then Bohm's interpretation appears as one among a family of possible candidates for an observer-free 'no collapse' interpretation. The relation of non-Booleanity to the irreducible indeterminism in a quantum world appears in a rather different light. While the probabilities are irreducible on the lattice $\mathscr{L}$, they are reducible to measures over 2-valued homomorphisms on the determinate sublattices of $\mathscr{L}$. If the preferred determinate observable R is a discrete coarse-grained observable, then irreducible indeterminism arises because the actual values of R, and hence the actual properties in a quantum world, undergo a stochastic temporal evolution, constrained by the (deterministic) dynamical evolution of the determinate sublattice in $\mathscr{L}$. If R is continuous, as in Bohm's choice of position in configuration space, then the actual properties evolve deterministically. But since measurements are essentially non-ideal in this case, we can never, in principle, use a measurement to fix the position of a system in configuration space precisely, and so our description of events in a quantum world is necessarily probabilistic.

The interpretative framework proposed here has been developed explicitly only for nonrelativistic quantum mechanics, and one might wonder whether a relativistic extension is possible. In Bohmian mechanics, the trajectory of a composite system in configuration space is not Lorentz-invariant. Consider, for example, two spin-$\frac{1}{2}$ particles in the singlet spin state that move in opposite directions away from a source, and measurements of the spin component in the z-direction on each of the particles. There will be a configuration of the particles for which a measurement of spin in the z-direction on the left particle yields the result 'up,' corresponding to the left particle moving in the 'up' direction. A subsequent measurement of spin in the z-direction on the right particle will then yield the result 'down,' corresponding to the right particle moving in the 'down' direction. This follows because the first measurement induces an evolution of the quantum state of the two-particle system to a linear superposition of states with effectively disjoint supports in configuration space. The position in configuration space is then effectively associated with just one of these states (in the sense that the 'empty wave' has a negligible effect on how this position changes in configuration space), so the measurement leads to an effective 'collapse' of the state to a state corresponding to just one of the product states in the singlet state (along the lines of the discussion of nonlocality in Bohm's theory in section 6.1). Since the particles are space-like separated in the 4-dimensional space–time manifold, whether the spin measurement on the left particle precedes the spin measurement on the right particle, or conversely, depends on the Lorentz frame relative to which the space–time coordinates are defined. There are initial configurations of the particles for which a spin measurement on the right particle, if performed before the spin measurement on the left particle, will also yield the result 'up,' in which case a subsequent spin measurement on the left particle will yield the result 'down.' So, whether the spin measurements result in the left particle moving 'up' and the right particle moving 'down' (in which case the spin components of the two particles are measured as $+\frac{1}{2}$ and $-\frac{1}{2}$, respectively), or whether the measurements result in the left particle moving 'down' and the right particle moving 'up' (in which case the spin components of the two particles are measured as $-\frac{1}{2}$ and $+\frac{1}{2}$, respectively), depends on the Lorentz frame. (For an elementary discussion, see Albert, 1992, pp. 156–61.)

Now, although the configuration space trajectory is not Lorentz invariant, the statistics of measurement outcomes will be the same in every Lorentz frame. It is impossible to exploit features of an individual configuration space trajectory in Bohmian mechanics to signal superluminally, because measurements depend on the dynamical evolution of the quantum state to a certain form (a linear superposition of states with effectively disjoint supports in configuration space), and however precise such measurements are, they can never, in principle, yield point positions in configuration space (because they never yield linear superpositions of states with supports that are literally disjoint in configuration space). So measurements always yield distributions in configuration space, which will evolve deterministically via the Bohmian equation of motion, by continuity, to conform to the dynamical evolution of the effectively 'collapsed' quantum state (the state with support in configuration space covering the range of configurations consistent with the actual coarse-grained pointer position that is the outcome of the measurement). Since we can never get to the individual Bohmian trajectories by any possible measurements, there is no empirical means of establishing our state of absolute motion in a quantum world. We are confined to the Born probabilities, and these will be the same in every Lorentz frame. In Bohmian mechanics, Lorentz invariance conflicts with Einsteinian realism in a quantum world, but the way in which Lorentz invariance is violated has no empirical consequences.

The story is a little different for an observer-free 'no collapse' interpretation with a fixed discrete determinate observable R. Events that are simultaneous in one inertial frame define a simultaneity-slice or 3-dimensional space-like hyperplane in the 4-dimensional Minkowski manifold of special relativity. Since a Lorentz transformation of the space–time coordinates induces a unitary transformation in Hilbert space, the quantum state in a relativistic quantum world is hyperplane-dependent, and so the determinate sublattice, which evolves with the quantum state, is hyperplane-dependent. By contrast, in the version of the modal interpretation discussed in section 6.2, the collection of determinate properties of a system S is obtained from the spectral representation of the density operator representing the reduced state of S and is not hyperplane-dependent (see Dieks, 1989b, 1993, 1994a and Clifton, 1996a). The reduced density operator generates the Born probabilities for S-measurements and is derived by 'tracing out' the environment E in the state W_{S+E} of $S + E$. Since these probabilities are local and do not depend on how events involving distant subsystems in E are represented – which might be different for different hyperplanes – the reduced density operator, and hence the collection of determinate properties it defines, is not hyperplane-dependent.

For an argument showing that a Lorentz invariant evolution of a system's possessed properties is impossible in the modal interpretations proposed by van Fraassen (1991), Kochen (1985), Dieks (1989a, 1994b), Clifton (1995b), and Vermaas and Dieks (1995), see Dickson and Clifton (1996).

While there has been some work on extending Bohm's theory to both fermion and boson fields (see Bell, 1987; Bohm and Hiley, 1993; Holland, 1993b; Valentini, 1996), how to construct a satisfactory relativistic extension or field-theoretic generalization of an observer-free 'no collapse' interpretation of quantum mechanics for a suitable preferred determinate observable (say a coarse-grained observable associated with a quasiclassical domain in something like the sense of Gell-Mann and Hartle) is an open question. What is now clear is that this is a task that cannot be avoided without giving up on quantum mechanics.

Appendix
Some mathematical machinery

The discussion in this appendix is intended to be self-contained for a reader with some minimal mathematical competence, but no particular background is assumed, beyond a passing familiarity with the basic concepts of vector spaces, complex numbers, and probability theory.

A.1 Hilbert space

The central interpretative problems of quantum mechanics, which reflect the projective geometry of Hilbert space as a representation space for the properties of quantum systems, appear most directly and simply in finite-dimensional Hilbert spaces. Two dimensions are sufficient to illustrate certain features of quantum mechanics associated with the non-Booleanity of the property structure (or the noncommutativity of the algebra of physical quantities), but a system whose states are represented on a 2-dimensional Hilbert space is 'almost classical,' in the sense that the lattice of subspaces representing the property structure of the system is embeddable into a Boolean algebra.[120] So we will need to consider higher dimensional Hilbert spaces of three, four, and eight dimensions to bring out what is conceptually puzzling about quantum mechanics.

In fact, an infinite-dimensional Hilbert space would be required to fully represent the changing properties of any real physical system, such as a particle moving through a magnetic field. Technical difficulties having to do with issues of continuity, convergence, and boundedness arise in the infinite-dimensional case that are best handled in a general treatment utilizing the resources of functional analysis. The Dirac notation introduced in section A.3, with its improper 'delta functions,' provides a handy bookkeeping device that avoids these technicalities and suffices for the purposes of this discussion.

An *n-dimensional Hilbert space* is just an n-dimensional Euclidean space, with vectors represented in terms of complex numbers instead of real numbers, and a scalar (or inner) product between vectors that can take complex numbers as values. In a 2-dimensional Euclidean space – a plane – we can represent unit vectors in the x- and

[120] See section 3.1.

y-directions of a Cartesian coordinate system by $\binom{1}{0}$ and $\binom{0}{1}$, where the number pairs represent Cartesian coordinates. Other vectors can be constructed by multiplying these vectors by real number coefficients, for example:

$$2\begin{pmatrix}1\\0\end{pmatrix} = \begin{pmatrix}2\\0\end{pmatrix},$$

or by adding vectors:

$$2\begin{pmatrix}1\\0\end{pmatrix} + \sqrt{5}\begin{pmatrix}0\\1\end{pmatrix} = \begin{pmatrix}2\\\sqrt{5}\end{pmatrix}$$

represents a vector with x-coordinate 2 and y-coordinate $\sqrt{5}$. In Hilbert space the coefficients can be complex numbers.

The length or magnitude (or 'norm') of a vector ψ in a Euclidean plane is defined (via Pythagoras' theorem) as the square root of the sum of the squares of the Cartesian components of the vectors and denoted by $\|\psi\|$. So the length of $\binom{2}{\sqrt{5}}$ is $\sqrt{(2^2 + 5)} = 3$ and the length of $\begin{pmatrix}1\\0\end{pmatrix}$ is 1.

The scalar product of two vectors is defined as the product of their lengths and the cosine of the angle θ between them. If the angles between the vectors and the x-direction are θ_1 and θ_2, then $\theta = \theta_1 - \theta_2$. So $\cos\theta = \cos(\theta_1 - \theta_2) = \cos\theta_1\cos\theta_2 + \sin\theta_1\sin\theta_2$. It follows that the scalar product of the vectors is just the sum of the product of their x-coordinates and the product of their y-coordinates, since each x-coordinate is equal to the product of the length of the vector and the cosine of θ_1 or θ_2, and each y-coordinate is equal to the product of the length of the vector and the sine of θ_1 or θ_2. (This generalizes to any number of dimensions.) For example, the scalar product of $\binom{1}{0}$ and $\binom{2}{\sqrt{5}}$ is $1{\cdot}2 + 0{\cdot}\sqrt{5} = 2$, and the scalar product of $\binom{1}{0}$ and $\binom{0}{1}$ is $1{\cdot}0 + 0{\cdot}1 = 0$, which reflects the orthogonality of the two basis vectors ($\cos(\pi/2) = 0$). The length of a vector can be expressed as the square root of the scalar product of a vector with itself (since the relevant angle θ is 0 in this case, and $\cos(0) = 0$). Note that in a general Hilbert space, the scalar product of two vectors can be a complex number.

Formally, a *Hilbert space* $\mathscr{H}$ is a linear space over the field of complex numbers. This means that if ψ and ϕ are elements – 'vectors' – of $\mathscr{H}$, then $\psi + \phi$ belongs to $\mathscr{H}$, and if c is any complex number then $c\psi$ belongs to $\mathscr{H}$. Vector addition is associative and commutative: $\psi + (\phi + \chi) = (\psi + \phi) + \chi$, and $\psi + \phi = \phi + \psi$. Multiplication by complex numbers is distributive (that is, $c(\psi + \phi) = c\psi + c\phi$ and $(c_1 + c_2)\psi = c_1\psi + c_2\,\psi$) and associative (that is, $c_1(c_2\psi) = (c_1c_2)\psi$).[121]

A *scalar product* is defined on $\mathscr{H}$ as a map from the Cartesian product, $\mathscr{H} \times \mathscr{H}$, into the complex numbers, satisfying certain conditions. The image (complex number) of an

[121] I use lower case Greek letters, with or without subscripts, to label vectors (elements of $\mathscr{H}$), for example, ψ, ϕ, α_1, β_i. I use lower case Roman letters to label (complex) numbers, for example a, b_1, c_i. In the Dirac notation (introduced in section A.3 below), I use any sort of label to distinguish vectors, for example, $|\psi\rangle$, $|a_i\rangle$, $|+\rangle$, $|\text{up}\rangle$.

ordered pair of vectors, ψ, ϕ, under the map is denoted by (ψ, ϕ). The conditions on the map are:

$(\psi, \phi) = (\phi, \psi)^*$, where '*' denotes the complex conjugate
$(\psi, \phi + \chi) = (\psi, \phi) + (\psi, \chi)$
$(\psi, c\phi) = c(\psi, \phi)$, for any complex number c
$(\psi, \psi) > 0$, unless $\psi = 0$ (the zero or null vector), in which case $(\psi, \psi) = 0$

It follows that

$$(c\psi, \phi) = (\phi, c\psi)^* = c^*(\phi, \psi)^* = c^*(\psi, \phi)$$

The *norm* or *length* of a vector ψ is denoted by $\|\psi\|$ and defined as $\|\psi\| = (\psi, \psi)^{1/2}$. The map on $\mathscr{H} \times \mathscr{H}$ that associates each pair of vectors ψ, ϕ with the norm of their difference $\|\psi - \phi\|$ is a metric or distance function on the Hilbert space $\mathscr{H}$. It follows from the properties of the scalar product that:

$$\|\psi - \phi\| \geq 0$$
$$\|\psi - \phi\| = 0 \text{ if and only if } \psi = \phi$$
$$\|\psi - \phi\| = \|\phi - \psi\|$$
$$\|\psi - \phi\| \leq \|\psi - \chi\| + \|\chi - \phi\|$$

The last 'triangle inequality' is equivalent to the inequality $\|\psi + \phi\| \leq \|\psi\| + \|\phi\|$, and this follows from the Schwarz inequality $|(\psi, \phi)| \leq \|\psi\| \, \|\phi\|$, where $|(\psi, \phi)|$ denotes the absolute value of the complex number (ψ, ϕ). So, the length or magnitude of a vector ψ can be taken as $\|\psi\|$, and the distance between points ψ and ϕ in $\mathscr{H}$ as $\|\psi - \phi\|$. It is consistent with this definition of the metric to take the condition $(\psi, \phi) = 0$ as the condition of orthogonality of ψ and ϕ (for example, Pythagoras's theorem holds for the relation between the length of a vector ψ and the lengths of any n mutually orthogonal vectors that sum to ψ). It follows that if

$$\psi = \begin{pmatrix} a \\ b \end{pmatrix} = a\begin{pmatrix} 1 \\ 0 \end{pmatrix} + b\begin{pmatrix} 0 \\ 1 \end{pmatrix}$$

and

$$\phi = \begin{pmatrix} c \\ d \end{pmatrix} = c\begin{pmatrix} 1 \\ 0 \end{pmatrix} + d\begin{pmatrix} 0 \\ 1 \end{pmatrix}$$

in a 2-dimensional Hilbert space $\mathscr{H}_2$, then

$$(\psi, \phi) = a^*c + b^*d$$

because $\begin{pmatrix} 1 \\ 0 \end{pmatrix}$ and $\begin{pmatrix} 0 \\ 1 \end{pmatrix}$ are orthogonal.

The dimension of a Hilbert space $\mathscr{H}$ is defined via the concept of linear independence. A set of n non-null vectors $\{\alpha_1, \alpha_2 \ldots, \alpha_n\}$ is *linearly independent* if and only if the only way a linear combination of the vectors, $\Sigma_i c_i \alpha_i$, can sum to the null vector is if all the coefficients c_i are equal to zero; that is $\Sigma_i c_i \alpha_i = 0$ only if $c_i = 0$ for all i, where the c_i are any complex numbers. If the n vectors are mutually orthogonal ($(\alpha_i, \alpha_j) = 0$ for $i \neq j$), they are linearly independent. For suppose $\Sigma_i c_i \alpha_i = 0$, then

$$(\alpha_j, \sum_i c_i \alpha_i) = (\alpha_j, 0) = 0, \text{ for } j = 1, 2, \ldots, n$$

But

$$(\alpha_j, \sum_i c_i \alpha_i) = c_j(\alpha_j, \alpha_j)$$

and so $c_j(\alpha_j, \alpha_j) = 0$, and hence $c_j = 0$ for all j.

The *dimension* of $\mathscr{H}$ is the maximum number of linearly independent vectors in $\mathscr{H}$. If $\mathscr{H}$ contains n linearly independent vectors, but every set of $n + 1$ vectors is linearly dependent, then $\mathscr{H}$ is n-dimensional. If there is no maximum number; that is, if there are arbitrarily many linearly independent vectors, then $\mathscr{H}$ is infinite-dimensional.

Since an orthogonal set $\{\alpha_1, \alpha_2, \ldots, \alpha_n\}$ is linearly independent in an n-dimensional Hilbert space $\mathscr{H}_n$, it follows that the set $\{\psi, \alpha_1, \alpha_2, \ldots, \alpha_n\}$ for any vector ψ is linearly dependent: there exist complex numbers $k_1, k_2, \ldots, k_n$, not all zero, such that

$$k\psi + k_1\alpha_1 + k_2\alpha_2 + \ \ldots \ + k_n\alpha_n = 0$$

or

$$\psi = \sum_i (-k_i/k)\alpha_i = \sum_i c_i \alpha_i$$

The condition of orthogonality requires that $(\alpha_i, \alpha_j) = 0$, for $i \neq j$. In addition, if each α_j if of unit length ($\|\alpha_j\|^2 = (\alpha_j, \alpha_j) = 1$), it follows that:

$$(\alpha_j, \psi) = (\alpha_j, \sum_i c_i \alpha_i) = c_j(\alpha_j, \alpha_j) = c_j$$

and so any vector ψ in $\mathscr{H}_n$ can be expressed in the form

$$\psi = \sum_i (\alpha_i, \psi)\alpha_i$$

with respect to a complete *orthonormal* set of vectors in $\mathscr{H}_n$ – an orthogonal set of unit vectors spanning the Hilbert space – just as a vector in a Euclidean space can be expressed in terms of its components relative to a Cartesian coordinate system or set of axes in the space. The orthogonal unit vectors $\{\alpha_i, i = 1, 2, \ldots, n\}$ are said to form a *basis* for the Hilbert space $\mathscr{H}_n$, and the vectors $(\alpha_i, \psi)\alpha_i$ are the projections of ψ along the directions α_i (see below).

In a 2-dimensional Euclidean space, for example, the vector $\psi = \binom{2}{5}$ can be expressed as:

$$\psi = 2\alpha_1 + 5\alpha_2$$
$$= 2\begin{pmatrix}1\\0\end{pmatrix} + 5\begin{pmatrix}0\\1\end{pmatrix}$$

where $2\binom{1}{0} = \binom{2}{0}$ is the projection of $\psi = \binom{2}{5}$ along the direction $\binom{1}{0}$ (the x-axis in the standard basis) and $5\binom{0}{1} = \binom{0}{5}$ is the projection of $\psi = \binom{2}{5}$ along the direction $\binom{0}{1}$ (the y-axis in the standard basis). Note that $(\alpha_1, \psi) = 2$ and $(\alpha_2, \psi) = 5$.

A *linear manifold* in $\mathscr{H}$ is a set of vectors such that $c\psi$ and $\psi + \phi$ belong to the set if ψ, ϕ belong to the set, and c is any complex number. Equivalently, a linear manifold contains all linear combinations of any finite subset of its elements. The linear manifold spanned by a set of vectors is the smallest linear manifold that includes the set: it contains all and only linear combinations of vectors from the set. A *subspace* $\mathscr{K}$ of a Hilbert space $\mathscr{H}$ is a *closed* linear manifold (a linear manifold containing all its limit points).[122] The subspaces form a lattice under the partial ordering defined by subspace inclusion. Evidently, a subspace $\mathscr{K} \subseteq \mathscr{H}$ is itself a Hilbert space, with dimension less than or equal to the dimension of $\mathscr{H}$.

An *operator* A on $\mathscr{H}$ is a function from a subset of $\mathscr{H}$ into $\mathscr{H}$. The operator A is *linear* if its domain is a linear manifold and

$$A(\psi + \phi) = A\psi + A\phi$$
$$A(c\psi) = cA\psi$$

for any vectors ψ, ϕ in $\mathscr{H}$, and any complex number c. It follows that the range of A is also a linear manifold.

The set of vectors in $\mathscr{H}$ orthogonal to all vectors in a subspace $\mathscr{K}$ is a subspace $\mathscr{K}^{\perp} = \mathscr{H} - \mathscr{K}$. Two subspaces $\mathscr{K}_1$ and $\mathscr{K}_2$ are orthogonal if and only if every vector in $\mathscr{K}_1$ is orthogonal to every vector in $\mathscr{K}_2$; that is, if and only if $\mathscr{K}_1 \subseteq \mathscr{K}_2{}^{\perp}$ or, equivalently, if $\mathscr{K}_2 \subseteq \mathscr{K}_1{}^{\perp}$. Note that two planes in a 3-dimensional Euclidean space that cut at right angles are not 'orthogonal' planes in this sense, because their intersection is a line containing vectors (all multiples of the same unit vector) that belong to both planes.

Every vector $\psi \in \mathscr{H}$ can be resolved in one and only one way into two components

$$\psi = \psi_1 + \psi_2 = P\psi + P^{\perp}\psi,$$

where $\psi_1 = P\psi \in \mathscr{K}$ is the projection of ψ onto a subspace $\mathscr{K}$ of $\mathscr{H}$, and $\psi_2 = P^{\perp}\psi \in \mathscr{K}^{\perp}$ is the projection of ψ onto the subspace $\mathscr{K}^{\perp}$.

The map $P: \mathscr{H} \to \mathscr{H}$, associating each vector in $\mathscr{H}$ with its projection in $\mathscr{K}$, is a linear operator defined everywhere on $\mathscr{H}$. It is obvious geometrically that P is *idempotent*:

$$P^2 = P$$

[122] A sequence of vectors $\{\psi_i\}$ in $\mathscr{H}_\infty$ converges to ψ, and ψ is the limit of the sequence, if the sequence of numbers $\{\|\psi - \psi_i\|\}$ converges to zero.

in the sense that $P(P\psi) = P\psi$, and *self-adjoint* or *Hermitian:*[123]

$$(P\psi, \phi) = (\psi, P\phi), \text{ for all } \psi, \phi \in \mathscr{H}.$$

To see this, note that

$$(P\psi, \phi) = (P\psi, P\phi + P^{\perp}\phi) = (P\psi, P\phi) + (P\psi, P^{\perp}\phi)$$

and

$$(\psi, P\phi) = (P\psi + P^{\perp}\psi, P\phi) = (P\psi, P\phi) + (P^{\perp}\psi, P\phi)$$

so $(P\psi, \phi) = (\psi, P\phi)$, since $(P\psi, P^{\perp}\phi) = (P^{\perp}\psi, P\phi) = 0$.

A *projection operator* may be defined as a self-adjoint, idempotent linear operator on $\mathscr{H}$, and it is then a theorem that each projection operator P corresponds to a unique subspace $\mathscr{K}$ that is its range.[124] It is an obvious consequence of the definition that:

$$(\psi, P\psi) = (\psi, P^2\psi) = (P\psi, P\psi) = \|P\psi\|^2,$$

that is, the square of the length of the projection of ψ onto $\mathscr{K}$ is $(\psi, P\psi)$. If $\psi = \Sigma_i(\alpha_i, \psi)\alpha_i$ and P_{α_j} is the projection operator onto the 1-dimensional subspace spanned by the vector α_j, then $P_{\alpha_j}\psi = (\alpha_j, \psi)\alpha_j$, so the length of the projection of ψ onto the direction α_j is $\|P_{\alpha_j}\psi\| = |(\alpha_j, \psi)|$.

If A is a self-adjoint operator in $\mathscr{H}_n$, there exist n nontrivial solutions to the *eigenvalue equation*

$$A\alpha = a\alpha$$

that is, n nonzero *eigenvectors* α_i (of unit length), determined up to multiplication by a common phase factor $e^{i\theta}$, and n corresponding *eigenvalues* a_i (besides the trivial solution $\alpha = 0$). The eigenvalues of A can be found as the solutions of the equation

$$|A_{ij} - aI_{ij}| = 0$$

where A_{ij} is an $n \times n$ matrix representing the operator A, I_{ij} is the unit matrix (with all off-diagonal elements 0 and all diagonal elements 1), and $|A_{ij} - aI_{ij}|$ is the determinant of the matrix $A_{ij} - aI_{ij}$, an nth order polynomial in a. By the fundamental theorem of algebra, this equation has n solutions, which are said to form the *spectrum* of the operator.

Each eigenvalue a_i is real, since A is self-adjoint:

$$a_i^*(\alpha_i, \alpha_i) = (a_i\alpha_i, \alpha_i) = (A\alpha_i, \alpha_i) = (\alpha_i, A\alpha_i) = (\alpha_i, a_i\alpha_i) = a_i(\alpha_i, \alpha_i)$$

Eigenvectors corresponding to different eigenvalues are orthogonal, because

$$a_i(\alpha_i, \alpha_j) = (a_i\alpha_i, \alpha_j) = (A\alpha_i, \alpha_j) = (\alpha_i, A\alpha_j) = (\alpha_i, a_j\alpha_j) = a_j(\alpha_i, \alpha_j)$$

123 The adjoint of an operator A is an operator $A^{\dagger}$ satisfying the condition $(A^{\dagger}\psi, \phi) = (\psi, A\phi)$, for all $\psi, \phi \in \mathscr{H}$.

124 Von Neumann (1955), Theorem 12.

and so $(\alpha_i, \alpha_j) = 0$ if $a_i \neq a_j$. The eigenvectors of A therefore form an orthogonal set, and since there are n orthogonal eigenvectors, these span $\mathcal{H}_n$, and the orthonormal set is said to be *complete*. Self-adjoint operators represent dynamical variables in quantum mechanics. For simple physical examples of self-adjoint operators and their associated eigenvectors, see section A.4 below.

The n eigenvalues need not all be distinct. If there are $k < n$ distinct eigenvalues of A, each eigenvalue a_i $(1 \leq i \leq k)$ corresponds to at most $m(i)$ linearly independent eigenvectors $\alpha_{i_1}, \alpha_{i_2}, \ldots, \alpha_{i_{m(i)}}$ that span a subspace $\mathcal{H}_{a_i}$ of dimension $m(i)$, the *eigenspace* of A corresponding to the eigenvalue a_i. The number $m(i)$, the *multiplicity* of the eigenvalue a_i, is the maximum number of linearly independent solutions to the eigenvalue equation $A\alpha = a_i\alpha$. An eigenvalue with multiplicity greater than 1 is said to be *degenerate*. If all the eigenvalues are nondegenerate, the spectrum is said to be *simple*. An operator A is *degenerate* or *nonmaximal* if A has one or more degenerate eigenvalues. For example, in a Hilbert space of more than two dimensions, a projection operator P is degenerate, with two distinct eigenvalues, 1 and 0, and two orthogonal eigenspaces: the eigenspace $\mathcal{K}$ that is the range of the operator corresponding to the eigenvalue 1, and eigenspace $\mathcal{K}^\perp$ corresponding to the eigenvalue 0. Every vector in $\mathcal{K}$ is an eigenvector of P with eigenvalue 1, and every vector in $\mathcal{K}^\perp$ is an eigenvector of P with eigenvalue 0.

A degenerate eigenvalue a_i determines a unique eigenspace $\mathcal{H}_{a_i}$, but since any vector in the subspace $\mathcal{H}_{a_i}$ is a solution to the eigenvalue equation, the eigenvalue a_i does not determine a unique set of $m(i)$ orthogonal vectors in $\mathcal{H}_{a_i}$. Every vector in $\mathcal{H}_{a_i}$ is representable in the form

$$\psi = \sum_{j=1}^{m(i)} (\alpha_{i_j}, \psi)\alpha_{i_j}$$

so that

$$A\psi = \sum_{j=1}^{m(i)} (\alpha_{i_j}, \psi)A\alpha_{i_j} = \sum_{j=1}^{m(i)} (\alpha_{i_j}, \psi)a_i\alpha_{i_j} = a_i \sum_{j=1}^{m(i)} (\alpha_{i_j}, \psi)\alpha_{i_j} = a_i\psi$$

A complete orthonormal set of eigenvectors of a degenerate operator A is a set of unit vectors $\{\alpha_{i_j}\}$ where, for each i, the vectors $\alpha_{i_1}, \alpha_{i_2}, \ldots, \alpha_{i_{m(i)}}$ are *any* set of $m(i)$ orthonormal vectors spanning the eigenspace $\mathcal{H}_{a_i}$.

In $\mathcal{H}_n$, then, there always exists a complete orthonormal set of eigenvectors for any self-adjoint operator A: a set of basis vectors $\{\alpha_i\}$ that is complete in the sense that any vector ψ in the space can be represented as a linear sum or superposition of the basis vectors:

$$\psi = \sum_{i=1}^{n} c_i\alpha_i = \sum_{i=1}^{n} (\alpha_i, \psi)\alpha_i,$$

where some of the coefficients $c_i = (\alpha_i, \psi)$ may be zero. If A is nondegenerate, the eigenvectors α_i in the basis are uniquely determined. If A is degenerate, the basis is not unique, but the set of eigenspaces of A is uniquely determined.

Every self-adjoint operator A in $\mathscr{H}_n$ can be represented uniquely in the form of a *spectral resolution* or *spectral representation*:

$$A = \sum_{i=1}^{k} a_i P_{a_i},$$

where the P_{a_i} are projection operators onto the k eigenspaces $\mathscr{K}_{a_i}$ determined by the k distinct eigenvalues a_i of A. The projection operators P_{a_i} define a unique *spectral measure*:

$$P_A(E) = \sum_{a_i \in S} P_{a_i},$$

a map from the field of Borel sets $\{E\}$ on the real line $\mathbb{R}$ onto a set of projection operators $\{P_A(E)\}$, satisfying the conditions:

$$P_A(\emptyset) = 0$$
$$P_A(\mathbb{R}) = I$$
$$P_A(E_1 \cap E_2) = P_A(E_1) \wedge P_A(E_2) = P_A(E_1) \cdot P_A(E_2)$$
$$P_A(E_1 \cup E_2) = P_A(E_1) \vee P_A(E_2) = P_A(E_1) + P_A(E_2) - P_A(E_1) \cdot P_A(E_2)$$

Here $\emptyset$ denotes the empty set, 0 denotes the null operator (the linear operator that maps every vector in $\mathscr{H}_n$ onto the null vector), and I denotes the unit operator (the operator that maps every vector in $\mathscr{H}_n$ onto itself).

The generalization of this result to $\mathscr{H}_\infty$ is the content of the spectral representation theorem. An infinite-dimensional Hilbert space is required to be separable and complete. *Separability* is the property that there exists a countable set in $\mathscr{H}_\infty$ that is *everywhere dense* in $\mathscr{H}_\infty$: every vector in $\mathscr{H}_\infty$ is the limit of a convergent sequence from the countable set or, equivalently, for any $\psi \in \mathscr{H}_\infty$ there is an element ψ_n in the countable set such that $\|\psi - \psi_n\| < \varepsilon$, for any positive real number ε. *Completeness* is the property that every *convergent sequence* in $\mathscr{H}_\infty$ converges to a limit point in $\mathscr{H}_\infty$ (where a convergent sequence is a sequence satisfying the Cauchy convergence criterion, that for each $\varepsilon > 0$ there exists an $N(\varepsilon)$ such that $\|\psi_m - \psi_n\| < \varepsilon$ if $m, n > N(\varepsilon)$). Separability guarantees the existence of countable sets of basis vectors in terms of which any vector can be represented as a linear sum. If $\{\alpha_i\}$ is a basis in $\mathscr{H}_\infty$, then $\psi = \Sigma_{i=1}^{\infty}(\alpha_i,\psi)\alpha_i$, in the sense that the sequence $\Sigma_{i=1}^{n}(\alpha_i,\psi)\alpha_i$ converges to ψ as $n \to \infty$. Every finite-dimensional Hilbert space is separable and complete.

In $\mathscr{H}_\infty$ a complete orthonormal set of solutions to the eigenvalue equation $A\alpha = a\alpha$ does not always exist for every self-adjoint operator A. While a countable basis always exists in any separable Hilbert space, the set of eigenvectors of an arbitrary self-adjoint operator A will not necessarily form such a basis: it might not be possible to represent every vector $\psi \in \mathscr{H}_\infty$ as a linear superposition of the vector solutions to the eigenvalue equation for A. In such a case, the spectrum of A is said to be (partly) continuous. The eigenvalues a_i for which the corresponding eigenvectors (or sets of eigenvectors, in the degenerate case) exist in $\mathscr{H}_\infty$ form the discrete spectrum of A.

A self-adjoint operator with a continuous spectrum is associated with a unique spectral measure $\{P_A(E)$: for all Borel subsets E of the real line $\mathbb{R}\}$, and a unique spectral representation in terms of this measure:

$$A = \int_{-\infty}^{\infty} a \, \mathrm{d}P_A(a),$$

where $P_A(a) = P_A((-\infty, a])$ and the integral is understood in the sense that

$$(\psi, A\psi) = (\psi, \int_{-\infty}^{\infty} a \, \mathrm{d}P_A(a)\psi) = \int_{-\infty}^{\infty} a \, \mathrm{d}(\psi, P_A(a)\psi)$$

Here $(-\infty, a]$ denotes the subset of the real line defined by the half-open interval from $-\infty$ to a.

A.2 Quantum states

Dynamical states in quantum mechanics are represented by rays in Hilbert space, and also by convex combinations of rays (see below). Dynamical variables, referred to as 'observables' in quantum mechanics, are represented by self-adjoint operators in Hilbert space.[125] The ray representing a quantum state assigns probabilities to ranges f values of observables according to the following rule, proposed by Born.[126]

In $\mathscr{H}_n$, if the spectrum of an observable A is simple (so A has n distinct eigenvalues a_i), the probability assigned to the value a_i of A by the ray spanned by the unit vector ψ is:

$$\mathrm{prob}_\psi(a_i) = |(\alpha_i, \psi)|^2 = \|P_{a_i}\psi\|^2$$

In other words, the probability $\mathrm{prob}_\psi(a_i)$ is the square of the length of the projection of ψ onto the eigenvector α_i corresponding to a_i.[127] If A is degenerate, then

$$\mathrm{prob}_\psi(a_i) = \|P_{a_i}\psi\|^2,$$

where P_{a_i} is the projection operator onto the $m(i)$-dimensional subspace $\mathscr{K}_{a_i}$. (Note that $P_{a_i} = \Sigma_{j=1}^{m(i)} P_{\alpha_{i_j}}$, where the $P_{\alpha_{i_j}}$ are the projection operators onto $m(i)$ mutually orthogonal 1-dimensional subspaces corresponding to *any* orthonormal set of vectors spanning $\mathscr{K}_{a_i}$.) In general, then, in $\mathscr{H}_n$:

[125] Or, more generally, by POV measures. See section 5.3.
[126] Born (1926).
[127] I use the same symbol for an observable and the operator representing the observable. So I refer, for example, to 'the spectrum of the observable A' rather than 'the spectrum of the operator representing the observable A.'

$$\text{prob}_\psi(a \in E) = \sum_{a_i \in E} \|P_{a_i}\psi\|^2 = \|P_A(E)\psi\|^2$$

To see that this is a probability assignment to the Borel subsets E of values of the observable A, note that $\|P_A(E)\psi\|^2$ is the square of the length of the projection of ψ onto the subspace $\mathscr{K}_A(E)$ corresponding to $P_A(E)$. Since ψ has unit length, $0 \le \|P_A(E)\|^2 \le 1$, and so $0 \le \text{prob}_\psi(a \in E) \le 1$, for all E. Also, $\text{prob}_\psi(a \in E) = 0$ if $E = \varnothing$, because $P_A(\varnothing) = 0$, and $\text{prob}_\psi(a \in E) = 1$ if $E = \mathbb{R}$, because $P_A(\mathbb{R}) = 1$. If E_1 and E_2 are disjoint sets in $\mathbb{R}$ $(E_1 \cap E_2 = 0)$, then $P_A(E_1 \cap E_2) = P_A(E_1) \cdot P_A(E_2) = 0$, and so:

$$\begin{aligned}
\text{prob}_\psi(a \in E_1 \text{ or } a \in E_2) &= \text{prob}_\psi(a \in (E_1 \cup E_2)) \\
&= \|P_A(E_1 \cup E_2)\psi\|^2 \\
&= (\psi, (P_A(E_1) + P_A(E_2))\psi) \\
&= (\psi, P_A(E_1)\psi) + (\psi, P_A(E_2)\psi) \\
&= \text{prob}_\psi(a \in E_1) + \text{prob}_\psi(a \in E_2)
\end{aligned}$$

In $\mathscr{H}_n$, the expectation value of A in the state represented by a vector ψ can be expressed as:

$$\begin{aligned}
\text{Exp}_\psi(A) &= \sum_{i=1}^{n} a_i \text{prob}_\psi(a_i) \\
&= \sum_{i=1}^{n} a_i (P_{a_i}\psi, P_{a_i}\psi) \\
&= \sum_{i=1}^{n} a_i (\psi, P_{a_i}\psi) \\
&= (\psi, \sum_{i=1}^{n} a_i P_{a_i}\psi) \\
&= (\psi, A\psi)
\end{aligned}$$

since $A = \Sigma_{i=1}^{n} a_i P_{a_i}$ in the spectral representation. Note that $(\psi, A\psi)$ is always real if A is self-adjoint, because $(\psi, A\psi) = (A\psi, \psi) = (\psi, A\psi)^*$.

The probabilities assigned by quantum states represented by unit vectors can be generalized to states represented by *convex sets* of unit vectors: sets of vectors $\psi_1, \psi_2, \ldots$ with weights $w_1, w_2, \ldots$ $(w_1 \ge 0, w_2 \ge 0, \ldots; w_1 + w_2 + \ldots = 1)$:

$$\text{prob}(a \in E) = \sum_i w_i \text{prob}_{\psi_i}(a \in E) = \sum_i w_i \|P_A(E)\psi_i\|^2$$

and

$$\text{Exp}(A) = \sum_i w_i \text{Exp}_{\psi_i}(A) = \sum_i w_i (\psi_i, A\psi_i)$$

Quantum states represented by individual rays or unit vectors in $\mathscr{H}$ are termed *pure states*. Convex sets of pure states are termed *mixed states* or *mixtures*.

It is convenient to express the probability assignments defined by pure states and mixed states in terms of the trace function of an operator. The *trace* of an operator A is defined as the sum $\Sigma_i(\phi_i,A\phi_i)$, where $\{\phi_i\}$ is any complete orthonormal set in $\mathscr{H}$. The trace is invariant under a change of basis:

$$\mathrm{tr}(A) = \sum_i(\phi_i,A\phi_i) = \sum_i(\psi_i,A\psi_i)$$

for any other complete orthonormal set $\{\psi_i\}$. If A is self-adjoint, $\mathrm{tr}(A)$ is real. If A is also *definite* ($(\psi,A\psi)\geq 0$, for all ψ), then $\mathrm{tr}(A)\geq 0$, and $\mathrm{tr}(A) = 0$ if and only if $A = 0$.

It can easily be verified that

$$\mathrm{tr}(cA) = c\,\mathrm{tr}(A)$$
$$\mathrm{tr}(A + B) = \mathrm{tr}(A) + \mathrm{tr}(B)$$

Moreover, the trace function is *cyclically invariant*:

$$\mathrm{tr}(AB) = \mathrm{tr}(BA), \text{ for } all\ A, B \text{ (even noncommuting } A, B),$$

and similarly:

$$\mathrm{tr}(ABC) = \mathrm{tr}(BCA) = \mathrm{tr}(CAB), \text{ etc.}$$

If P is a projection operator onto a k-dimensional subspace $\mathscr{K}$ in an n-dimensional Hilbert space $\mathscr{H}_n$, then $\mathrm{tr}(P) = k$. For

$$\mathrm{tr}(P) = \sum_{i=1}^{k}(\psi_i,P\psi_i) + \sum_{j=k+1}^{n}(\phi_j,P\phi_j)$$

where $\{\psi_1,\psi_2,\ldots,\psi_k\}$ is any orthonormal set of vectors spanning $\mathscr{K}$, and $\{\phi_{k+1},\phi_{k+2},\ldots,\phi_n\}$ is an orthonormal set spanning $\mathscr{K}^\perp$. Hence

$$\mathrm{tr}(P) = \sum_{i=1}^{k}(\psi_i,\psi_i) + 0 = k$$

From the definition of the trace, it follows that:

$$\mathrm{Exp}_\psi(A) = (\psi,A\psi) = (\psi,AP_\psi\psi) = \mathrm{tr}(AP_\psi) = \mathrm{tr}(P_\psi A)$$

where P_ψ is the projection operator onto ψ. This follows because $\mathrm{tr}(AP_\psi) = \Sigma_i(\phi_i,AP_\psi\phi_i)$ is invariant under a change of basis, and so we can choose as a basis any complete orthonormal set $\{\phi_i\}$ with $\phi_i = \psi$ (in which case $\mathrm{tr}(AP_\psi) = (\psi,AP_\psi\psi) + \Sigma_{i=2}(\phi_i,AP_\psi\phi_i) = (\psi,AP_\psi\psi) + 0$). The probability assigned to the range E of an observable A by a pure state represented by a vector ψ can therefore be expressed in terms of the trace as:

$$\mathrm{prob}_\psi(a \in E) = \|P_A(E)\psi\|^2 = (\psi,P_A(E)\psi) = \mathrm{Exp}_\psi(P_A(E)) = \mathrm{tr}(P_\psi P_A(E))$$

where $P_A(E)$ is the projecti on operator in the spectral measure of A corresponding to the range E of A.

The expectation value of A in a mixed state represented by the convex set of vectors ψ_i with weights w_i is therefore:

$$\begin{aligned}\sum_i^{\mathrm{i}} w_i \mathrm{Exp}_{\psi_i}(A) &= \sum_i^{\mathrm{i}} w_i(\psi_i, A\psi_i)\\ &= \sum_i^{\mathrm{i}} w_i \mathrm{tr}(P_{\psi_i} A)\\ &= \mathrm{tr}((\sum_i^{\mathrm{i}} w_i P_{\psi_i})A)\end{aligned}$$

The operator $W = \Sigma_i w_i P_{\psi_i}$, introduced by von Neumann as the *density operator*[128] representing the mixed state, is self-adjoint and definite, because each P_{ψ_i} is self-adjoint and definite and each $w_i \geq 0$. Also, $\mathrm{tr}(W) = \Sigma_i w_i = 1$, because $\mathrm{tr}(P_{\psi_i}) = 1$. Evidently, the projection operators onto the 1-dimensional subspaces are the density operators of the pure states. Since density operators are self-adjoint with unit trace, a necessary and sufficient condition that a density operator W is the density operator of a pure state is that W is idempotent ($W^2 = W$).

The probability assignments defined by a pure or mixed state are completely characterized by the density operator according to the rule:

$$\mathrm{prob}_W(a \in E) = \mathrm{tr}(WP_A(E))$$

and

$$\mathrm{Exp}_W(A) = \mathrm{tr}(WA),$$

where $W = P_\psi$ in the case of a pure state. Note that $\mathrm{prob}_W(a \in E) = \mathrm{Exp}_W(P_A(E))$.

If W is nondegenerate and has a discrete spectrum of eigenvalues w_i and corresponding eigenvectors ω_i, then the spectral representation of W is

$$W = \sum_i w_i P_{w_i} = \sum_i w_i P_{\omega_i},$$

where the P_{ω_i} are the projection operators onto the vectors ω_i, and each $w_i \geq 0$ (since W is definite). In this case, W can be interpreted as representing a mixture of mutually orthogonal pure states represented by the unit vectors ω_i, with weights w_i. The decomposition into orthogonal states is unique (by the spectral theorem), but W will also be decomposable into mixtures of non-orthogonal pure states.

If some eigenvalues of W are degenerate, then

$$W = \sum_i w_i P_{w_i}$$

and the degenerate eigenvalues correspond to mutually orthogonal multi-dimensional

[128] Von Neumann (1955, p. 319). Von Neumann uses the term 'statistical operator.'

subspaces. The subspace $\mathscr{K}_{w_i}$ corresponding to the degenerate eigenvalue w_i is the set of all solutions to the eigenvalue equation $W\omega = w_i\omega$. In this case it is not possible to represent W as a mixture of a unique set of orthogonal pure states. For example, if a mixture is formed from an orthogonal set of k pure states, represented by the unit vectors $\psi_1, \psi_2, \ldots, \psi_k$, with equal weights (corresponding to the density operator

$$W = \tfrac{1}{k}(P_{\psi_1} + P_{\psi_2} + \ldots + P_{\psi_k}) = \tfrac{1}{k}\,P,$$

where P is the projection operator onto the subspace $\mathscr{K}$ spanned by the k vectors, with a single n-fold degenerate eigenvalue $1/k$), then any equal-weight mixture of k orthogonal pure states spanning the subspace $\mathscr{K}$ would be represented by the same density operator W, and would yield the same statistics for all observables:

$$W = \tfrac{1}{k}\,(P_{\psi_1} + P_{\psi_2} + \ldots + P_{\psi_k}) = \tfrac{1}{k}\,(P_{\varphi_1} + P_{\varphi_2} + \ldots + P_{\varphi_k}) = \tfrac{1}{k}\,P$$

Pure states, represented by idempotent density operators, are homogeneous in the sense that no idempotent density operator is expressible as a convex sum of two or more different density operators. That is, if W is idempotent and

$$W = p_1 W_1 + p_2 W_2 \quad (p_1 + p_2 = 1;\, p_1 > 0,\, p_2 > 0)$$

then $W = W_1 = W_2$.

To see this, note that W^2 can be expressed as:

$$\begin{aligned} W^2 &= p_1{}^2 W_1{}^2 + p_2{}^2 W_2{}^2 + p_1 p_2\,(W_1 W_2 + W_2 W_1) \\ &= p_1{}^2 W_1{}^2 + p_2{}^2 W_2{}^2 + p_1 p_2[W_1{}^2 + W_2{}^2 - (W_1 - W_2)^2] \\ &= p_1(p_1 + p_2)W_1{}^2 + p_2(p_1 + p_2)W_2{}^2 - p_1 p_2(W_1 - W_2)^2 \\ &= p_1 W_1{}^2 + p_2 W_2{}^2 - p_1 p_2(W_1 - W_2)^2 \end{aligned}$$

So $\quad W - W^2 = p_1(W_1 - W_1{}^2) + p_2(W_2 - W_2{}^2) + p_1 p_2(W_1 - W_2)^2 = 0.\quad$ But $W_1 - W_1{}^2$, $W_2 - W_2{}^2$, and $(W_1 - W_2)^2$ are all definite, from which it follows that they are all equal to the null operator, in particular:

$$(W_1 - W_2)^2 = 0$$

and so

$$W_1 - W_2 = 0$$

since $W_1 - W_2$ is self-adjoint. So $W_1 = W_2$, and hence (since $p_1 + p_2 = 1$), $W = W_1 = W_2$.[129]

So a pure state represented by an idempotent density operator is pure in the sense that it is homogeneous with respect to the set of states represented by density operators, but a mixed state represents a mixture only in the sense that it is nonhomogeneous, not in the sense that it represents a definite mixture of pure states. An 'ignorance interpretation' of mixed states is not appropriate.

[129] This proof is due to London and Bauer (1939, pp. 31, 32).

The probability assigned to a range of values E of an observable A by a pure or mixed quantum state according to the Born rule, expressed in terms of the trace of the density operator representing the state:

$$\mathrm{prob}_W(a \in E) = \mathrm{tr}(WP_A(E))$$

can be extended to ranges of values of several mutually *compatible* observables (observables represented by *commuting* operators).

Two operators in $\mathscr{H}_n$, A_1 and A_2, are said to *commute* if and only if[130]

$$A_1 A_2 = A_2 A_1$$

It can be shown that a self-adjoint operator A_1 commutes with a self-adjoint operator A_2, if and only if each $P_{A_1}(E)$ in the spectral measure of A_1 commutes with A_2, and each $P_{A_2}(E)$ in the spectral measure of A_2 commutes with A_1; in fact, if and only if $P_{A_1}(E)$ commutes with $P_{A_2}(F)$, for all Borel sets E and F.[131] Also, any function of A_1 commutes with any function of A_2 if A_1 and A_2 commute. In fact, if A_1 and A_2 commute, there exists a self-adjoint operator B such that:

$$A_1 = g_1(B)$$
$$A_2 = g_2(B).$$

A self-adjoint operator A representing an observable is said to be a function, $g(B)$, of a self-adjoint operator B, if there exists a map $g\colon \mathbb{R} \to \mathbb{R}$ such that $A = \int g(r)\mathrm{d}P_B(r)$.[132] It follows that if $A = g(B)$, then:

$$\mathrm{prob}_W(a \in E) = \mathrm{prob}_W(g(B) \in E) = \mathrm{prob}_W(b \in g^{-1}(E))$$

for all quantum states (represented by density operators W), and all Borel sets E. The implicit assumption here is that two observables, A_1 and A_2, are equivalent if and only if

$$\mathrm{prob}_W(a_1 \in E) = \mathrm{prob}_W(a_2 \in E)$$

for all W and all E; that is, that the set of quantum states $\{W\}$ is complete.

Gleason's theorem (1957) demonstrates that the set of pure and mixed states is complete, in the sense that all possible probability measures μ definable on the lattice $\mathscr{L}$ of projection operators of a Hilbert space are generated by the density operators of pure or mixed states by the generalized Born rule:

$$\mu(P) = \mathrm{tr}(WP)$$

for all $P \in \mathscr{L}$. So the Born rule turns out to be unique for the probability measures definable on the propositional structure of a quantum mechanical system characterized

[130] In $\mathscr{H}_\infty$ the different domains of definition of the two operators have to be taken into account, if both operators are not defined everywhere. In $\mathscr{H}_N$ all linear operators are defined everywhere.

[131] Von Neumann (1955, p. 171).

[132] Von Neumann (1955, footnote 94, p. 145).

by the projective geometry of Hilbert space (where a probability measure is a map from $\mathscr{L}$ onto the interval (0,1) satisfying the usual conditions for a probability measure on each Boolean sublattice of $\mathscr{L}$).

Gleason's theorem does not preclude the possibility of extending the set of states by introducing additional 'hidden variables.' In such an extension or modification of quantum mechanics, two observables that are equivalent for all quantum states might no longer be equivalent for the extended set of states. As shown in chapter 3, what Gleason's theorem excludes is a class of hidden variable reconstructions of the quantum statistics satisfying a certain constraint.

If $A_1, A_2, \ldots, A_n$ are a set of mutually compatible observables, represented by mutually commuting self-adjoint operators, then the probability assigned by a pure state represented by a vector ψ to ranges of values E_i of the observables A_i is given by:

$$prob_\psi((a_1 \in E_1)\&(a_2 \in E_2) \ldots \&(a_n \in E_n)) = \|P_{A_1}(E_1)P_{A_2}(E_2) \ldots P_{A_n}(E_n)\psi\|^2.$$

In the general case of a pure or mixed state represented by the density operator W:[133]

$$prob_W((a_1 \in E_1)\&(a_2 \in E_2) \ldots \&(a_n \in E_n)) = \mathrm{tr}(WP_{A_1}(E_1)P_{A_2}(E_2) \ldots P_{A_n}(E_n))$$

This follows because if the observables A_i are mutually compatible, then $A_i = g_i(B)$, for some observable B. So

$$\begin{aligned}
&\mathrm{prob}_W((a_1 \in E_1)\&(a_2 \in E_2) \ldots \&(a_n \in E_n)) \\
&\quad = \mathrm{prob}_W((b \in g_1^{-1}(E_1))\&(b \in g_2^{-1}(E_2)) \ldots \&(b \in g_n^{-1}(E_n))) \\
&\quad = \mathrm{prob}_W(b \in \bigcap_i g_i^{-i}(E_i)) \\
&\quad = \mathrm{tr}(WP_B(\bigcap_i g_i^{-1}(E_i)) \\
&\quad = \mathrm{tr}(W\prod_i P_B(g_i^{-1}(E_i)) \\
&\quad = \mathrm{tr}(W\prod_i P_{A_i}(E_i))
\end{aligned}$$

by the properties of the spectral measure.

The dynamical evolution of quantum states in Hilbert space is defined by Schrödinger's time-dependent equation of motion

$$i\hbar\frac{d\psi(t)}{dt} = H\psi(t)$$

for the vector representative $\psi(t)$ of a pure state. This evolution is characterized by a *unitary transformation* of the state, defined by a *unitary operator* $U(t) = e^{-iHt/\hbar}$:

$$\psi(t) = U(t)\psi$$

133 Here the symbol a_i is a variable denoting a general value of the observable A_i, not a name for the ith eigenvalue of an observable A. So $a_i \in E$ is to be read: the value of the observable A_i lies in the range E.

where ψ represents the state at $t = 0$. A unitary operator U satisfies the condition:

$$U^*U = UU^* = I,$$

so $U^* = U^{-1}$. It follows that a unitary transformation preserves the lengths of vectors $(U\psi, U\psi) = (U^{-1}U\psi, \psi) = (\psi, \psi)$.

The unitary operators $U(t)$, $-\infty < t < \infty$, characterizing the dynamical evolution of a quantum system, define a one-parameter group of transformations on Hilbert space. Since $P_\psi \phi = (\psi, \phi)\psi$, for any vector ϕ:

$$\begin{aligned} P_{U(t)\psi}\phi &= (U(t)\psi, \phi)U(t)\psi \\ &= (\psi, U^{-1}(t)\phi)U(t)\psi \\ &= U(t)(\psi, U^{-1}(t)\phi)\psi \\ &= U(t)P_\psi U^{-1}(t)\phi \end{aligned}$$

and so the dynamical evolution of a general quantum state represented by a density operator $W(t)$ is given by:

$$W(t) = U(t)WU^{-1}(t)$$

A.3 The Dirac notation

There is an alternative notation for Hilbert space vectors and operators, the Dirac notation, that is often more useful than the von Neumann notation introduced above. I shall exploit the resources of both notations, but for the most part I shall use the Dirac notation.

A linear functional on $\mathscr{H}$ is a complex-valued map f on $\mathscr{H}$ such that $f(c\psi) = cf(\psi)$, for any complex number c, and $f(\psi + \phi) = f(\psi) + f(\phi)$. The set of continuous linear functionals on $\mathscr{H}$ forms a Hilbert space, called the dual space to $\mathscr{H}$. A 1–1 correspondence can be defined between $\mathscr{H}$ and its dual by associating a vector ψ in $\mathscr{H}$ with the linear functional that maps any vector ϕ in $\mathscr{H}$ onto the scalar product (ψ,ϕ).

Dirac denotes the vectors of the dual space by $\langle\ |$ and refers to them as 'bra' vectors. He refers to the vectors $|\ \rangle$ of $\mathscr{H}$ as 'ket' vectors. The scalar product of two kets $|\psi\rangle$ and $|\phi\rangle$ in $\mathscr{H}$, denoted by the 'bra(c)ket' $\langle\psi|\phi\rangle$ instead of (ψ, ϕ), can then be understood as the value of the linear functional $\langle\psi|$ at the vector $|\phi\rangle$ (that is, $\langle\phi|\phi\rangle = \langle\psi(|\phi\rangle)\rangle$).

In the Dirac notation, eigenvectors can be denoted by the index of the value of the corresponding eigenvalue. For example, in the case of an operator with two possible eigenvalues, $+$ and $-$, the eigenvectors can be denoted as $|+\rangle$ and $|-\rangle$, respectively. This is convenient, but the main advantage of the Dirac notation is the transparent representation of operators, so that the action of an operator on a vector can be read directly from the representation. In the Diract notation, $|\psi\rangle$ represents a vector in $\mathscr{H}$, $\langle\psi|\phi\rangle$ represents the scalar product of two vectors (a complex number), and $|\psi\rangle\langle\phi|$ represents an operator (the operator that maps the vector $|\chi\rangle$ onto the vector $|\psi\rangle\langle\phi|\cdot|\chi\rangle = |\psi\rangle\langle\phi|\chi\rangle = \langle\phi|\chi\rangle\,|\psi\rangle$).

The projection operator onto an eigenvector $|a_i\rangle$ of the operator A is represented in the Dirac notation as the operator $|a_i\rangle\langle a_i|$. The projection of $|\psi\rangle$ onto $|a_i\rangle$ is represented as:

$$|a_i\rangle\langle a_i|\psi\rangle = \langle a_i|\psi\rangle|a_i\rangle.$$

If the eigenvectors form a complete set in $\mathscr{H}_n$, then:

$$\sum_{i=1}^{n} |a_i\rangle\langle a_i| = I$$

and

$$|\psi\rangle = \sum_{i=1}^{n} |a_i\rangle\langle a_i|\psi\rangle = \sum_{i=1}^{n} \langle a_i|\psi\rangle|a_i\rangle = \sum_{i=1}^{n} c_i|a_i\rangle \qquad \text{(A.1)}$$

The spectral representation of A can be expressed transparently as:

$$A = A{\cdot}I = A\sum_{i=1}^{n} |a_i\rangle\langle a_i| = \sum_{i=1}^{n} A|a_i\rangle\langle a_i| = \sum_{i=1}^{n} a_i|a_i\rangle\langle a_i|$$

Observables with continuous spectra, like position, do not have normalizable eigenvectors. Dirac introduces position 'eigenvectors' $|q'\rangle$ as convenient mathematical fictions, defined formally by the eigenvalue equation:

$$Q|q'\rangle = q'|q'\rangle,$$

where Q represents the (1-dimensional) position operator. The kets $|q'\rangle$ satisfy an orthogonality condition expressed in terms of Dirac's 'delta function' (another mathematical fiction) as:

$$\langle q|q'\rangle = \delta(q - q'),$$

where $\delta(q)$ is defined by the conditions:

(i) $\displaystyle\int_{-\infty}^{\infty} dq\delta(q) = 1$

(ii) $\delta(q) = 0$ if $q \neq 0$

It follows that:

$$\int_{-\infty}^{\infty} dqf(q)\delta(q) = f(0)$$

or

$$\int_{-\infty}^{\infty} \mathrm{d}q f(q)\delta(q-q') = f(q')$$

(The corresponding relation in the discrete case is:

$$\sum_{j=1}^{n} c_j \delta_{ij} = c_i$$

where δ_{ij} is the Kronecker delta, defined by $\delta_{ij} = 0$ if $i \neq j$ and $\delta_{ij} = 1$ if $i = j$.)

The unit operator can now be expressed in terms of these kets as:

$$I = \int \mathrm{d}q |q\rangle\langle q|$$

and so a quantum state $|\psi\rangle$ can be represented formally as a continuous superposition of position eigenvectors:

$$\begin{aligned} |\psi\rangle &= \int \mathrm{d}q |q\rangle\langle q|\psi\rangle \\ &= \int \mathrm{d}q \psi(q)|q\rangle, \end{aligned} \tag{A.2}$$

where $\psi(q) = \langle q|\psi\rangle$ is the 'wave function' in position space. Equation (A.2) is the continuous analogue of equation (A.1), the expression for the representation of $|\psi\rangle$ as a superposition of eigenstates of an observable A with a discrete spectrum.

Similarly, $|\psi\rangle$ can be expressed formally as a continuous superposition of momentum 'eigenvectors' $|p\rangle$ of the momentum operator P conjugate to Q:

$$|\psi\rangle = \int \mathrm{d}p \psi(p)|p\rangle,$$

where $\langle p|p'\rangle = \delta(p-p')$ and $\psi(p) = \langle p|\psi\rangle$ is the wave function in momentum space. In the position representation, the momentum eigenvectors are:

$$\langle q|p\rangle = \frac{1}{\sqrt{2\pi}} \mathrm{e}^{\mathrm{i}pq}$$

for all values of p and q, and so the two representations are related by the equation:

$$\begin{aligned} \psi(p) &= \langle p|\psi\rangle \\ &= \int \mathrm{d}q \langle p|q\rangle\langle q|\psi\rangle \\ &= \frac{1}{\sqrt{2\pi}} \int \mathrm{d}q \mathrm{e}^{-\mathrm{i}pq} \psi(q) \end{aligned}$$

Just as $|c_i|^2 = |\langle a_i|\psi\rangle|^2$ represents the probability of obtaining the value a_i as the outcome of a measurement of the discrete observable A, so $|\psi(q)|^2 \mathrm{d}q$ represents the probability of finding the value of position in the range $[q,q + \mathrm{d}q]$ in a position measurement, and $|\psi(p)|^2\mathrm{d}p$ represents the probability of finding the value of momentum in the range $[p,p + \mathrm{d}p]$ in a momentum measurement.

The Dirac notation provides a handy mathematical bookkeeping device for keeping track of relationships between vectors and operators in Hilbert space. I present below a number of identities that are useful at various points in the discussion.

In the von Neumann notation, to show that:

$$P_{\alpha_i}WP_{\alpha_i} = (\alpha_i, W\alpha_i)P_{\alpha_i}$$

note that, for any vector θ, $P_\psi\theta = (\psi,\theta)\psi$, and so:

$$\begin{aligned} P_{\alpha_i}WP_{\alpha_i}\theta &= P_{\alpha_i}W(\alpha_i,\theta)\alpha_i \\ &= (\alpha_i,\theta)P_{\alpha_i}W\alpha_i \\ &= (\alpha_i,\theta)(\alpha_i,\ W\alpha_i)\alpha_i \\ &= (\alpha_i, W\alpha_i)P_{\alpha_i}\theta \end{aligned}$$

In the Dirac notation, we see directly that:

$$|\alpha_i\rangle\langle\alpha_i|W|\alpha_i\rangle\langle\alpha_i| = \langle\alpha_i|W|\alpha_i\rangle|\alpha_i\rangle\langle\alpha_i|$$

Notice that $\mathrm{tr}(WP_\psi) = (\psi, W\psi)$, for any vector ψ. To see this, take a basis $\{\phi_i\}$ with $\phi_1 = \psi$ to compute the trace. Then:

$$P_{\alpha_i}WP_{\alpha_i} = (\alpha_i, W\alpha_i)P_{\alpha_i} = \mathrm{tr}(WP_{\alpha_i})P_{\alpha_i}$$

In the von Neumann notation, to show that:

$$\mathrm{tr}(P_\psi P_{\alpha_i}) = (\psi,\ P_{\alpha_i}\psi) = \|P_{\alpha_i}\psi\|^2$$

note first that:

$$\mathrm{tr}(P_\psi P_{\alpha_i}) = \mathrm{tr}(P_{\alpha_i}P_\psi) = (\psi,\ P_{\alpha_i}\psi)$$

(taking a basis $\{\phi_i\}$ with $\phi_1 = \psi$ to compute the trace). And:

$$\begin{aligned} (\psi,\ P_{\alpha_i}\psi) &= (\psi,(\alpha_i,\psi)\alpha_i) \\ &= (\alpha_i,\psi)(\psi,\alpha_i) \\ &= |(\alpha_i,\psi)|^2 \end{aligned}$$

In the Dirac notation:

$$\mathrm{tr}(|\psi\rangle\langle\psi|\alpha_i\rangle\langle\alpha_i|) = \langle\psi|\psi\rangle\langle\psi|\alpha_i\rangle\langle\alpha_i|\psi\rangle = |\langle\alpha_i|\psi\rangle|^2$$

But the analogous result:

$$\mathrm{tr}(P_\psi P_{a_i}) = (\psi,P_{a_i}\psi) = \|P_{a_i}\psi\|^2,$$

where P_{a_i} is a projection operator onto a multi-dimensional subspace, is more transparent in the von Neumann notation:

$$\begin{aligned}\mathrm{tr}(P_\psi P_{a_i}) &= \mathrm{tr}(P_{a_i} P_\psi)\\ &= (\psi, P_{a_i}\psi)\\ &= (\psi, P_{a_i}{}^2\psi)\\ &= (P_{a_i}\psi, P_{a_i}\psi)\\ &= \|P_{a_i}\psi\|^2\end{aligned}$$

In the von Neumann notation:

$$P_\phi P_\psi P_\phi = |(\phi,\psi)|^2 P_\phi$$

because, for any vector θ:

$$\begin{aligned}P_\phi P_\psi P_\phi \theta &= P_\phi P_\psi (\phi,\theta)\phi\\ &= P_\phi (\phi,\theta)(\psi,\phi)\psi\\ &= (\phi,\theta)(\psi,\phi)(\phi,\psi)\phi\\ &= |(\phi,\psi)|^2 P_\phi \theta\end{aligned}$$

In the Dirac notation, the inference is immediate:

$$|\phi\rangle\langle\phi|\psi\rangle\langle\psi|\phi\rangle\langle\phi| = |\langle\phi|\psi\rangle|^2|\phi\rangle\langle\phi|$$

If $\psi' = \psi/\|\psi\|$, then $P_\psi/\|\psi\|^2 = P_{\psi'}$ because for any vector θ:

$$\frac{P_\psi}{\|\psi\|^2}\theta = \frac{(\psi,\theta)\psi}{\|\psi\|^2} = (\psi',\theta)\psi' = P_{\psi'}\theta$$

In the Dirac notation, this follows immediately:

$$\frac{|\psi\rangle\langle\psi|}{\|\psi\|^2} = |\frac{\psi}{\|\psi\|}\rangle\langle\frac{\psi}{\|\psi\|}| = |\psi'\rangle\langle\psi'|$$

The Dirac notation is particularly useful in the representation of states and observables of composite systems, although it is sometimes more perspicuous to use the resources of both notations.

A.4 Spin

The simplest physical examples of quantum mechanical dynamical variables or observables involve *spin*, a vector quantity ***S*** (in real space) whose components relative to a Cartesian coordinate system, S_x, S_y, S_z, satisfy commutation relations similar to those derived from the Poisson brackets for the components of angular momentum, J_x, J_y, J_z:

$$[S_x, S_y] = \mathrm{i}\,\hbar\, S_z$$

and cyclic permutations, or

$$[S_x, S_y] = iS_z$$

in units in which $\hbar = 1$. Spin is a purely quantum mechanical quantity representing internal degrees of freedom of a system.

It turns out (after solving the eigenvalue equation) that the eigenvalues of $S^2 = S_x^2 + S_y^2 + S_z^2$ can take the values $s(s+1)$ for integral or half-integral values of s. Each eigenvalue is degenerate, associated with a $(2s+1)$-dimensional eigenspace of S^2. For each value of s (that is, each eigenvalue of S^2), each of the components of $\boldsymbol{S}$ can take any of the $2s+1$ values $\{-s, -s+1, \ldots, s-1, s\}$. So, for $s = \frac{1}{2}$, the eigenvalues of S_z (or S_x, S_y) can take the values $-\frac{1}{2}, +\frac{1}{2}$. For $s = 1$, the eigenvalues of S_z (or S_x, S_y) can take the values $-1, 0, +1$. The case $s = \frac{1}{2}$ is referred to as a 'spin-$\frac{1}{2}$' system, and the case $s = 1$ as a 'spin-1' system: these are systems for which the state lies in the 2-dimensional or 3-dimensional S^2-eigenspace, respectively, corresponding to the eigenvalues $\frac{3}{4}$ and 2 for S^2.

In the spin-$\frac{1}{2}$ case, it is usual to introduce the so-called *Pauli operators*, $\sigma_x, \sigma_y, \sigma_z$ in the 2-dimensional eigenspace of S^2, where

$$\sigma_x = 2S_x$$
$$\sigma_y = 2S_y$$
$$\sigma_z = 2S_z$$

in this subspace. Then the eigenvalues of σ_x, σ_y, σ_z are ± 1.

If we represent the two eigenvectors of σ_z as $|\sigma_z = +1\rangle$ and $|\sigma_z = -1\rangle$, abbreviated to $|z+\rangle$ and $|z-\rangle$, then

$$\sigma_z|z+\rangle = |z+\rangle$$
$$\sigma_z|z-\rangle = -|z-\rangle$$

The matrix representatives of the Pauli operators in the σ_z-basis, $|z+\rangle = \begin{pmatrix}1\\0\end{pmatrix}$, $|z-\rangle = \begin{pmatrix}0\\1\end{pmatrix}$, are:

$$\sigma_x = \begin{pmatrix}0 & 1\\ 1 & 0\end{pmatrix}$$
$$\sigma_y = \begin{pmatrix}0 & -i\\ i & 0\end{pmatrix}$$
$$\sigma_z = \begin{pmatrix}1 & 0\\ 0 & -1\end{pmatrix}$$

It is easily checked that these matrices satisfy the commutation relations for $\sigma_x, \sigma_y, \sigma_z$ ($[\sigma_x, \sigma_y] = 2i\sigma_z$, and cyclic permutations). Note that $\sigma_x^2 + \sigma_y^2 + \sigma_z^2 = 4(S_x^2 + S_y^2 + S_z^2) = 3$, so $S^2 = \frac{3}{4} I$.

The eigenvectors of σ_x and σ_y in the σ_z-basis are found by solving the eigenvalue equations for the operators:

$$|x+\rangle = \tfrac{1}{\sqrt{2}}\begin{pmatrix}1\\1\end{pmatrix},\ |x-\rangle = \tfrac{1}{\sqrt{2}}\begin{pmatrix}1\\-1\end{pmatrix}$$
$$|y+\rangle = \tfrac{1}{\sqrt{2}}\begin{pmatrix}1\\i\end{pmatrix},\ |y-\rangle = \tfrac{1}{\sqrt{2}}\begin{pmatrix}1\\-i\end{pmatrix}$$

The projection operators onto these vectors are:[134]

$$P_{z+} = |z+\rangle\langle z+| = \begin{pmatrix}1\\0\end{pmatrix}(1,0) = \begin{pmatrix}1&0\\0&0\end{pmatrix}$$
$$P_{z-} = |z-\rangle\langle z-| = \begin{pmatrix}0\\1\end{pmatrix}(0,1) = \begin{pmatrix}0&0\\0&1\end{pmatrix}$$
$$P_{x+} = |x+\rangle\langle x+| = \tfrac{1}{2}\begin{pmatrix}1\\1\end{pmatrix}(1,1) = \tfrac{1}{2}\begin{pmatrix}1&1\\1&1\end{pmatrix}$$
$$P_{x-} = |x-\rangle\langle x-| = \tfrac{1}{2}\begin{pmatrix}1\\-1\end{pmatrix}(1,-1) = \tfrac{1}{2}\begin{pmatrix}1&-1\\-1&1\end{pmatrix}$$
$$P_{y+} = |y+\rangle\langle y+| = \tfrac{1}{2}\begin{pmatrix}1\\i\end{pmatrix}(1,-i) = \tfrac{1}{2}\begin{pmatrix}1&-i\\i&1\end{pmatrix}$$
$$P_{y-} = |y-\rangle\langle y-| = \tfrac{1}{2}\begin{pmatrix}1\\-i\end{pmatrix}(1,i) = \tfrac{1}{2}\begin{pmatrix}1&i\\-i&1\end{pmatrix}$$

The spectral representations of σ_x, σ_y, and σ_z are:

$$\sigma_x = P_{x+} - P_{x-}$$
$$\sigma_y = P_{y+} - P_{y-}$$
$$\sigma_z = P_{z+} - P_{z-}$$

and since:

$$P_{x+} + P_{x-} = I$$
$$P_{y+} + P_{y-} = I$$
$$P_{z+} + P_{z-} = I$$

it follows that

$$P_{x+} = \tfrac{1}{2}(I + \sigma_x),\ P_{x-} = \tfrac{1}{2}(I - \sigma_x)$$

and similarly for σ_y, σ_z. This relation between the spin component projection operators and the Pauli operators is used in section 3.4.

In the spin-1 case, the spin operators S_x, S_y, S_z have the following matrix representations in the basis $|\sigma_z = -1\rangle$, $|\sigma_z = 0\rangle$, $|\sigma_z = +1\rangle$:

[134] I write P_{z+} for the projection operator onto the vector $|z+\rangle$, rather than $P_{|z+\rangle}$. In general, I use the symbol P_a to denote the projection operator onto the eigenspace corresponding to the eigenvalue a of an operator A. If this eigenspace is 1-dimensional, $P_a = |a\rangle\langle a|$ is the projection operator onto the unit vector $|a\rangle$. If a vector is denoted in the Dirac notation by $|\psi\rangle$, I write P_ψ for the projection operator onto $|\psi\rangle$, rather than $P_{|\psi\rangle}$.

$$S_x = \tfrac{1}{\sqrt{2}}\begin{pmatrix} 0 & 1 & 0 \\ 1 & 0 & 1 \\ 0 & 1 & 0 \end{pmatrix}$$

$$S_y = \tfrac{1}{\sqrt{2}}\begin{pmatrix} 0 & i & 0 \\ -i & 0 & i \\ 0 & -i & 0 \end{pmatrix}$$

$$S_z = \begin{pmatrix} -1 & 0 & 0 \\ 0 & 0 & 0 \\ 0 & 0 & 1 \end{pmatrix}$$

Each of these operators has three eigenvalues, $-1, 0, +1$. The operators S_x^2, S_y^2, S_z^2 each have two eigenvalues, 0 and 1, and are therefore degenerate or nonmaximal in the 3-dimensional spin-1 eigenspace. For S_x^2, the 0-eigenvalue corresponds to the eigenvector $|S_z = 0\rangle$ and the 1-eigenvalue corresponds to the plane spanned by the eigenvectors $|S_x = -1\rangle$ and $|S_x + 1\rangle$, and similarly for S_y^2 and S_z^2.

It is easy to see that:

$$|S_x = 0\rangle = \tfrac{1}{\sqrt{2}}\begin{pmatrix} 1 \\ 0 \\ -1 \end{pmatrix}$$

$$|S_y = 0\rangle = \tfrac{1}{\sqrt{2}}\begin{pmatrix} 1 \\ 0 \\ 1 \end{pmatrix}$$

$$|S_z = 0\rangle = \begin{pmatrix} 0 \\ 1 \\ 0 \end{pmatrix}$$

so the three 0-eigenstates of S_x, S_y, S_z form an orthogonal basis. It follows that, although S_x, S_y, S_z do not commute pairwise, S_x^2, S_y^2, S_z^2 do commute pairwise. Since the vectors $|S_y = 0\rangle$ and $|S_z = 0\rangle$ are orthogonal to $|S_x = 0\rangle$, the 0-eigenvector of S_x, they must span the ± 1-eigenplane of S_x. In fact:

$$|S_y = 0\rangle = \tfrac{1}{\sqrt{2}}\left(|S_x = -1\rangle - |S_x = +1\rangle\right)$$
$$|S_z = 0\rangle = \tfrac{1}{\sqrt{2}}\left(|S_x = -1\rangle + |S_x = +1\rangle\right)$$

that is, the orthogonal pair $|S_y = 0\rangle$, $|S_z = 0\rangle$ is obtained from the orthogonal pair $|S_x = -1\rangle$, $|S_x = +1\rangle$ by rotating through $\dfrac{\pi}{2}$ clockwise about $|S_x = 0\rangle$. Similarly, $|S_x = 0\rangle$, $|S_z = 0\rangle$ span the ± 1-eigenplane of S_y, and $|S_x = 0\rangle$, $|S_y = 0\rangle$ span the ± 1-eigenplane of S_z. So the 1-eigenplanes of S_x^2, S_y^2, S_z^2 intersect at right angles in the rays spanned by the vectors $|S_x = 0\rangle$, $|S_y = 0\rangle$, $|S_z = 0\rangle$, and hence the projection operators onto these planes commute. But these projection operators are just the operators S_x^2, S_y^2, S_z^2. In the spectral representation:

$$S_x^2 = 1 \cdot P_{x=0}^{\perp} + 0 \cdot P_{x=0}$$
$$S_y^2 = 1 \cdot P_{y=0}^{\perp} + 0 \cdot P_{y=0}$$
$$S_z^2 = 1 \cdot P_{z=0}^{\perp} + 0 \cdot P_{z=0}$$

where $P_{x=0} = |S_x = 0\rangle\langle S_x = 0|$ is the projection operator onto the ray spanned by the unit vector $|S_x = 0\rangle$, and $P^{\perp}_{x=0}$ is the projection operator onto the orthogonal plane, and similarly for $P_{y=0}$, $P_{z=0}$.

Notice that these 1-eigenplanes are not 'orthogonal' in the formal sense, because they contain a common 1-dimensional subspace or ray – that is, not every vector in one of these planes is orthogonal to *every* vector in another of these planes. But they are 'orthogonal except for an overlap,' and this characterizes the compatibility of these subspaces as elements of the lattice of Hilbert space subspaces, equivalently the commutativity of the corresponding operators.

Of course, it can easily be checked directly from the matrix representations of S_x, S_y, S_z that S_x^2, S_y^2, S_z^2 commute pairwise. Similarly, it is easy to check that

$$P_{x=0} = |S_x = 0\rangle\langle S_x = 0| = \tfrac{1}{2}\begin{pmatrix} 1 & 0 & -1 \\ 0 & 0 & 0 \\ -1 & 0 & 1 \end{pmatrix}$$

$$P_{y=0} = |S_y = 0\rangle\langle S_y = 0| = \tfrac{1}{2}\begin{pmatrix} 1 & 0 & 1 \\ 0 & 0 & 0 \\ 1 & 0 & 1 \end{pmatrix}$$

$$P_{z=0} = |S_z = 0\rangle\langle S_z = 0| = \begin{pmatrix} 0 & 0 & 0 \\ 0 & 1 & 0 \\ 0 & 0 & 0 \end{pmatrix}$$

and hence that

$$P^{\perp}_{x=0} = \begin{pmatrix} 1 & 0 & 0 \\ 0 & 1 & 0 \\ 0 & 0 & 1 \end{pmatrix} - \tfrac{1}{2}\begin{pmatrix} 1 & 0 & -1 \\ 0 & 0 & 0 \\ -1 & 0 & 1 \end{pmatrix} = \tfrac{1}{2}\begin{pmatrix} 1 & 0 & 1 \\ 0 & 2 & 0 \\ 1 & 0 & 1 \end{pmatrix}$$

$$P^{\perp}_{y=0} = \begin{pmatrix} 1 & 0 & 0 \\ 0 & 1 & 0 \\ 0 & 0 & 1 \end{pmatrix} - \tfrac{1}{2}\begin{pmatrix} 1 & 0 & 1 \\ 0 & 0 & 0 \\ 1 & 0 & 1 \end{pmatrix} = \tfrac{1}{2}\begin{pmatrix} 1 & 0 & -1 \\ 0 & 2 & 0 \\ -1 & 0 & 1 \end{pmatrix}$$

$$P^{\perp}_{z=0} = \begin{pmatrix} 1 & 0 & 0 \\ 0 & 1 & 0 \\ 0 & 0 & 1 \end{pmatrix} - \begin{pmatrix} 0 & 0 & 0 \\ 0 & 1 & 0 \\ 0 & 0 & 0 \end{pmatrix} = \begin{pmatrix} 1 & 0 & 0 \\ 0 & 0 & 0 \\ 0 & 0 & 1 \end{pmatrix}$$

Consider the operator

$$H = aS_x^2 + bS_y^2 + cS_z^2$$

in the 3-dimensional spin-1 eigenspace, with distinct values for a, b, and c. This H can be written in the spectral representation as:

$$\begin{aligned} H &= aP^{\perp}_{x=0} + bP^{\perp}_{y=0} + cP^{\perp}_{z=0} \\ &= a(P_{y=0} + P_{z=0}) + b(P_{x=0} + P_{z=0}) + c(P_{x=0} + P_{y=0}) \\ &= (b + c)P_{x=0} + (a + c)P_{y=0} + (a + b)P_{z=0} \end{aligned}$$

that is, H is a maximal operator in the spin-1 eigenspace, with eigenvalues $b + c$, $a + c$, $a + b$. In the basis $|S_x = 0\rangle$, $|S_y = 0\rangle$, $|S_z = 0\rangle$, the matrix representation of H is:

$$H = \begin{pmatrix} b + c & 0 & 0 \\ 0 & a + c & 0 \\ 0 & 0 & a + b \end{pmatrix}$$

Clearly, the operators S_x^2, S_y^2, S_z^2 are all functions of H, because $P_{x=0}$, $P_{y=0}$, $P_{z=0}$ are functions of H: when $H = b + c$, $P_{x=0} = 1$, $P_{y=0} = 0$, $P_{z=0} = 0$, so $S_x^2 = P_{x=0}^{\perp} = 0$, $S_y^2 = P_{y=0}^{\perp} = 1$, and $S_z^2 = P_{z=0}^{\perp} = 1$. Similarly, when $H = a + c$, $S_x^2 = 1$, $S_y^2 = 0$, $S_z^2 = 1$, and when $H = a + b$, $S_x^2 = 1$, $S_y^2 = 1$, $S_z^2 = 0$.

A.5 Composite systems

An 'internal' discussion of measurement processes in quantum mechanics, in terms of the dynamics of interacting systems, requires the analysis of composite systems – systems composed of subsystems. Suppose we have two systems, S associated with a Hilbert space $\mathscr{H}_S$, and M associated with a Hilbert space $\mathscr{H}_M$. Here S might represent the system measured and M the measuring instrument, but the following discussion is quite general. It applies to the spin states of a pair of spin-$\frac{1}{2}$ particles, say, or even to a single spin-$\frac{1}{2}$ particle, where the particle's position and spin are represented by operators on different Hilbert spaces.

States of the combined system, $S + M$, are represented by density operators on the tensor product Hilbert space, $\mathscr{H}_S \otimes \mathscr{H}_M$. If $\{\alpha_i, i = 1, 2, \ldots, n\}$ is a complete orthonormal basis for $\mathscr{H}_S$ and $\{\rho_j, j = 1, 2, \ldots, m\}$ is a complete orthonormal basis for $\mathscr{H}_M$, then $\{\alpha_i \otimes \rho_j, i = 1, 2, \ldots, n; j\, 1, 2, \ldots, m\}$ is a complete orthonormal basis for $\mathscr{H}_S \otimes \mathscr{H}_M$. So a general unit vector in $\mathscr{H}_S \otimes \mathscr{H}_M$ can be represented as a linear superposition of the form:

$$\psi = \sum_{ij} c_{ij} \alpha_i \otimes \rho_j = \sum_{ij} (\alpha_i \otimes \rho_j, \psi) \alpha_i \otimes \rho_j,$$

where the scalar product of $\psi \otimes \phi$ and $\chi \otimes \eta$ is defined by:

$$(\psi \otimes \phi,\ \chi \otimes \eta) \equiv (\psi, \chi)(\phi, \eta)$$

In the Dirac notation, a basis vector of $\mathscr{H}_S \otimes \mathscr{H}_M$ is represented as $|a_i\rangle|r_j\rangle$ (without the tensor products symbol '$\otimes$'), or as $|a_i, r_j\rangle$, or even as $|i\rangle|j\rangle$ or $|i,j\rangle$, if the bases are understood as given, where the order indicates to which Hilbert spaces the indices refer. The scalar product of $|\psi\rangle|\phi\rangle$ and $|\chi\rangle|\eta\rangle$ is represented as:

$$\langle \phi, \psi | \chi, \eta \rangle \equiv \langle \psi | \chi \rangle \langle \phi | \eta \rangle$$

A density operator W, representing a pure or mixed state of $S + M$, assigns probabilities to ranges of values of S-observables and M-observables, as well as $(S + M)$-observables. There is a unique density operator W_S, called the *reduced density operator* for the system S, representing the *reduced state* of S, that assigns the same probabilities to ranges of values of S-observables as W. Similarly, there is a unique density operator W_M, the reduced density operator for M, representing the reduced state of M, that assigns the same probabilities to ranges of values of M-observables as W. Equivalently, W_S and W_M define the same expectation values as W for all S-observables and all M-observables, respectively.

The reduced density operators are therefore defined by the conditions:

$$\mathrm{tr}_S(W_S Q_S) = \mathrm{tr}(W(Q_S \otimes I_M)), \text{ for all self-adjoint operators } Q_S \text{ of } \mathscr{H}_S$$
$$\mathrm{tr}_M(W_M Q_M) = \mathrm{tr}(W(I_S \otimes Q_M)), \text{ for all self-adjoint operators } Q_M \text{ of } \mathscr{H}_M,$$

where I_S and I_M are the unit operators in $\mathscr{H}_S$ and $\mathscr{H}_M$, respectively, and tr_S, tr_M, tr refer to the trace function in $\mathscr{H}_S$, $\mathscr{H}_M$, and $\mathscr{H}_S \otimes \mathscr{H}_M$, respectively. The density operator W_S of S is uniquely determined by the density operator W of $S + M$: if $\mathrm{tr}_S(W_S Q_S) = \mathrm{tr}_S(W'_S Q_S) = \mathrm{tr}(W(Q_S \otimes I_M)$ for all self-adjoint operators Q_S in $\mathscr{H}_S$, then $W_S = W'_S$. Similarly, W_M is uniquely determined by W.

Suppose the state of $S + M$ is a pure state represented by the vector $|\psi\rangle = \Sigma c_{ij}|a_i\rangle|r_j\rangle$, so that the density operator W is the projection operator $P_\psi = |\psi\rangle\langle\psi|$. To find W_S in this case, consider the expectation value of an $(S + M)$-observable represented by an operator of the form $Q_S \otimes I_M$:

$$\begin{aligned}
\mathrm{tr}(P_\psi(Q_S \otimes I_M)) &= \langle\psi|Q_S \otimes I_M|\psi\rangle \\
&= \sum_{ii'jj'} \langle r_{j'}|\langle a_{i'}|c^*_{i'j'}Q_S \otimes I_M c_{ij}|a_i\rangle|r_j\rangle \\
&= \sum_{ii'jj'} c_{ij}c^*_{i'j'}\langle a_{i'}|Q_S|a_i\rangle\langle r_{j'}|r_j\rangle \\
&= \sum_{ii'j} c_{ij}c^*_{i'j}\langle a_{i'}|Q_S|a_i\rangle \\
&= \sum_{ii'} u_{ii'}\langle a_{i'}|Q_S|a_i\rangle,
\end{aligned}$$

where $u_{ii'} = \Sigma_j c_{ij}c^*_{i'j}$.

Since $\langle a_k|a_i\rangle = \delta_{ki}$, this can be expressed as:

$$\begin{aligned}
\mathrm{tr}(P_\psi(Q_S \otimes I_M)) &= \sum_{ii'} u_{ii'} \sum_k \delta_{ki}\langle a_{i'}|Q_S|a_k\rangle \\
&= \sum_{ii'} u_{ii'} \sum_k \langle a_k|a_i\rangle\langle a_{i'}|Q_S|a_k\rangle \\
&= \sum_k \langle a_k|\sum_{ii'} u_{ii'}|a_i\rangle\langle a_{i'}|Q_S|a_k\rangle \\
&= \sum_k \langle a_k|W_S Q_S|a_k\rangle \\
&= \mathrm{tr}_S(W_S Q_S)
\end{aligned}$$

where $W_S = \Sigma_{ii'} u_{ii'}|a_i\rangle\langle a_{i'}| = \Sigma_{ii'j} c_{ij}c^*_{i'j}|a_i\rangle\langle a_{i'}| = \Sigma_h\langle r_h|P_\psi|r_h\rangle = \mathrm{tr}_M(P_\psi)$.

So $W_S = \mathrm{tr}_M(P_\psi)$, the *partial trace* of P_ψ over the Hilbert space $\mathscr{H}_M$. Similarly, $W_M = \Sigma_{jj'} v_{jj'}|r_j\rangle\langle r_{j'}| = \Sigma_k\langle a_k|P_\psi|a_k\rangle = \mathrm{tr}_S(P_\psi)$, the partial trace of P_ψ over the Hilbert space $\mathscr{H}_S$.

A general density operator W in $\mathscr{H}_S \otimes \mathscr{H}_M$ can be expressed as:

$$\begin{aligned}W &= IWI\\ &= \sum_{ij}|a_i\rangle\langle a_i| \otimes |r_j\rangle\langle r_j|\cdot W\cdot \sum_{i'j'}|a_{i'}\rangle\langle a_{i'}| \otimes |r_{j'}\rangle\langle r_{j'}|\\ &= \sum_{ii'jj'} (\langle a_i|\langle r_j|W|a_{i'}\rangle|r_{j'}\rangle)|a_i\rangle\langle a_{i'}| \otimes |r_j\rangle\langle r_{j'}|\\ &= \sum_{ii'jj'} W_{ij,i'j'}|a_i\rangle\langle a_{i'}| \otimes |r_j\rangle\langle r_{j'}|,\end{aligned}$$

where I represents the unit operator in $\mathcal{H}_S \otimes \mathcal{H}_M$, and $W_{ij,i'j'} = \langle a_i|\langle r_j|W|a_{i'}\rangle|r_{j'}\rangle$ is a matrix element of W. The reduced density operator W_S is the partial trace of W over $\mathcal{H}_M$:

$$\begin{aligned}W_S &= \mathrm{tr}_M(W)\\ &= \sum_h \langle r_h|\left(\sum_{ii'jj'} W_{ij,i'j'}|a_i\rangle\langle a_{i'}| \otimes |r_j\rangle\langle r_{j'}|\right)|r_h\rangle\\ &= \sum_{ii'}\left(\sum_h W_{ih,i'h}\right)|a_i\rangle\langle a_{i'}|\end{aligned}$$

and similarly W_M is the partial trace of W over $\mathcal{H}_S$:

$$W_M = \sum_{jj'}\left(\sum_k W_{kj,kj'}\right)|r_j\rangle\langle r_{j'}|$$

While the states of subsystems of a composite system are uniquely determined by the state of the composite system via the reduced density operators, the state of a composite system is not, in general, uniquely determined by the states of its subsystems.

As a simple example, consider a pure state of $S + M$ represented by the unit vector:

$$|\psi\rangle = \sum_i c_i|a_i\rangle|r_i\rangle,$$

where the $|a_i\rangle$ are eigenstates of some observable A of S and the $|r_i\rangle$ are eigenstates of some observable R of M. This is an 'entangled' state: a state that cannot be decomposed into a tensor product of an S-state and an M-state. The density operator of this state is the projection operator onto $|\psi\rangle$:

$$\begin{aligned}|\psi\rangle\langle\psi| &= \sum_{ij} c_i c^*_j |a_i\rangle\langle a_j| \otimes |r_i\rangle\langle r_j|\\ &= \sum_i |c_i|^2 |a_i\rangle\langle a_i| \otimes |r_i\rangle\langle r_i| + \sum_{i\neq j} c_i c^*_j |a_i\rangle\langle a_j| \otimes |r_i\rangle\langle r_j|\end{aligned}$$

The reduced density operators W_S and W_M are:

$$W_S = \sum_h \langle r_h|\psi\rangle\langle\psi|r_h\rangle = \sum_i |c_i|^2 |a_i\rangle\langle a_i| = \sum_i |c_i|^2 P_{a_i}$$

$$W_M = \sum_k \langle a_k|\psi\rangle\langle\psi|a_k\rangle = \sum_i |c_i|^2 |r_i\rangle\langle r_i| = \sum_i |c_i|^2 P_{r_i}$$

So the states of S and M are mixed states, and this is generally the case if the state of $S+M$ is a pure state that cannot be represented by a vector of the form $\psi \otimes \phi$. But W_S and W_M are also the reduced density operators of the *mixed state*:

$$W = \sum_i^{\mathrm{i}} |c_i|^2 |a_i\rangle\langle a_i| \otimes |r_i\rangle\langle r_i| = \sum_i^{\mathrm{i}} |c_i|^2 P_{a_i} \otimes P_{r_i}$$

because

$$|\psi\rangle\langle\psi| = W + \sum_{i \neq j} c_i c^*{}_j |a_i\rangle\langle a_j| \otimes |r_i\rangle\langle r_j| = W + X$$

and the additional 'interference term' X contributes zero to the partial traces.

The state of a composite system is uniquely determined by the states of its subsystems if and only if at least one of the subsystem states is pure.[135] If the state of a composite system W is a mixed state that can be decomposed (non-uniquely) into some mixture of pure or mixed states W_i, then the reduced states of the subsystems are also mixed states that can be similarly decomposed into the same mixtures of the reduced states of the W_i.

Since the mixed state $W = \Sigma_i |c_i|^2 P_{a_i} \otimes P_{r_i}$ and the pure state $P_\psi = |\psi\rangle\langle\psi|$, where $|\psi\rangle = \Sigma_i |a_i\rangle|r_i\rangle$, determine the same reduced states for S and M, W and P_ψ are indistinguishable by the measurements of any S-observable alone or any M-observable alone.[136] The states W and P_ψ are also indistinguishable by the measurement statistics of any $(S+M)$-observable of the form $A' \otimes R'$, where A' commutes with A and R' commutes with R. So W and P_ψ can be distinguished only in terms of the measurement statistics of observables of the form $B \otimes T$, where B is incompatible with A and T is incompatible with R, or global observables of $S+M$ that are not representable in the form $Q_S \otimes Q_R$ but rather as sums of such tensor products. This was pointed out by Furry (1936) in a discussion of the Einstein–Podolsky–Rosen argument.

According to the biorthogonal decomposition theorem (or Schmidt decomposition theorem), any pure state $|\psi\rangle$ of a system $S+M$ can be expressed in the form:

$$|\psi\rangle = \sum c_i |u_i\rangle|v_i\rangle$$

for some complete orthonormal set of vectors $\{|u_i\rangle\}$ in $\mathscr{H}_S$ and some complete orthornormal set $\{|v_i\rangle\}$ in $\mathscr{H}_M$. The decomposition is unique if and only if the $|c_i|^2$ are all distinct. The particular basis sets $\{|u_i\rangle\}$ and $\{|v_i\rangle\}$ of $\mathscr{H}_S$ and $\mathscr{H}_M$ are referred to as the Schmidt bases. In an ideal measurement, the state $|\psi(t)\rangle$ of $S+M$ evolves to a biorthogonal decomposition in the Schmidt bases defined by the eigenstates of the pointer observable R of M and the measured observable A of S. Evidently, the Schmidt bases are also the bases that diagonalize the reduced density matrices of S and M. The biorthogonal decomposition theorem is exploited by the modal interpretation dis-

135 Von Neumann (1955, pp. 426–9).

136 This also follows from the observation that

$$W = \sum_h |a_h\rangle\langle a_h| \otimes I_M \cdot |\psi\rangle\langle\psi \langle \cdot |a_h\rangle\langle a_h| \otimes I_M = \sum_k I_S \otimes |r_k\rangle\langle r_k| \cdot |\psi\rangle\langle\psi| \cdot I_S \otimes |r_k\rangle\langle r_k|$$

cussed in section 6.2, and is also relevant to the discussion of environmental 'monitoring' in section 5.4.

Suppose $S = S_1$ and $M = S_2$ represent two spin-$\frac{1}{2}$ particles. Ignoring spatial motion, the spin states of S_1 and S_2 are represented on 2-dimensional Hilbert spaces, $\mathcal{H}_1$ and $\mathcal{H}_2$. It turns out that in this case the spin quantum number, s, of the composite two-particle system, $S_1 + S_2$, can be either 0 or 1. If $s = 0$, the z-component of the total spin of the composite system, S_z, must also take the value 0. The state corresponding to this 0-eigenvalue of S_z is called the *singlet state* and takes the form of a biorthogonal decomposition:

$$|\psi\rangle = \tfrac{1}{\sqrt{2}}|\sigma_z = +1\rangle_1|\sigma_z = -1\rangle_2 - \tfrac{1}{\sqrt{2}}|\sigma_z = -1\rangle_1|\sigma_z = +1\rangle_2$$

or, in abbreviated form:

$$|\psi\rangle = \tfrac{1}{\sqrt{2}}|z+\rangle|z-\rangle - \tfrac{1}{\sqrt{2}}|z-\rangle|z+\rangle,$$

where the order of the two vectors in a product indicates the Hilbert space to which the vector belongs ($\mathcal{H}_1$ or $\mathcal{H}_2$).

This biorthogonal decomposition is non-unique, because the squares of the (real) coefficients are both equal to $\frac{1}{2}$. In fact, the singlet state is invariant under a change of basis. If we transform the σ_z-basis in $\mathcal{H}_1$ and $\mathcal{H}_2$ to, say, the σ_x-basis:

$$|z+\rangle = \tfrac{1}{\sqrt{2}}|x+\rangle - \tfrac{1}{\sqrt{2}}|x-\rangle$$
$$|z-\rangle = \tfrac{1}{\sqrt{2}}|x+\rangle + \tfrac{1}{\sqrt{2}}|x-\rangle$$

the singlet state retains the same form:

$$|\psi\rangle = \tfrac{1}{\sqrt{2}}|x+\rangle|x-\rangle - \tfrac{1}{\sqrt{2}}|x-\rangle|x+\rangle$$

In this case, the reduced states of S_1 and S_2 are represented by the statistical operators:

$$W_1 = \tfrac{1}{2}P^1_{z+} + \tfrac{1}{2}P^1_{z-} = \tfrac{1}{2}P^1_{x+} + \tfrac{1}{2}P^1_{x-} = \mathrm{I}_1$$
$$W_2 = \tfrac{1}{2}P^2_{z+} + \tfrac{1}{2}P^2_{z-} = \tfrac{1}{2}P^2_{x+} + \tfrac{1}{2}P^2_{x-} = \mathrm{I}_2$$

where I_1 and I_2 are the unit operators in the Hilbert spaces $\mathcal{H}_1$ and $\mathcal{H}_2$, respectively.

If $s = 1$, S_z can take the values $-1, 0, +1$. The three eigenstates are referred to as *triplet states* and take the form:

$$|z-\rangle|z-\rangle$$
$$\tfrac{1}{\sqrt{2}}|z+\rangle|z-\rangle + \tfrac{1}{\sqrt{2}}|z-\rangle|z+\rangle$$
$$|z+\rangle|z+\rangle$$

Notice that these states are all orthogonal to the singlet state, as we would expect, and that transforming from the σ_z-basis to the σ_x-basis transforms $\frac{1}{\sqrt{2}}|z+\rangle|z-\rangle + \frac{1}{\sqrt{2}}|z-\rangle|z+\rangle$ to $\frac{1}{\sqrt{2}}|x+\rangle|x+\rangle - \frac{1}{\sqrt{2}}|x-\rangle|x-\rangle$, not to $\frac{1}{\sqrt{2}}|x+\rangle|x-\rangle + \frac{1}{\sqrt{2}}|x-\rangle|x+\rangle$.

Bibliography

Y. Aharonov, J. Anandan, and L. Vaidman (1993), 'Meaning of the Wave Function,' *Physical Review A* **47**, 4616–26.

D.Z Albert (1986), 'How to Take a Photograph of Another Everett World,' in D.M. Greenberger (ed.), *New Techniques and Ideas in Quantum Measurement Theory* (Annals of the New York Academy of Sciences **480**) (New York: New York Academy of Sciences), pp. 498–502.

D.Z Albert (1987), 'A Quantum-Mechanical Automaton,' *Philosophy of Science* **54**, 577–85.

D.Z Albert (1992), *Quantum Mechanics and Experience* (Cambridge, Massachusetts: Harvard University Press).

D.Z Albert and J.A. Barrett (1995), 'On What it Takes to be a World,' *Topoi* **14**, 35–7.

D.Z Albert and B. Loewer (1988), 'Interpreting the Many-Worlds Interpretation,' *Synthese* **77**, 195–213.

D.Z Albert and B. Loewer (1989), 'Two No-Collapse Interpretations of Quantum Mechanics,' *Nous* **23**, 169–86.

D.Z Albert and B. Loewer (1990). 'Wanted Dead or Alive: Two Attempts to Solve Schrödinger's Paradox,' in A. Fine, M. Forbes, and L. Wessels (eds.), *PSA 1990*, Volume 1 (East Lansing, Michigan: Philosophy of Science Association), pp. 277–85.

D.Z Albert and B. Loewer (1991), 'Some Alleged Solutions to the Measurement Problem,' *Synthese* **88**, 87–98.

D.Z Albert and B. Loewer (1993), 'Non-Ideal Measurements,' *Foundations of Physics Letters* **6**, 297–305.

D.Z Albert and B. Loewer (1995), 'Tails of Schrödinger's Cat,' in R. Clifton (ed.), *Perspectives on Quantum Reality* (Dordrecht: Kluwer), pp. 81–92.

D.Z Albert and H. Putnam (1995), 'Further Adventures of Wigner's Friend,' *Topoi* **14**, 17–22.

A. Albrecht (1992), 'Investigating Decoherence in a Simple System,' *Physical Review D* **46**, 5504–20.

F. Arntzenius (1990), 'Kochen's Interpretation of Quantum Mechanics,' in A. Fine, M. Forbes, and L. Wessels (eds.), *PSA 1990*, Volume 1 (East Lansing, Michigan: Philosophy of Science Association), pp. 241–9.

F. Arntzenius (1992), 'Apparatus Independence in Proofs of Non-Locality,' *Foundations of Physics Letters* **5**, 517–25.

F. Arntzenius (1993), 'How to Discover that the Real is Unreal,' *Erkenntnis* **38**, 191–202.

F. Arntzenius (1994), 'Relativistic Hidden Variable Theories?' *Erkenntnis* **41**, 207–31.

A. Aspect, P. Grangier, and G. Roger (1981), 'Experimental Tests of Realistic Local Theories via Bell's Theorem,' *Physical Review Letters* **47**, 460–7.

A. Aspect, P. Grangier, and G. Roger (1982), 'Experimental Realization of EPR

Gedankenexperiment: A New Violation of Bell's Inequalities,' *Physical Review Letters* **49**, 91–4.

A. Aspect, J. Dalibard, and G. Roger (1982), 'Experimental Tests of Bell's Inequalities Using Time-Varying Analyzers,' *Physical Review Letters* **49**, 1804–7.

G. Bacciagaluppi (1995), 'Kochen–Specker Theorem in the Modal Interpretation of Quantum Mechanics,' *International Journal of Theoretical Physics* **34**, 1205–16.

G. Bacciagaluppi (1996), *Topics in the Modal Interpretation of Quantum Mechanics*, PhD dissertation, University of Cambridge.

G. Bacciagaluppi (1997) 'Review of J. Bub, *Interpreting the Quantum World*,' *Nuncius, Annali di Storia della Scienza* XII, no. 2, 606–10.

G. Bacciagaluppi and M. Dickson (1997), 'Dynamics for Density-Operator Interpretations of Quantum Theory: 9711048 (The Quantum Physics e-print Archive).

G. Bacciagaluppi and M. Dickson (1999), 'Modal Interpretations with Dynamics,' forthcoming.

G. Bacciagaluppi, A. Elby, and M. Hemmo (1996), 'Observers' Beliefs in the Modal Interpretation of Quantum Theory,' manuscript.

G. Bacciagaluppi and M. Hemmo (1996), 'Modal Interpretations, Decoherence and Measurements,' *Studies in the History and Philosophy of Modern Physics* **27B**, 239–77.

J. Baggott (1992), *The Meaning of Quantum Theory* (Oxford: Oxford University Press).

L.E. Ballentine (1970), 'The Statistical Interpretation of Quantum Mechanics,' *Reviews of Modern Physics* **42**, 358–81.

L.E. Ballentine (1987), 'Resource Letter IQM-2: Foundations of Quantum Mechanics Since the Bell Inequalities,' *American Journal of Physics* **55**, 785–92.

L.E. Ballentine (1990), *Quantum Mechanics* (Englewood Cliffs, New Jersey: Prentice Hall).

J.A. Barrett (1992), *Quantum Mechanics Without the Collapse Postulate*, Doctoral thesis, Philosophy Department, Columbia University.

J.A. Barrett (1994), 'The Suggestive Properties of Quantum Mechanics Without the Collapse Postulate,' *Erkenntnis* **41**, 233–52.

J.A. Barrett (1995a), 'The Distribution Postulate in Bohm's Theory,' *Topoi* **14**, 45–54.

J.A. Barrett (1995b), 'The Single-Mind and Many-Minds Versions of Quantum Mechanics,' *Erkenntnis* **42**, 89–105.

J.A. Barrett (1996a), 'On Everett's Formulation of Quantum Mechanics,' *The Monist*, forthcoming.

J.A. Barrett (1996b), 'Empirical Adequacy and the Availability of Reliable Records in Quantum Mechanics,' *Philosophy of Science*, **63**, 49–64.

F.J. Belinfante (1973), *A Survey of Hidden-Variables Theories* (Oxford: Pergamon Press).

J.L. Bell and R.K. Clifton (1995), 'QuasiBoolean Algebras and Definite Properties in Quantum Mechanics,' *International Journal of Theoretical Physics* **34**, 2409–21.

J.L. Bell and M. Hallett (1982), 'Logic, Quantum Logic and Empiricism,' *Philosophy of Science* **49**, 355–79.

J.L. Bell and M. Machover (1977), *A Course in Mathematical Logic* (Amsterdam: North-Holland).

J.L. Bell and A.B. Slomsen (1969), *Models and Ultraproducts* (Amsterdam: North Holland).

J.S. Bell (1966), 'On the Problem of Hidden Variables in Quantum Mechanics,' *Reviews of Modern Physics* **38**, 447–75. Reprinted in Bell (1987).

J.S. Bell (1964), 'On the Einstein–Podolsky–Rosen Paradox,' *Physics* **1**, 195–200. Reprinted in Bell (1987).

J.S. Bell (1987), *Speakable and Unspeakable in Quantum Mechanics* (Cambridge: Cambridge University Press).

J.S. Bell (1990), 'Against "Measurement",' in A. Miller (ed.), *Sixty-Two Years of Uncertainty: Historical, Philosophical and Physical Inquiries into the Foundations of Quantum Mechanics* (New York: Plenum), pp. 17–31.

J.S. Bell and M. Nauenberg (1966), 'The Moral Aspect of Quantum Mechanics,' in A. De Shalit, H. Feshbach, and L. van Hove (eds.), *Preludes in Theoretical Physics* (Amsterdam: North Holland), pp. 279–86. Reprinted in Bell (1987), pp. 22–8.

M. Beller and A. Fine (1994), 'Bohr's Response to EPR,' in J. Faye and H.J. Folse (eds.), *Niels Bohr and Contemporary Philosophy* (Dordrecht: Kluwer).

E.G. Beltrametti and G. Casinelli (1981), *The Logic of Quantum Mechanics* (Reading, Massachussetts: Addison-Wesley).

K. Berndl, D. Dürr, S. Goldstein, G. Peruzzi, and N. Zanghì (1993), 'Existence of Trajectories for Bohmian Mechanics,' *International Journal of Theoretical Physics* **32**, 2245–51.

G. Birkhoff (1966), *Lattice Theory*, 3rd edition (American Mathematical Society Colloquium, vol. 25) (Providence, Rhode Island: American Mathematical Society).

G. Birkhoff and J. von Neumann (1936), 'The Logic of Quantum Mechanics,' *Annals of Mathematics* **37**, 823–43.

D. Bohm (1951), *Quantum Theory* (Englewood Cliffs, New Jersey: Prentice-Hall).

D. Bohm (1952a), 'A Suggested Interpretation of Quantum Theory in Terms of "Hidden Variables",' Parts I and II, *Physical Review* **85**, 166–79, 180–93.

D. Bohm (1952b), 'Reply to a Criticism of a Causal Re-Interpretation of the Quantum Theory,' *Physical Review* **87**, 389–90.

D. Bohm (1953a), 'Comments on an Article of Takabayasi Concerning the Formulation of Quantum Mechanics with Classical Pictures,' *Progress of Theoretical Physics* **9**, 273–87.

D. Bohm (1953b), 'Comments on a Letter Concerning the Causal Interpretation of the Quantum Theory,' *Physical Review* **89**, 319–20.

D. Bohm (1953c), 'Proof That Probability Density Approaches $|\psi|^2$ in Causal Interpretation of the Quantum Theory,' *Physical Review* **89**, 458–66.

D. Bohm (1962), 'Hidden Variables in the Quantum Theory,' in D.R. Bates (ed.), *Quantum Theory*, vol. 3 (New York: Academic Press), pp. 345–87.

D. Bohm (1971), 'Quantum Theory as an Indication of a New Order in Physics,' in B. d'Espagnat (1971), pp. 412–69.

D. Bohm and J. Bub (1966a), 'A Proposed Solution of the Measurement Problem in Quantum Mechanics by a Hidden Variable Theory,' *Reviews of Modern Physics* **38**, 453–69.

D. Bohm and J. Bub (1966b), 'A Refutation of the Proof by Jauch and Piron that Hidden Variables Can Be Excluded in Quantum Mechanics,' *Reviews of Modern Physics* **38**, 470–5.

D. Bohm and J. Bub (1968), 'On Hidden Variables – A Reply to Comments by Jauch and Piron and by Gudder,' *Reviews of Modern Physics* **40**, 235–6.

D. Bohm and B.J. Hiley (1984), 'Measurement Understood Through the Quantum Potential Approach,' *Foundations of Physics* **14**, 255–74.

D. Bohm and B.J. Hiley (1985), 'Unbroken Quantum Realism, from Microscopic to Macroscopic Levels,' *Physical Review Letters* **55**, 2511–14.

D. Bohm and B.J. Hiley (1989), 'Non-Locality and Locality in the Stochastic Interpretation of Quantum Mechanics,' *Physics Reports* **172**, 93–122.

D. Bohm and B.J. Hiley (1993), *The Undivided Universe: An Ontological Interpretation of Quantum Theory* (London: Routledge).

D. Bohm, B.J. Hiley, and P.N. Kaloyerou (1987), 'An Ontological Basis for the Quantum Theory,' *Physics Reports* **144**, 321–75.

D. Bohm, R. Schiller, and J. Tiomno (1955), 'A Causal Interpretation of the Pauli Equations,' *Supplemento al Nuovo Cimento* **1**, 48–66.

N. Bohr (1934), *Atomic Theory and the Description of Nature* (Cambridge: Cambridge University Press).

N. Bohr (1935), 'Can Quantum-Mechanical Description of Physical Reality be Considered Complete?' *Physical Review* **38**, 696–702.

N. Bohr (1939), 'The Causality Problem in Modern Physics,' in *New Theories in Physics* (Paris: International Institute of Intellectual Cooperation), pp. 11–45.

N. Bohr (1948), 'On the Notions of Causality and Complementarity,' *Dialectica* **2**, 312–19.

N. Bohr (1949), 'Discussion with Einstein on Epistemological Problems in Modern Physics,' in P.A. Schilpp (1949), pp. 201–41.

N. Bohr (1961), *Atomic Physics and Human Knowledge* (New York: Science Editions).

N. Bohr (1963), *Essays 1958–1962 on Atomic Physics and Human Knowledge* (New York: Random House).

M. Born (1926), 'Zur Quantenmechanik der Stossvorgange,' *Zeitschrift für Physik* **37**, 863–7.

M. Born (ed.) (1971), *The Born–Einstein Letters* (London: Walker and Co.).

H.R. Brown (1986), 'The Insolubility Proof of the Quantum Measurement Problem,' *Foundations of Physics* **16**, 857–70.

H.R. Brown (1992), 'Bell's Other Theorem and its Connections with Nonlocality. Part I,' in A. van der Merwe, F. Selleri, and G. Tarozzi (1992), pp. 104–16.

H.R. Brown, C. Dewdney, and G. Horton (1995), 'Bohm Particles and their Detection in the Light of Neutron Interferometry,' *Foundations of Physics* **25**, 329–47.

H.R. Brown, A. Elby, and R. Weingard (1996), 'Cause and Effect in the Pilot-Wave Interpretation of Quantum Mechanics,' in J. Cushing, A. Fine, and S. Goldstein (eds.), *Bohmian Mechanics and Quantum Theory: An Appraisal* (Dordrecht: Kluwer), pp. 309–19.

J. Bub (1968), 'The Daneri–Loinger–Prosperi Quantum Theory of Measurement,' *Il Nuovo Cimento* **57B**, 503–20.

J. Bub (1973), 'On the Possibility of a Phase-Space Reconstruction of the Quantum Statistics: A Refutation of the Bell–Wigner Locality Argument,' *Foundations of Physics* **3**, 29–44.

J. Bub (1974), *The Interpretation of Quantum Mechanics* (Dordrecht: Reidel).

J. Bub (1976), 'Hidden Variables and Locality,' *Foundations of Physics* **6**, 511–26.

J. Bub (1979), 'The Measurement Problem of Quantum Mechanics,' in G. Toraldo di Francia (ed.), *Problems in the Foundations of Physics* (International School of Physics: Enrico Fermi, LXXII Course) (Dordrecht: Reidel), pp. 71–124.

J. Bub (1989), 'On Bohr's Response to EPR: A Quantum Logical Analysis,' *Foundations of Physics* **19**, 793–805.

J. Bub (1990), 'On Bohr's Response to EPR: II,' *Foundations of Physics* **20**, 929–41.

J. Bub (1992a), 'Quantum Mechanics Without the Projection Postulate,' *Foundations of Physics* **22**, 737–54.

J. Bub (1992b) 'Quantum Mechanics as a Theory of "Beables",' in A. van der Merwe, F. Selleri, and G. Tarozzi (1992), pp. 117–24.

J. Bub (1993a), 'Measurement and Objectivity in Quantum Mechanics,' in H.D. Doebner, W. Scherer, and F. Schroeck, Jr. (eds.), *Classical and Quantum Systems – Foundations and Symmetries* (Proceedings of the II. International Wigner Symposium, Goslar, Germany) (Singapore: World Scientific), pp. 9–18.

J. Bub (1993b), 'Measurement: It Ain't Over Till It's Over,' *Foundations of Physics Letters* **6**, 21–35.

J. Bub (1993c), 'Non-Ideal Measurements,' in P. Busch, P. Lahti, and P. Mittelstaedt (1993), pp. 125–36.

J. Bub (1994a), 'On the Structure of Quantal Proposition Systems,' *Foundations of Physics* **24**, 1261–80.

J. Bub (1994b), 'How to Interpret Quantum Mechanics,' *Erkenntnis* **41**, 253–73.

J. Bub (1995a), 'Interference, Noncommutativity, and Determinateness in Quantum Mechanics,' *Topoi* **14**, 39–43.

J. Bub (1995b), 'Why Not Take All Observables As Beables?' in D.M. Greenberger and A. Zeilinger (1995), pp. 761–7.

J. Bub (1995c), 'Maximal Structures of Determinate Propositions in Quantum Mechanics,' *International Journal of Theoretical Physics* **34**, 1–10.

J. Bub (1996), 'Schrödinger's Cat and Other Entanglements of Quantum Mechanics,' in J. Earman and J. Norton (eds.), *The Cosmos of Science* (Pittsburgh: University of Pittsburgh Press).

J. Bub and H. Brown (1986), 'Curious Properties of Quantum Ensembles which have been both Pre- and Post-Selected,' *Physical Review Letters* **56**, 2337–40.

J. Bub and R. Clifton (1996), 'A Uniqueness Theorem for "No Collapse" Interpretations of Quantum Mechanics,' *Studies in the History and Philosophy of Modern Physics*, **27**, 181–219.

J. Bub, R. Clifton, and B. Monton (1996), 'The Bare Theory Has No Clothes,' in G. Hellman and R. Healey (1997).

J. Bub and W. Demopoulos (1974), 'The Interpretation of Quantum Mechanics,' in R.S. Cohen and M. Wartofsky (eds.), *Boston Studies in the Philosophy of Science Vol. XIII* (Dordrecht: Reidel), pp. 92–122.

P. Busch (1985), 'Indeterminacy Relations and Simultaneous Measurements in Quantum Theory,' *International Journal of Theoretical Physics* **24**, 63–92.

P. Busch, G. Cassinelli, P. Lahti (1990), 'On the Quantum Theory of Sequential Measurements,' *Foundations of Physics* **20**, 757–75.

P. Busch, M. Grabowski, and P.J. Lahti (1995), *Operational Quantum Physics* (Berlin: Springer-Verlag).

P. Busch and P.J. Lahti (1990), 'Some Remarks on Unsharp Quantum Measurements, Quantum Non-Demolition, and All That,' *Annalen der Physik* **47**, 369–82.

P. Busch, P.J. Lahti, and P. Mittelstaedt (1991), *The Quantum Theory of Measurement* (Berlin: Springer).

P. Busch, P. Lahti, and P. Mittelstaedt (eds.) (1993), *Symposium on the Foundations of Modern Physics 1993* (Singapore: World Scientific).

P. Busch and F.E. Schroeck, Jr. (1989), 'On the Reality of Spin and Helicity,' *Foundations of Physics* **19**, 807–72.

J. Butterfield (1989), '*A Space–Time Approach to the Bell Inequality.*' in J.T. Cushing and E. McMullin (1989), pp. 114–44.

J. Butterfield (1994), 'Outcome Dependence and Stochastic Einstein Locality,' in D. Prawitz and D. Westerstahl (eds.), *Logic and Philosophy of Science in Uppsala* (Dordrecht: Kluwer), pp. 385–424.

A. Cabello, J.M. Estebaranz, and G. García-Alcaine (1996a), 'Bell–Kochen–Specker Theorem: A Proof with 18 Vectors,' *Physics Letters A* **212**, 183–7.

A Cabello, J.M. Estebaranz, and G. García-Alcaine (1996b), 'New Variants of the Bell–Kochen–Specker Theorem,' *Physics Letters, A* **218**, 115–18.

A. Cabello and G. García-Alcaine (1996), 'Bell–Kochen–Specker Theorem for any Finite Dimension $n \geq 3$,' *Journal of Physics A: Math. Gen.* **29**, 1025–36.

G. Cassinelli, E. DeVito, and P. Lahti (1994), 'Properties of the Range of a State Operator,' *Reports on Mathematical Physics* **43**, 211–24.

G. Cassinelli and P. Lahti (1993), 'The Copenhagen Variant of the Modal Interpretation and the Quantum Theory of Measurement,' *Foundations of Physics Letters* **6**, 533–44.

G. Casssinelli and P. Lahti (1995), 'Quantum Theory of Measurement and the Modal Interpretation of Quantum Mechanics,' *International Journal of Theoretical Physics* **34**, 1271–81.

J.F. Clauser and M.A. Horne (1974), 'Experimental Consequences of Objective Local Theories,' *Physical Review D* **10**, 526–35.

J.F. Clauser and A. Shimony (1978), 'Bell's Theorem: Experimental Tests and Implications,' *Reports on Progress in Physics* **41**, 1881–927.

E. Clavadetscher-Seeberger (1983), *Eine Partielle Praedikatenlogik* (Zürich: Eidgenoessische Technische Hochschule).

R. Clifton (1993), 'Getting Contextual and Nonlocal Elements of Reality the Easy Way,' *American Journal of Physics* **61**, 443–7.

R. Clifton (1995a), 'The Irrelevance of the Triorthogonal Uniqueness Theorem to Modal Interpretations of Quantum Mechanics,' in K.V. Laurikainen et al. (eds.), *Symposium on the Foundations of Modern Physics 1994 – 70 Years of Matter Waves* (Paris: Editions Frontiers), pp. 45–60.

R. Clifton (1995b), 'Independently Motivating the Kochen–Dieks Modal Interpretation of Quantum Mechanics,' *British Journal for the Philosophy of Science* **46**, 33–57.

R. Clifton (1995c), 'Making Sense of the Kochen–Dieks "No Collapse" Interpretation of Quantum Mechanics Independent of the Measurement Problem,' in D. Greenberger and A. Zeilinger (1995), pp. 570–8.

R. Clifton (1995d), 'Why Modal Interpretations of Quantum Mechanics Must Abandon Classical Reasoning About the Physical Properties,' *International Journal of Theoretical Physics* **34**, 1302–12.

R. Clifton (1996a), 'Is There Hyperplane Dependence in the Modal Interpretation?' manuscript.

R. Clifton (1996b), 'The Properties of Modal Interpretations of Quantum Mechanics,' *British Journal for the Philosophy of Science*, **47**, 371–98.

R. Clifton (1999), 'Beables in Algebraic Quantum Mechanics,' in J. Butterfield and C. Pagonis (eds.), *From Physics to Philosophy*, (Cambridge: Cambridge University Press).

R. Clifton and H. Halvorson (1999), 'Maximal Subalgebras of Beables,' forthcoming.

R. Clifton and J. Zimba (1998), 'Valuations on Functionally Closed Sets of Quantum-Mechanical Observables and Von Neumann's No-Hidden-Variables Theorem,' in Dennis Dieks and Pieter Vermaas (eds.), *The Modal Interpretation of Quantum Mechanics*, The University of Western Ontario Series in Philosophy of Science (Dordrecht: Kluwer).

R.K. Clifton, M.L.G. Redhead, and J.N. Butterfield (1991a), 'Generalization of the Greenberger–Horne–Zeilinger Algebraic Proof of Nonlocality.' *Foundations of Physics* **21**, 149–84.

R.K. Clifton, M.L.G. Redhead, and J.N. Butterfield (1991b), 'A Second Look at a Recent Algebraic Proof of Nonlocality,' *Foundations of Physics Letters* **4**, 395–403.

J.T. Cushing (1994), *Quantum Mechanics: Historical Contingency and the Copenhagen Hegemony* (Chicago: University of Chicago Press).

J.T. Cushing (1996), 'What Measurement Problem?' in R. Clifton (ed.), *Perspectives on Quantum Reality* (Dordrecht: Kluwer), pp. 167–81.

J.T. Cushing, A. Fine, and S. Goldstein (eds.) (1996), *Bohmian Mechanics and Quantum Theory: An Appraisal, Boston Studies in the Philosophy of Science Vol.* 184 (Dordrecht: Kluwer).

J.T. Cushing and E. McMullin (eds.) (1989), *Philosophical Consequences of Quantum Theory: Reflections on Bell's Inequality* (Notre Dame, Illinois: University of Notre Dame Press).

A. Daneri, A. Loinger, and G.M. Prosperi (1962), 'Quantum Theory of Measurement and Ergodicity Conditions,' *Nuclear Physics* **33**, 297–319.

A. Daneri, A. Loinger, and G.M. Prosperi (1966), 'Further Remarks on the Relations Between Statistical Mechanics and Quantum Theory of Measurement,' *Il Nuovo Cimento* **44B**, 119–28.

E.B. Davies (1976), *Quantum Theory of Open Systems* (London: Academic Press).

W. Demopoulos (1976), 'The Possibility Structure of Physical Systems,' in W. Harper and C.A. Hooker (eds.), *Foundations and Philosophy of Statistical Theories in the Physical Sciences* (Dordrecht: Reidel), pp. 55–80.

W. Demopoulos (1980), 'Locality and the Algebraic Structure of Quantum Mechanics,' in P. Suppes (ed.), *Studies in the Foundations of Quantum Mechanics* (East Lansing, Michigan: Philosophy of Science Association), pp. 119–44.

B. d'Espagnat (ed.) (1971), *Foundations of Quantum Mechanics* (International School of Physics: Enrico Fermi, XLIX Course, New York: Academic Press).

B. d'Espagnat (1976), *Conceptual Foundations of Quantum Mechanics* (Reading, Massachusetts: Benjamin).

B. d'Espagnat (1995), *Veiled Reality* (Reading, Massachussetts: Addison-Wesley).

D. Deutsch (1985a), 'Quantum Theory, the Church–Turing Principle and the Universal Quantum Computer,' *Proceedings of the Royal Society (London)*, **A400**, 97–117.

D. Deutsch (1985b), 'Quantum Theory as a Universal Physical Theory,' *International Journal of Theoretical Physics* **24**, 1–41.

C. Dewdney (1992), 'Constraints on Quantum Hidden Variables and the Bohm Theory,' *Journal of Physics A: Math. Gen.* **25**, 3615–26.

C. Dewdney, L. Hardy, and E.J. Squires (1993), 'How Late Measurements of Quantum Trajectories Can Fool a Detector,' *Physics Letters A* **184**, 6–11.

C. Dewdney, P.R. Holland, and A. Kyprianidis (1986), 'What Happens in a Spin Measurement?' *Physics Letters A* **119**, 259–67.

C. Dewdney, P.R. Holland, A. Kyprianidis, and J.P. Vigier (1988), 'Spin and Nonlocality in Quantum Mechanics,' *Nature* **336**, 536–44.

C. Dewdney, G. Horton, M.M. Lam, Z. Malik, and M. Schmidt (1992), 'Wave–Particle Dualism and the Interpretation of Quantum Mechanics,' *Foundations of Physics* **22**, 1217–65.

B.S. DeWitt and N. Graham (1973), *The Many-Worlds Interpretation of Quantum Mechanics* (Princeton: Princeton University Press).

M. Dickson (1994), 'Wavefunction Tails in the Modal Interpretation,' in D. Hull, M. Forbes, and R. Burian (eds.), *PSA 1994*, Volume 1 (East Lansing, Michigan: Philosophy of Science Association), pp. 366–76.

M. Dickson (1995), 'Is There *Really* No Projection Postulate in the Modal Interpretation?' *British Journal for the Philosophy of Science* **46**, 197–218.

M. Dickson and R. Clifton (1996), 'Lorentz-Invariance in the Modal Interpretation,' in D. Dieks and P. Vermaas (eds.), *The Modal Interpretation of Quantum Mechanics*, (Dordrecht: Kluwer), forthcoming.

D. Dieks (1988), 'The Formalism of Quantum Theory: An Objective Description of Reality?' *Annalen der Physik* **7**, 174–90.

D. Dieks (1989a), 'Quantum Mechanics Without the Projection Postulate and Its Realistic Interpretation,' *Foundations of Physics* **19**, 1397–423.

D. Dieks (1989b), 'Resolution of the Measurement Problem Through Decoherence of the Quantum State,' *Physics Letters A* **142**, 439–46.

D. Dieks (1993), 'The Modal Interpretation of Quantum Mechanics and Some of Its Relativistic Aspects,' *International Journal of Theoretical Physics* **32**, 2363–75.

D. Dieks (1994a), 'Objectification, Measurement and Classical Limit According to the Modal Interpretation of Quantum Mechanics,' in P. Busch, P. Lahti, and P. Mittelstaedt (1993), pp. 160–7.

D. Dieks (1994b), 'Modal Interpretation of Quantum Mechanics, Measurements, and Macroscopic Behavior,' *Physical Review A* **49**, 2290–300.

D. Dieks (1996), 'Preferred Factorizations and the Modal Interpretation of Quantum Mechanics,' in G. Hellman and R. Healey (1997).

P.A.M. Dirac (1958), *Quantum Mechanics*, 4th edition (Oxford: Clarendon Press).

D. Dürr, W. Fusseder, S. Goldstein, and N. Zanghì (1993), 'Comment on "Surrealistic Bohm Trajectories",' *Zeitschrift für Naturforschung* **48a**, 1261–2.

D. Dürr, S. Goldstein, and N. Zanghì (1990), 'On a Realistic Theory for Quantum Physics,' in S. Albeverio, G. Casati, U. Cattaneo, D. Merlini, and R. Moresi (eds.), *Stochastic Processes, Physics, and Geometry* (Singapore: World Scientific), pp. 374–91.

D. Dürr, S. Goldstein, and N. Zanghì (1992a), 'Quantum Equilibrium and the Origin of Absolute Uncertainty,' *Journal of Statistical Physics* **67**, 843–907.

D. Dürr, S. Goldstein, and N. Zanghì (1992b), 'Quantum Chaos, Classical Randomness, and Bohmian Mechanics,' *Journal of Statistical Physics* **68**, 259–70.

D. Dürr, S. Goldstein, and N. Zanghì (1992c), 'Quantum Mechanics, Randomness, and Deterministic Reality,' *Physics Letters A* **172**, 6–12.

D. Dürr, S. Goldstein, and N. Zanghì (1993), 'A Global Equilibrium as the Foundation of Quantum Randomness,' *Foundations of Physics* **23**, 721–38.

A. Einstein (1948), 'Quantenmechanik und Wirklichkeit,' *Dialectica* **2**, 320–4. Translated as 'Quantum Mechanics and Reality,' in M. Born (1971), pp. 168–72.

A. Einstein, B. Podolsky, and N. Rosen (1935), 'Can Quantum-Mechanical Description of Physical Reality be Considered Complete?' *Physical Review* **47**, 777–80.

A. Elby (1992), 'Bells' *Other* Theorem and Its Connection with Nonlocality, Part 2,' in A. van der Merwe, F. Selleri, and G. Tarozzi (1992), pp. 184–93.

A. Elby (1993a), 'Why "Modal" Interpretations of Quantum Mechanics Don't Solve the Measurement Problem,' *Foundations of Physics Letters* **6**, 5–19.

A. Elby (1993b), 'Decoherence and Zurek's Existential Interpretation of Quantum Mechanics,' in P. Busch, P. Lahti, and P. Mittelstaedt (1993), pp. 168–81.

A. Elby (1994), 'The "Decoherence" Approach to the Measurement Problem in Quantum Mechanics,' in D. Hull, M. Forbes, and R. Burian (eds.), *PSA 1994*, Volume 1 (East Lansing, Michigan: Philosophy of Science Association), pp. 355–65.

A. Elby and J. Bub (1994), 'Triorthogonal Uniqueness Theorem and its Relevance to the Interpretation of Quantum Mechanics,' *Physical Review A* **49**, 4213–16.

A. Elby and M.R. Jones (1992), 'Weakening the Locality Conditions in Algebraic Nonlocality Proofs,' *Physics Letters A* **171**, 11–16.

B-G. Englert, M.O. Scully, G. Süssmann, and H. Walther (1993a), 'Surrealistic Bohm Trajectories,' *Zeitschrift für Naturforschung* **47a**, 1175–86.

B-G. Englert, M.O. Scully, G. Süssmann, and H. Walther (1993b), 'Reply to Comment on "Surrealistic Bohm Trajectories",' *Zeitschrift für Naturforschung* **48a**, 1263–4.

T. Epstein (1952), 'The Causal Interpretation of Quantum Mechanics,' *Physical Review* **89**, 319.

T. Epstein (1953), 'The Causal Interpretation of Quantum Mechanics,' *Physical Review* **91**, 985.

H. Everett (1957), 'Relative State Formulations of Quantum Mechanics,' *Reviews of Modern Physics* **29**, 454–62.

H. Everett (1973), 'The Theory of the Universal Wave Function,' in B.S. DeWitt and N. Graham (1973), pp. 3–140.

J. Faye (1991), *Niels Bohr: His Heritage and Legacy* (Dordrecht: Kluwer).

J. Faye and H.J. Folse (eds.) (1994) *Niels Bohr and Contemporary Philosophy* (Dordrecht: Kluwer).

M.H. Fehrs and A. Shimony (1974), 'Approximate Measurement in Quantum Mechanics I,' *Physical Review D* **9**, 2317–20.

A. Fine (1970), 'Insolubility of the Quantum Measurement Problem,' *Physical Review D* **2**, 2783–7.

A. Fine (1973), 'Probability and the Interpretation of Quantum Mechanics,' *British Journal for the Philosophy of Science* **24**, 1–37.

A. Fine (1974), 'On the Completeness of Quantum Theory, *Synthese* **29**, 257–89. Reprinted in P. Suppes (1976), pp. 259–81.

A. Fine (1981), 'Correlations and Physical Locality,' in P. Asquith and R. Giere (eds.), *PSA 1980*, Volume 2 (East Lansing, Michigan: Philosophy of Science Association), pp. 535–62.

A. Fine (1982a), 'Joint Distributions, Quantum Correlations, and Commuting Observables,' *Journal of Mathematical Physics* **23**, 1306–10.

A. Fine (1982b), 'Hidden Variables, Joint Probability, and the Bell Inequalities,' *Physical Review Letters* **48**, 291–5.

A. Fine (1982c), 'Antinomies of Entanglement: The Puzzling Case of the Tangled Statistics.' *Journal of Philosophy* **79**, 733–47.

A. Fine (1982d), 'Some Local Models for Correlation Experiments,' *Synthese* **50**, 279–94.

A. Fine (1986), *The Shaky Game: Einstein, Realism, and the Quantum Theory* (Chicago: University of Chicago Press).

A. Fine (1989a), 'Do Correlations Have to be Explained?' in J.T. Cushing and E. McMullin (1989), pp. 174–94.

A. Fine (1989b), 'Correlations in Efficiency: Testing the Bell Inequalities,' *Foundations of Physics* **19**, 453–78.

G. Fleming (1966), 'A Manifestly Covariant Description of Arbitrary Dynamical Variables in Relativistic Quantum Mechanics,' *Journal of Mathematical Physics* **7**, 1959–81.

G. Fleming (1988), 'Hyperplane Dependent Quantized Fields and Lorentz Invariance,' in H.R. Brown and R. Harre (eds.), *Philosophical Foundations of Quantum Field Theory* (Oxford: Oxford University Press), pp. 93–115.

G. Fleming (1989), 'Lorentz Invariant State Reduction and Localization,' in A. Fine and J. Leplin (eds.), *PSA 1988*, Volume 2 (East Lansing, Michigan: Philosophy of Science Association), pp. 112–16.

G. Fleming (1996), 'Just How Radical is Hyperplane Dependence?' in R. Clifton (ed.), *Perspectives on Quantum Reality* (Dordrecht: Kluwer), pp. 11–27.

R. Folman (1994), 'A Search for Hidden Variables in the Domain of High Energy Physics,' *Foundations of Physics Letters* **7**, 191–200.

R. Folman (1995), 'A Search for a Non-Exponential Distribution of the Tau Lepton Decay Time,' *OPAL Physics Note PN173*, European Organization for Nuclear Research. (See also OPAL Collaboration (1996).)

H. Folse (1985), *The Philosophy of Niels Bohr* (Amsterdam: North-Holland).

W.H. Furry (1936), 'Note on the Quantum Mechanical Theory of Measurement,' *Physical Review* **49**, 393–9.

M. Gell-Mann and J.B. Hartle (1990), 'Quantum Mechanics in the Light of Quantum Cosmology,' in S. Kobayashi, H. Ezawa, Y. Murayama, and S. Nomura (eds.), *Proceedings of the 3rd International Symposium on the Foundations of Quantum Mechanics in the Light of New Technology* (Tokyo: Physical Society of Japan), pp. 321–43.

M. Gell-Mann and J.B. Hartle (1991a), 'Quantum Mechanics in the Light of Quantum Cosmology,' in W.H. Zurek (ed.), *Complexity, Entropy and the Physics of Information, Santa Fe Institute Studies in the Science of Complexity,* Vol. VIII (Redwood City, California: Addison-Wesley), pp. 425–58.

M. Gell-Mann and J. Hartle (1991b), 'Alternative Decohering Histories in Quantum Mechanics,' in K.K. Phua and Y. Yamaguchi (eds.), *Proceedings of the 25th International Conference on High Energy Physics, Singapore, August 2–8, 1990* (Singapore: World Scientific).

R. Geroch (1984), 'The Everett Interpretation,' *Noûs* **18**, 617–33.

G.C. Ghirardi (1992), 'Bell's Requirements for a "Serious Theory",' in A. van der Merwe, F. Selleri, and G. Tarrozi (1992), pp. 228–43.

G.C. Ghirardi, R. Grassi, and P. Pearle (1990), 'Relativistic Dynamical Reduction Models: General Framework and Examples,' *Foundations of Physics* **20**, 1271–316.

G.C. Ghirardi, R. Grassi, and P. Pearle (1991), 'Relativistic Dynamical Reduction Models and Nonlocality,' in P. Lahti and P. Mittelstaedt (eds.), *Symposium on the Foundations of Modern Physics 1990* (Singapore: World Scientific), pp. 109–23.

G.C. Ghirardi, P. Pearle, and A. Rimini (1990), 'Markov Processes in Hilbert Space and Continuous Spontaneous Localization of Systems of Identical Particles,' *Physical Review A* **42**, 78–89.

G.C. Ghirardi and A. Rimini (1990), 'Old and New Ideas in the Theory of Quantum Measurement,' in A. Miller (ed.), *Sixty-Two Years of Uncertainty* (New York: Plenum), pp. 167–91.

G.C. Ghirardi, A. Rimini, and T. Weber (1986), 'Unified Dynamics for Microscopic and Macroscopic Systems,' *Physical Review D* **34**, 470–91.

A.M. Gleason (1957), 'Measures on the Closed Sub-Spaces of Hilbert Spaces,' *Journal of Mathematics and Mechanics* **6**, 885–93.

S. Goldstein (1987), 'Stochastic Mechanics and Quantum Theory,' *Journal of Statistical Physics* **47**, 645–67.

D.M. Greenberger, M.A. Horne, and A. Zeilinger (1989), 'Going Beyond Bell's Theorem,' in M. Kafatos (1989), pp. 69–72.

D.M. Greenberger and A. Zeilinger (eds.) (1995), *Fundamental Problems in Quantum Theory* (Annals of the New York Academy of Sciences **775**) (New York: New York Academy of Sciences).

R. Griffiths (1984), 'Consistent Histories and the Interpretation of Quantum Mechanics,' *Journal of Statistical Physics* **36**, 219–72.

R. Griffiths (1987), 'Correlation in Separated Quantum Systems: A Consistent History Analysis of the EPR Problem,' *American Journal of Physics* **55**, 11–17.

A. Hajek and J. Bub (1992), 'EPR,' *Foundations of Physics* **22**, 313–32.

O. Halpern (1952), 'A Proposed Re-Interpretation of Quantum Mechanics,' *Physical Review* **87**, 389.

L. Hardy (1992), 'Quantum Mechanics, Local Realistic Theories, and Lorentz-Invariant Theories,' *Physical Review Letters* **68**, 2981–4.

J. Hartle (1968), 'Quantum Mechanics of Individual Systems,' *American Journal of Physics* **36**, 704–12.

R. Healey (1989), *The Philosophy of Quantum Mechanics* (Cambridge: Cambridge University Press).

R. Healey (1993a), 'Why Error-Prone Quantum Measurements Have Outcomes,' *Foundations of Physics Letters* **6**, 37–54.

R. Healey (1993b), 'Measurement and Quantum Indeterminateness,' *Foundations of Physics Letters* **6**, 307–16.

R. Healey (1995), 'Dissipating the Quantum Measurement Problem,' *Topoi* **14**, 55–65.

R. Healey (1996), '"Modal" Interpretations, Decoherence and the Quantum Measurement Problem,' in G. Hellman and R. Healey (1997).

W. Heisenberg (1967), 'Quantum Theory and its Interpretation,' in S. Rozental (ed.), *Niels Bohr: His Life and Work as Seen by his Friends and Colleagues* (New York: Wiley Interscience), pp. 94–108.

G. Hellmann (1982), 'Stochastic Einstein-Locality and the Bell Theorems,' *Synthese* **53**, 461–504.

G. Hellman (1987), 'EPR, Bell, and Collapse: A Route Around "Stochastic" Hidden Variables,' *Philosophy of Science* **54**, 639–57.

G. Hellman and R. Healey (eds.) (1997), *Quantum Measurement, Decoherence and Modal Interpretations.* Forthcoming as a volume in *Minnesota Studies in the Philosophy of Science*, University of Minnesota Press.

K. Hepp (1972), 'Quantum Theory of Measurement and Macroscopic Observables,' *Helvetica Physica Acta* **45**, 237–48.

P. Heywood and M. Redhead (1983), 'Nonlocality and the Kochen–Specker Paradox,' *Foundations of Physics* **13**, 481–99.

W.V.D. Hodge, *The Theory and Application of Harmonic Integrals* (Cambridge: Cambridge University Press).

D.G. Holdsworth and C.A. Hooker (1983), 'A Critical Survey of Quantum Logic,' *Scientia Special Issue: Logic in the 20th Century*, 127–246.

P.R. Holland (1988), 'Causal Interpretation of a System of Two Spin-1/2 Particles,' *Physics Reports* **169**, 293–327.

P.R. Holland (1993a), 'The de Broglie–Bohm Theory of Motion and Quantum Field Theory,' *Physics Reports* **224**, 95–150.

P.R. Holland (1993b), *The Quantum Theory of Motion* (Cambridge: Cambridge University Press).

P.R. Holland and J.P. Vigier (1988), 'The Quantum Potential and Signalling in the Einstein–Podolsky–Rosen Experiment,' *Foundations of Physics* **18**, 741–9.

D. Home and F. Selleri (1991), 'Bell's Theorem and the EPR Paradox,' *La Rivista del Nuovo Cimento* **14**, 1–95.

J. Horgan (1993), 'The Artist, the Physicist, and the Waterfall,' *Scientific American* **268**, no. 2 (February), 30.

D. Howard (1985), 'Einstein on Locality and Separability,' *Studies in History and Philosophy of Science* **16**, 171–201.

D. Howard (1989). 'Holism, Separability, and the Metaphysical Implications of Bell's Inequality,' in J.T. Cushing and E. McMullin (1989), pp. 224–53.

R.I.G. Hughes (1981), 'Quantum Logic,' *Scientific American* **243**(10), 202–13.

R.I.G. Hughes (1987), *The Structure and Interpretation of Quantum Mechanics* (Cambridge: Cambridge University Press).

M. Jammer (1966), *The Conceptual Development of Quantum Mechanics* (New York: McGraw-Hill).

M. Jammer (1974), *The Philosophy of Quantum Mechanics* (New York: Wiley).

J.P. Jarrett (1984), 'On the Physical Significance of the Locality Conditions in the Bell Arguments,' *Noûs* **18**, 569–89.

J.P. Jarrett (1989), 'Bell's Theorem: A Guide to the Implications,' in J.T. Cushing and E. McMullin (1989), pp. 60–79.

J.M. Jauch (1964), 'The Problem of Measurement in Quantum Mechanics,' *Helvetica Physica Acta* **37**, 293–316.

J.M. Jauch (1968), *Foundations of Quantum Mechanics* (Reading, Massachusetts: Addison-Wesley).

J.M. Jauch (1971), 'Foundations of Quantum Mechanics,' in B. d'Espagnat (1971), pp. 20–35.

J.M. Jauch and C. Piron (1963), 'Can Hidden Variables be Excluded in Quantum Mechanics?' *Helvetica Physica Acta* **36**, 827–37.

J.M. Jauch and C. Piron (1969), 'On the Structure of Quantal Proposition Systems,' *Helvetica Physica Acta* **42**, 842–8.

M. Jones (1991), 'Some Difficulties for Clifton, Redhead, and Butterfield's Recent Proof of Nonlocality,' *Foundations of Physics Letters* **4**, 385–94.

M. Jones and R. Clifton (1993), 'Against Experimental Metaphysics,' in P.A. French, T.E. Euling, Jr., and H.K. Wettstein (eds.), *Mid-West Studies in Philosophy, vol. XVIII: Philosophy of Science*, (Notre Dame, Indiana: University of Notre Dame Press), pp. 297–316.

E. Joos (1986), 'Quantum Theory and the Appearance of a Classical World,' in *New Techniques and Ideas in Quantum Measurement Theory* (Annals of the New York Academy of Science **480**), p. 242.

E. Joos and H.D. Zeh (1985), 'The Emergence of Classical Properties Through Interaction with the Environment,' *Zeitschrift für Physik* **B59**, 223–43.

R. Jost (1976), 'Measures on the Finite Dimensional Subspaces of a Hilbert Space: Remarks on a Theorem by A.M. Gleason,' in E.H. Lieb, B. Simon, and S. Wightman (eds.), *Studies in Mathematical Physics: Essays in Honour of Valentine Bargmann* (Princeton: Princeton University Press), pp. 221–45.

M. Kafatos (ed.) (1989), *Bell's Theorem, Quantum Theory and Conceptions of the Universe* (Dordrecht: Kluwer).

J.B. Keller (1953), 'Bohm's Interpretation of the Quantum Theory in Terms of "Hidden" Variables,' *Physical Review* **89**, 1040–1.

Kelvin, Lord (1901), 'Nineteenth Century Clouds Over the Dynamical Theory of Heat and Light,' *Philosophical Magazine* **2**, 1–40.

M. Kernaghan (1994), 'Bell–Kochen–Specker Theorem with 20 Vectors,' *Journal of Physics A* **27**, L829.

M. Kernaghan and A. Peres (1995), 'Kochen–Specker Theorem for Eight-Dimensional Space,' *Physics Letters A* **198**, 1–5.

S. Kochen (1985), 'A New Interpretation of Quantum Mechanics,' in P. Lahti and P. Mittelstaedt (eds.), *Symposium on the Foundations of Modern Physics* (Singapore: World Scientific), pp. 151–69.

S. Kochen and E.P. Specker (1965), 'Logical Structures Arising in Quantum Theory,' *Symposium on the Theory of Models* (Amsterdam: North Holland).

S. Kochen and E.P. Specker (1967), 'The Problem of Hidden Variables in Quantum Mechanics,' *Journal of Mathematics and Mechanics* **17**, 59–87.

H. Krips (1987), *The Metaphysics of Quantum Theory* (Oxford: Clarendon Press).
I. Lakatos (1976), *Proofs and Refutations: The Logic of Mathematical Discovery* (Cambridge: Cambridge University Press).
P. Lewis (1995), 'GRW and the Tails Problem,' *Topoi* **14**, 23–33.
M. Lockwood (1992), *Mind, Brain, and the Quantum: The Compound I* (Oxford: Blackwell).
F. London and E. Bauer (1939), *La Theorie de l'Observation en Mecanique Quantique* (Paris: Hermann).
G. Lüders (1951), 'Über die Zustandsanderung durch den Messprozess,' *Annalen der Physik* **8**, 322–8.
G. Mackey (1963), *Mathematical Foundations of Quantum Mechanics* (New York: Benjamin).
M.J. Maczynski (1971), 'Boolean Properties of Observables in Axiomatic Quantum Mechanics,' *Reports on Mathematical Physics* **2**, 135–50.
H. Margenau (1936), 'Quantum-Mechanical Description,' *Physical Review* **49**, 240–2.
H. Margenau (1963a), 'Measurements and Quantum States. Parts I and II,' *Philosophy of Science* **30**, 1–16, 138–57.
H. Margenau (1963b), 'Measurements in Quantum Mechanics,' *Annals of Physics* **23**, 469–85.
T.W. Marshall, E. Santos, and F. Selleri (1983), 'Local Realism Has Not Been Refuted by Atomic Cascade Experiments,' *Physics Letters* **98A**, 5–9.
T. Maudlin (1992), 'Bell's Inequality, Information Transmission, and Prism Models' in D. Hull, M. Forbes, and K. Okruhlik (eds.), *PSA 1992*, Volume 1 (East Lansing, Michigan: Philosophy of Science Association), pp. 404–17.
T. Maudlin (1994), *Quantum Non-Locality and Relativity* (Oxford: Blackwell).
T. Maudlin (1995a), 'Why Bohm's Theory Solves the Measurement Problem,' *Philosophy of Science* **62**, 479–83.
T. Maudlin (1995b), 'Three Measurement Problems,' *Topoi* **14**, 7–15.
N.D. Mermin (1981a), 'Bringing Home the Atomic World: Quantum Mysteries for Anybody,' *American Journal of Physics* **49**(10), 940–3.
N.D. Mermin (1981b), 'Quantum Mysteries for Anyone,' *The Journal of Philosophy* **78**, 397–408.
N.D. Mermin (1985), 'Is The Moon There When Nobody Looks? Reality and the Quantum Theory,' *Physics Today* **38**(4), 38–47.
N.D. Mermin (1989), 'Can You Help Your Team Tonight by Watching on TV? More Experimental Metaphysics from Einstein, Podolsky, and Rosen,' in J.T. Cushing and E. McMullin (1989), pp. 38–49.
N.D. Mermin (1990a), 'Quantum Mysteries Revisited,' *American Journal of Physics* **58**, 731–4.
N.D. Mermin (1990b), 'What's Wrong With These Elements of Reality?' *Physics Today* **43**(6), 9–11.
N.D. Mermin (1990c), 'Simple Unified Form for the Major Unified No-Hidden-Variables Theorems,' *Physical Review Letters* **65**, 3373–6.
N.D. Mermin (1990d), *Boojums All The Way Through* (Cambridge: Cambridge University Press).
N.D. Mermin (1993), 'Hidden Variables and the Two Theorems of John Bell,' *Reviews of Modern Physics* **65**, 803–15.
N.D. Mermin (1995), 'Limits to Quantum Mechanics as a Source of Magic Tricks: Retrodiction and the Bell–Kochen–Specker Theorem,' *Physical Review Letters* **74**, 831–4.
E. Nelson (1966), 'Derivation of Schrödinger Equation from Newtonian Mechanics,' *Physical Review* **150**, 1079–85.

E. Nelson (1967), *Dynamical Theories of Brownian Motion* (Princeton: Princeton University Press).

R. Omnès (1990), 'From Hilbert Space to Common Sense: A Synthesis of Recent Progress in the Interpretation of Quantum Mechanics,' *Annals of Physics (N.Y.)* **20**(1), 354–447.

R. Omnès (1992), 'Consistent Interpretations of Quantum Mechanics,' *Reviews of Modern Physics* **64**, 339–82.

R, Omnès (1994), *The Interpretation of Quantum Mechanics* (Princeton: Princeton University Press).

OPAL Collaboration (1996), 'Test of the Exponential Decay Law at Short Decay Times Using Tau Leptons,' *Physics Letters* **B368**, 244–50.

C. Pagonis and R. Clifton (1995), 'Unremarkable Contextualism: Dispositions in the Bohm Theory,' *Foundations of Physics* **25**, 281–96.

C. Papaliolios (1967), 'Experimental Test of a Hidden-Variable Quantum Theory,' *Physical Review Letters* **18**, 622–5.

W. Pauli (1933), 'Die Allgemeinen Prinzipien der Wellenmechanik,' in H. Geiger and K. Scheel (eds.), *Handbuch der Physik*, 2nd edition, Volume 24 (Berlin: Springer), pp. 83–272. English translation: *General Principles of Quantum Mechanics* (Berlin: Springer, 1980).

W. Pauli (1948), 'Editorial on the Concept of Complementarity,' *Dialectica* **2**(3/4), 307–11.

J.P. Paz, S. Habib, and W. Zurek (1993), 'Reduction of the Wave Packet: Preferred Observable and Decoherence Time Scale,' *Physical Review* **D47**, 488.

J.P. Paz and W.H. Zurek (1993), 'Environment-Induced Decoherence, Classicality, and Consistency of Quantum Histories,' *Physical Review* **D48**, 2728–38.

P.M. Pearle (1986), 'Stochastic Dynamical Reduction Theories and Superluminal Communication,' *Physical Review D* **33**, 2240–52.

P. Pearle (1989), 'Combining Stochastic Dynamical State-Vector Reduction with Spontaneous Localization,' *Physical Review A* **39**, 2277–89.

P. Pearle (1990), 'Toward a Relativistic Theory of Statevector Reduction,' in A. Miller (ed.), *Sixty-Two Years of Uncertainty* (New York: Plenum), pp. 193–214.

P. Pearle (1996), 'Wavefunction Collapse Models with Nonwhite Noise,' in R. Clifton (ed.), *Perspectives on Quantum Reality* (Dordrecht: Kluwer), pp. 93–109.

R. Penrose (1989), *The Emperor's New Mind* (Oxford: Oxford University Press).

R. Penrose (1994a), 'On Bell Non-Locality Without Probabilities: Some Curious Geometry, in J. Ellis and A. Amati (eds.), *Quantum Reflections* (in honour of J.S. Bell) (Cambridge: Cambridge University Press).

R. Penrose (1994b), *Shadows of the Mind* (Oxford: Oxford University Press).

A. Peres (1990), 'Incompatible Results of Quantum Measurements,' *Physics Letters A* **151**, 107–8.

A. Peres (1991), 'Two Simple Proofs of the Kochen–Specker Theorem,' *Journal of Physics A: Math. Gen.* **24**, L175–L178.

A. Peres (1993), *Quantum Theory: Concepts and Methods* (Dordrecht: Kluwer).

A. Peres (1996), 'Generalized Kochen–Specker Theorem,' *Foundations of Physics*, forthcoming.

A. Peres and A. Ron (1988), 'Cryptodeterminism and Quantum Theory,' in A. van der Merwe, F. Selleri, and G. Tarozzi (eds.), *Microphysical Reality and Quantum Formalism*, Volume 2 (Dordrecht: Kluwer), p. 115.

A. Petersen (1963), 'The Philosophy of Niels Bohr,' *Bulletin of the Atomic Scientists* **19**(7), 8–14.

I. Pitowsky (1989a), *Quantum Probability – Quantum Logic* (Berlin: Springer-Verlag).

I. Pitowsky (1989b), 'From George Boole to John Bell – The Origins of Bell's Inequality,' in M. Kafatos (1989), pp. 37–49.

K. Popper (1982), *Quantum Theory and the Schism in Physics* (London: Hutchinson).

K. Przibram (ed.) (1967), *Letters on Wave Mechanics* (New York: Philosophical Library).

H. Putnam (1969), 'Is Logic Empirical?' *Boston Studies in the Philosophy of Science* **5**, 216–41.

H. Putnam (1974), 'How To Think Quantum Logically,' *Synthese* **29**, 55–61.

H. Putnam (1981), 'Quantum Mechanics and the Observer,' *Erkenntnis* **16**, 193–219.

A.I.M. Rae (1986), *Quantum Physics: Illusion or Reality?* (Cambridge: Cambridge University Press).

M.L.G. Redhead (1986), 'Relativity and Quantum Mechanics – Conflict or Peaceful Coexistence,' *Annals of the New York Academy of Sciences* **480**, 14–20.

M.L.G. Redhead (1987), *Incompleteness, Nonlocality, and Realism* (Oxford: Clarendon Press).

M.L.G. Redhead (1989), 'Nonfactorizability, Stochastic Causality, and Passion-at-a-Distance,' in J.T. Cushing and E. McMullin (1989), pp. 145–53.

L. Rosenfeld (1965), 'The Measuring Process in Quantum Mechanics,' *Supplement of the Progress of Theoretical Physics* (Commemoration Issue for the 30th Anniversary of the Meson Theory by Dr H. Yukawa), 222–31.

S. Saunders (1992), 'Decoherence, Relative States, and Evolutionary Adaptation,' *Foundations of Physics* **23**, 1553–85.

S. Saunders (1995), 'Time, Quantum Mechanics, and Decoherence,' *Synthese* **102**, 235–66.

E. Scheibe (1973), *The Logical Analysis of Quantum Mechanics* (Oxford: Pergamon).

P.A. Schilpp (ed.) (1949), *Albert Einstein: Philosopher-Scientist*, (La Salle, Illinois: Open Court).

E. Schrödinger (1935a), 'Discussion of Probability Relations Between Separated Systems,' *Proceedings of the Cambridge Philosophical Society* **31**, 553–63.

E. Schrödinger (1935b), Die Gegenwärtige Situation in der Quantenmechanik,' *Naturwissenschaften* **23**, 807–12, 824–8, 844–9. Reprinted as 'The Present Situation in Quantum Mechanics,' in J.A. Wheeler and W.H. Zurek (1983), pp. 152–67.

F. Selleri (ed.) (1971), *Quantum Mechanics Versus Local Realism: The Einstein–Podolsky–Rosen Paradox* (New York: Plenum).

F. Selleri (1990), *Quantum Paradoxes and Physical Reality* (Dordrecht: Kluwer).

F. Selleri and G. Tarozzi (1980), 'Is Clauser and Horne's Factorizability a Necessary Requirement for a Probabilistic Local Theory?' *Lettere al Nuovo Cimento* **29**, 533–6.

W.D. Sharp and N. Shanks (1985), 'Fine's Prism Models for Quantum Correlation Statistics,' *Philosophy of Science* **52**, 538–64.

A. Shimony (1971), 'Experimental Test of Local Hidden Variable Theories,' in B. d'Espagnat (1971), pp. 182–94.

A. Shimony (1974), 'Approximate Measurement in Quantum Mechanics II,' *Physical Review D* **9**, 2321–3.

A. Shimony (1978), 'Metaphysical Problems in the Foundations of Quantum Mechanics,' *International Philosophical Quarterly* **18**, 3–17.

A. Shimony (1981), 'Critique of the Papers of Fine and Suppes,' in P. Asquith and R. Giere (eds.), *PSA 1980*, Volume 2 (East Lansing, Michigan: Philosophy of Science Association), pp. 572–80.

A. Shimony (1984a), 'Controllable and Uncontrollable Non-Locality,' in S. Kamefuchi *et al.* (eds.), *Foundations of Quantum Mechanics in the Light of New Technology* (Tokyo: Physical Society of Japan), pp. 225–30.

A. Shimony (1984b), 'Contextual Hidden Variables and Bell's Inequalities,' *British Journal*

for the Philosophy of Science **35**, 24–45.
A. Shimony (1986), 'Events and Processes in the Quantum World,' in R. Penrose and C.J. Isham (eds.), *Quantum Processes in Space and Time* (Oxford: Clarendon Press), pp. 182–203.
A. Shimony (1989), 'Search for a Worldview Which Can Accommodate our Knowledge of Microphysics,' in J.T. Cushing and E. McMullin (1989), pp. 25–37.
A. Siegel and N. Wiener (1956), '"Theory of Measurement" in Differential-Space Quantum Theory,' *Physical Review* **101**, 429–32.
E. Specker (1960), 'Die Logik Nicht Gleichzeitig Entscheidbarer Aussagen,' *Dialectica* **14**, 239–46.
E.J. Squires (1993), 'A Local Hidden-Variable Theory that, FAPP, Agrees with Quantum Theory,' *Physics Letters A* **178**, 22–6.
A. Stairs (1983), 'On the Logic of Pairs of Quantum Systems,' *Synthese* **56**, 437–60.
H.P. Stapp, (1975), 'Bell's Theorem and World Process,' *Il Nuovo Cimento* **29B**, 270–6.
H.P. Stapp (1977), 'Are Superluminal Connections Necessary?' *Il Nuovo Cimento* **40B**, 191–204.
H. Stapp (1985), 'Bell's Theorem and the Foundations of Quantum Physics,' *American Journal of Physics* **53**, 306–17.
H. Stapp (1988), 'Einstein Locality, EPR Locality, and the Significance for Science of the Nonlocal Character of Quantum Theory,' in A. van der Merwe, F. Selleri, and G. Tarozzi (eds.), *Microphysical Reality and Quantum Formalism*, Volume 2 (Dordrecht: Kluwer), pp. 36–78.
H. Stein and A. Shimony (1971), 'Limitations on Measurement,' in B. d'Espagnat (1971), pp. 56–76.
A. Stone (1994), 'Does the Bohm Theory Solve the Measurement Problem?' *Philosophy of Science* **61**, 250–66.
M.H. Stone (1936), 'The Theory of Representations for Boolean Algebras,' *Transactions of the American Mathematical Society* **40**, 37–111.
P. Suppes (1976), *Logic and Probability in Quantum Mechanics* (Dordrecht: Reidel).
P. Suppes (ed.) (1980), *Studies in the Foundations of Quantum Mechanics* (East Lansing, Michigan: Philosophy of Science Association).
P. Suppes and M. Zanotti (1976), 'On the Determinism of Hidden Variable Theories with Strict Correlation and Conditional Statistical Independence of Observables,' in P. Suppes (1976), pp. 445–55.
G. Svetlichny, M. Redhead, H. Brown, and J. Butterfield (1988), 'Do the Bell Inequalities Require the Existence of Joint Probability Distributions?' *Philosophy of Science* **55**, 387–401.
K. Svozil (1993), *Randomness and Undecidability in Physics* (Singapore: World Scientific).
K. Svozil (1995), 'A Constructivist Manifesto for the Physical Sciences,' in W.D. Schimanovich, E. Köhler, and F. Stadler (eds.), *The Foundational Debate, Complexity and Constructivity in Mathematics and Physics*, (Dordrecht: Kluwer), pp. 65–88.
K. Svozil and J. Tkadlec (1996), 'Greechie Diagrams, Nonexistence of Measures in Quantum Logics and Kochen–Specker Type Constructions,' manuscript.
A.R. Swift and R. Wright (1980), 'Generalized Stern–Gerlach Experiments and the Observability of Arbitrary Spin Operators,' *Journal of Mathematical Physics* **21**, 77–82.
T. Takabayasi (1952), 'On the Formulation of Quantum Mechanics Associated with Classical Pictures,' *Progress of Theoretical Physics* **8**, 143–82.
P. Teller (1986), 'Relativity, Wholeness, and Quantum Mechanics,' *British Journal for the Philosophy of Science* **37**, 71–81.

P. Teller (1989), 'Relativity, Relational Wholeness, and Bell's Inequality,' in J.T. Cushing and E. McMullin (1989), pp. 208–23.
J.H. Tutsch (1968), 'Collapse Time for the Bohm–Bub Hidden Variable Theory,' *Reviews of Modern Physics* **40**, 232–4.
J.H. Tutsch (1969), 'Simultaneous Measurement in the Bohm–Bub Hidden-Variable Theory,' *Physical Review* **183**, 1116–31.
J.H. Tutsch (1971), 'Mathematics of the Measurement Problem in Quantum Mechanics,' *Journal of Mathematical Physics* **12**, 1711–18.
A. Valentini (1991a), Signal-Locality, Uncertainty, and the Subquantum H-Theorem I,' *Physics Letters A* **156**, 5–11.
A. Valentini (1991b), 'Signal-Locality, Uncertainty, and the Subquantum H-Theorem II,' *Physical Letters A* **158**, 1–8.
A. Valentini (1996), *On the Pilot-Wave Theory of Classical, Quantum and Subquantum Physics* (Berlin: Springer).
A. van der Merwe, F. Selleri, and G. Tarozzi (eds.) (1992), *Bell's Theorem and the Foundations of Modern Physics* (Singapore: World Scientific).
B. van Fraassen (1973), 'A Semantic Analysis of Quantum Logic,' in C.A. Hooker (ed.), *Contemporary Research in the Foundations and Philosophy of Quantum Theory* (Dordrecht: Reidel), pp. 80–113.
B. van Fraassen (1974), 'The Einstein–Podolsky–Rosen Paradox,' *Synthese* **29**, 291–309.
B. van Fraassen (1979), 'Hidden Variables and the Modal Interpretation of Quantum Statistics,' *Synthese* **42**, 155–65.
B. van Fraassen (1981), 'A Modal Interpretation of Quantum Mechanics,' in E. Beltrametti and B. van Fraassen (eds.), *Current Issues in Quantum Logic* (Singapore: World Scientific), pp. 229–58.
B. van Fraassen (1982), 'The Charybdis of Realism: Epistemological Implications of Bell's Inequality,' *Synthese* **52**, 25–38. Reprinted with addenda in J.T. Cushing and E. McMullin (1989), pp. 97–113.
B. van Fraassen (1991), *Quantum Mechanics: An Empiricist View* (Oxford: Oxford University Press).
P.E. Vermaas (1996a), 'Unique Transition Probabilities in the Modal Interpretation,' *Studies in the History and Philosophy of Modern Physics* **27**, 133–59.
P.E. Vermaas (1996b), 'Expanding the Property Ascriptions in the Modal Interpretation of Quantum Mechanics, in G. Hellman and R. Healey (eds.) *Minnesota Studies in the Philosophy of Science* (Minneapolis: University of Minnesota Press), forthcoming.
P.E. Vermaas (1996c), 'A No-Go Theorem for Joint Property Ascriptions in the Modal Interpretation of Quantum Mechanics,' manuscript.
P. Vermaas and D. Dieks (1995), 'The Modal Interpretation of Quantum Mechanics and Its Generalization to Density Operators,' *Foundations of Physics* **25**, 145–58.
J.C. Vink (1993), 'Quantum Mechanics in Terms of Discrete Beables,' *Physical Review A* **48**, 1808–18.
J. von Neumann (1939), untitled remarks following an address by Bohr, in *New Theories in Physics* (Paris: International Institute of Intellectual Cooperation), pp. 30–8.
J. von Neumann (1954), 'Unsolved Problems in Mathematics,' unpublished address to the International Mathematical Congress, Amsterdam, Sept. 2, 1954. Typescript, von Neumann Archives, Library of Congress, Washington, DC, pp. 21, 22.
J. von Neumann (1955), *Mathematical Foundations of Quantum Mechanics* (Princeton: Princeton University Press). (Translated from *Mathematische Grundlagen der Quantenmechanik*, Springer, Berlin, 1932.)

J.A. Wheeler (1983), 'Law Without Law,' in J.A. Wheeler and W.H. Zurek (1983), pp. 183–213.

J.A. Wheeler and W.H. Zurek (eds.) (1983), *Quantum Theory and Measurement* (Princeton: Princeton University Press).

J.H.M. Whiteman (1961), *The Mystical Life* (London: Allen and Unwin).

J.H.M. Whiteman (1967), *Philosophy of Space and Time* (London: Allen and Unwin).

G.J. Whitrow (1961), *The Natural Philosophy of Time* (London: Thomas Nelson and Sons).

N. Wiener and A. Siegel (1953), 'A New Form for the Statistical Postulate of Quantum Mechanics,' *Physical Review* **91**, 1551–60.

N. Wiener and A. Siegel (1955), 'The Differential-Space Theory of Quantum Systems,' *Il Nuovo Cimento, Supplemento al volume II, Serie X*, 982–1003.

E.P. Wigner (1962), 'Remarks on the Mind-Body Question,' in I.J. Good (ed.), *The Scientist Speculates* (London: Heinemann), pp. 284–301.

E.P. Wigner (1970), 'On Hidden Variables and Quantum Mechanical Probabilities,' *American Journal of Physics* **38**, 1005–9.

J. Zimba and R. Penrose (1993), 'On Bell Non-Locality Without Probabilities: More Curious Geometry,' *Studies in the History and Philosophy of Modern Physics* **24**, 697–720.

W.H. Zurek (1981), 'Pointer Basis of Quantum Apparatus: Into What Mixture Does the Wave Packet Collapse?' *Physical Review* **D24**, 1516–25.

W.H. Zurek (1982), 'Environment-Induced Superselection Rules,' *Physical Review* **D26**, 1862–80.

W.H. Zurek (1991), 'Decoherence and the Transition from Quantum to Classical,' *Physics Today* **44**, 36–44.

W.H. Zurek (1993a), 'Negotiating the Tricky Border Between Quantum and Classical,' *Physics Today* **46**(4), 13–15, 81–90.

W.H. Zurek (1993b), 'Preferred States, Predictability, Classicality, and the Environment-Induced Decoherence,' *Progress in Theoretical Physics* **89**, 281–312.

W.H. Zurek, S. Habib, and J.P. Paz (1993), 'Coherent States via Decoherence,' *Physical Review Letters* **70**, 1187–90.

W.H. Zurek and J.P. Paz (1993), 'Decoherence, Chaos, and the Quantum and the Classical,' in P. Busch, P. Lahti, and P. Mittelstaedt (1993), pp. 452–72.

Index

ed in the United States
Bookmasters